Lecture Notes in Computer Science 1457

Edited by G. Goos, J. Hartmanis and J. van Leeuwen

Springer

Berlin
Heidelberg
New York
Barcelona
Budapest
Hong Kong
London
Milan
Paris
Singapore
Tokyo

A. Ferreira   J. Rolim
H. Simon   S.-H. Teng   (Eds.)

Solving Irregularly Structured Problems in Parallel

5th International Symposium, IRREGULAR'98
Berkeley, California, USA, August 9-11, 1998
Proceedings

 Springer

Volume Editors

Afonso Ferreira
INRIA, CNRS
2004 Route des Lucioles, F-06902 SophiaAntipolis, France
E-mail: afonso.ferreira@sophia.inria.fr

José Rolim
University of Geneva, Computer Science Center
24, Rue Général Dufour, CH-1211 Geneva, Switzerland
E-mail: jose.rolim@cui.unige.ch

Horst Simon
University of California, Lawrence Berkeley National Laboratory
One Cyclotron Road, Berkeley, CA 94720, USA
E-mail: hdsimon@lbl.gov

Shang-Hua Teng
University of Illinois at Urbana-Champaign, Department of Computer Science
1304 West Springfield Avenue, Urbana, IL 61801-2987, USA
E-mail: steng@cs.uiuc.edu

Cataloging-in-Publication data applied for

Die Deutsche Bibliothek - CIP-Einheitsaufnahme

Solving irregularly structured problems in parallel : 5th
international symposium, Irregular '98, Berkeley, California, USA,
August 9 - 11 1998 ; proceedings / A. Ferreira ... (ed.). - Berlin ;
Heidelberg ; New York ; Barcelona ; Budapest ; Hong Kong ;
London ; Milan ; Paris ; Singapore ; Tokyo : Springer, 1998
   (Lecture notes in computer science ; Vol. 1457)
   ISBN 3-540-64809-7

CR Subject Classification (1991): F.2, D.1.3, C.1.2, B.2.6, D.4, G.1-2

ISSN 0302-9743
ISBN 3-540-64809-7 Springer-Verlag Berlin Heidelberg New York

© Springer-Verlag Berlin Heidelberg 1998
Printed in Germany

Typesetting: Camera-ready by author
SPIN 10638253   06/3142 – 5 4 3 2 1 0   Printed on acid-free paper

Preface

The Fifth International Symposium on Solving Irregularly Structured Problems in Parallel (IRREGULAR'98) was scheduled for August 9–11, 1998, in NERSC, Lawrence Berkeley National Laboratory, Berkeley, California, USA.

IRREGULAR focuses on algorithmic, applicational, and system aspects arising in the development of efficient parallel solutions to irregularly structured problems. It aims, in particular, at fostering cooperation among practitioners and theoreticians in the field. *IRREGULAR'98*, is the fifth in the series, after Geneva, Lyon, Santa Barbara, and Paderborn. IRREGULAR was promoted and coordinated by the Symposium Chairs, Afonso Ferreira (ENS Lyon) and José Rolim (Geneva).

This volume contains all contributed papers accepted for presentation at the '98 Symposium, together with abstracts of the invited lectures by Gary Miller (CMU), John Gilbert (Xerox PARC), Keshav Pingali (Cornell), Bruce Hendrickson (Sandia National Laboratory), and M. Valero (Technical U, Barcelona). In addition to the traditional contributed and invited talks, this year IRREGULAR has four mini-symposia: (1) *Principles and Practice of Preconditioned Iterative Methods*, organized by Edmond Chow (Lawrence Livermore National Laboratory), (2) *Mesh Generation*, organized by Kwok Ko (Stanford), (3) *Partitioning and Dynamic Load Balancing*, organized by Andrew Sohn (NJIT), and (4) *Information Retrieval*, organized by Hongyuan Zha (Penn State).

The contributed papers were selected out of several dozen submissions received in response to the call for papers. The selection of the contributed papers was based on originality, quality, and relevance to the symposium. Considerable effort was devoted to the evaluation of the submissions by the program committee and a number of other referees. Extensive feedback was provided to authors as a result, which we hope has proven helpful to them. We wish to thank the program committee and the referees as well as the authors of all submissions.

The final program represents an interesting cross section of research efforts in the area of irregular parallelism. By introducing mini-symposia to the conference, we hope to expand the scope of this conference especially by attracting papers on parallel processing for information organization and internet applications.

IRREGULAR'98 could not have happened without the dedicated work of our local organizers Roberta Boucher and Kare Zukor, from Lawrence Berkeley National Laboratory, whose sponsorship of the symposium is gratefully acknowledged.

May 1998   Horst Simon and Shang-Hua Teng

Program Committee

Contents

Invited Talk

Regular Talks

Minisymposium Talks

Combinatorial Preconditioning for Sparse Linear Systems

John R. Gilbert[*]

Xerox Palo Alto Research Center

Abstract. A key ingredient in the solution of a large, sparse system of linear equations by an iterative method like conjugate gradients is a *preconditioner*, which is in a sense an approximation to the matrix of coefficients. Ideally, the iterative method converges much faster on the preconditioned system at the extra cost of one solve against the preconditioner per iteration.

We survey a little-known technique for preconditioning sparse linear systems, called *support-graph* preconditioning, that borrows some combinatorial tools from sparse Gaussian elimination. Support-graph preconditioning was introduced by Vaidya and extended by Gremban, Miller, and Zagha. We extend the technique further and use it to analyze existing preconditioners based on incomplete factorization and on multilevel diagonal scaling. In the end, we argue that support-graph preconditioning is a ripe field for further research.

Introduction

We present new applications of a little-known technique for constructing and analyzing preconditioners called *support-graph preconditioning*. The technique was first proposed and used by Pravin Vaidya, who described it in a 1991 workshop talk [7]. Vaidya used the technique to design a family of novel preconditioners. Later, Gremban, Miller, and Zagha [2, 3] extended the technique and used it to construct another family of preconditioners. Our full paper explains the technique, extends it further, and uses it to analyze two classes of *known* preconditioners for model problems. Specifically, we use the extended technique to analyze certain modified-incomplete-Cholesky (MICC) preconditioners (see [5]) and multilevel-diagonal-scaling (MDS) preconditioners (see [1], for example).

Support-Graph Preconditioners

The support-graph technique analyzes a preconditioner B for a positive definite matrix A by giving bounds on the condition number of the preconditioned matrix

[*] This work is joint with Marshall Bern (Xerox PARC), Bruce Hendrickson (Sandia National Labs), Nhat Nguyen (Stanford University), and Sivan Toledo (Xerox PARC and Tel Aviv University). It was supported in part by DARPA contract number DABT63-95-C-0087 and by NSF contract number ASC-96-26298.

$B^{-1}A$. To bound the condition number, we bound the two extreme eigenvalues of the preconditioned matrix separately.

The largest eigenvalue of $B^{-1}A$ is bounded above by a quantity called the *support* of the pair (A, B), written $\sigma(A, B)$. (The smallest eigenvalue is similarly bounded below by $1/\sigma(B, A)$.) Intuitively, if A and B represent resistive networks, the support is the number of copies of B required to give at least as much conductance as A between every pair of nodes. Formally, $\sigma(A, B)$ is the smallest number τ such that $\tau B - A$ is positive semidefinite.

The support is estimated by combinatorial techniques: the graph of A is split into simple pieces (typically single edges), each of which is supported by a simple piece (typically one or a few simple paths) of the graph of B. Often this allows us to reduce a complex problem to many problems with simple structures.

Discussion of Results

Vaidya proposed two classes of preconditioners. In both, the graph of the preconditioner B is a subgraph of that of A; the preconditioner is formed by dropping some nonzero matrix entries (and modifying the diagonal). Vaidya's first class, based on a maximum-weight spanning tree, guarantees a condition-number bound of $O(n^2)$ for any $n \times n$ sparse M-matrix. These preconditioners are easy to generalize to sparse symmetric diagonally dominant matrices. They can be constructed and factored at insignificant cost using relatively simple graph algorithms.

Vaidya's second class of preconditioners is based on maximum spanning trees augmented with a few extra edges. They can be constructed at insignificant cost using a slightly more complex algorithm than the first class. The cost of factoring these preconditioners depends on how many edges are added to the spanning tree. Vaidya proposes that the factorization cost be balanced with the iteration costs, and he provides balancing guidelines for some classes of matrices. This class of preconditioners guarantees that the work in the linear solver is bounded by $O(n^{1.75})$ for any sparse M-matrix, and by $O(n^{1.2})$ for M-matrices whose underlying graphs are planar. No bounds are known for matrices arising from finite-element or finite-difference discretizations in 3D.

The strengths of Vaidya's preconditioners, especially of his second class, are that they are general, easy to construct and provide good condition-number bounds. For example, the work required to solve a model Poisson problem in 2D using Vaidya's preconditioner is $O(n^{1.2})$, but $O(n^{1.25})$ work is required for a solver based on a modified incomplete Cholesky factorization. Coupled with the facts that Vaidya's preconditioners are guaranteed to work well on irregular problems, and that the only numerical assumption they make is that the matrix is an M-matrix, these are impressive results.

The main weaknesses of Vaidya's preconditioners are that they require a high-quality direct solver to factor the preconditioner, that balancing the preconditioner-factorization costs and the iteration costs may be a nontrivial task, and that they are not guaranteed to parallelize well.

Gremban, Miller, and Zagha proposed a more flexible framework than Vaidya's (and coined the support-graph terminology). They do not require the preconditioner B to be a subgraph of A; indeed, they allow B to augment A with both extra vertices and extra edges. They proposed a family of multilevel preconditioners in which B is a tree of logarithmic diameter whose leaves are the vertices of A. The tree is based on a hierarchical partitioning of the matrix A. The cost of preconditioning in every iteration is small, and the preconditioners are guaranteed to parallelize. The condition number of the preconditioned system is similar, for model problems, to the condition numbers offered by modified incomplete factorizations.

Even on model problems, none of these preconditioners offer convergence rates as good as those of other multilevel preconditioners, such as multigrid preconditioners. On the other hand, the Gremban/Miller/Zagha preconditioners are guaranteed to parallelize well, so they are likely to be preferable to incomplete factorization on some architectures.

We add to the support-graph framework some tools for handling mixed-sign entries in the preconditioner, and for simplifying the analysis of preconditioners whose structures are not based on trees. Then the support-graph technique can be used to analyze some existing preconditioners. For example, we show that the condition number for MICC for a model Laplace problem in 2D with either Dirichlet or Neumann or mixed boundary conditions is $O(n^{1/2})$. The bound for a 3D model problem is $O(n^{1/3})$, but this only holds in the Dirichlet case and in some mixed boundary-condition cases. To the best of our knowledge, these are the first such bounds for MICC without diagonal perturbation. The proofs are relatively simple, requiring only background in elementary linear algebra. Therefore, we believe that these proofs are suitable for classroom teaching.

Our condition-number bound for a multilevel diagonal scaling preconditioner in one dimension is $O(\log n)$, which is tight or nearly so. This is the first polylogarithmic condition-number bound to emerge from a support-graph analysis. However, this bound seems difficult to generalize to more realistic problems. The analysis does provides some purely algebraic insight as to how this preconditioner works; it also suggests a construction for more general problems that we have not yet been able to analyze.

References

1. William D. Gropp Barry F. Smith, Petter E. Bjorstad. *Domain Decomposition: Parallel Multilevel Methods for Elliptic Partial Differential Equations.* Cambridge University Press, 1996.
2. K.D. Gremban, G.L. Miller, and M. Zagha. Performance evaluation of a parallel preconditioner. In *9th International Parallel Processing Symposium*, pages 65–69, Santa Barbara, April 1995. IEEE.
3. Keith D. Gremban. *Combinatorial Preconditioners for Sparse, Symmetric, Diagonally Dominant Linear Systems.* PhD thesis, School of Computer Science, Carnegie Mellon University, October 1996. Technical Report CMU-CS-96-123.

4. Stephen Guattery. Graph embedding techniques for bounding condition numbers of incomplete factor preconditioners. Technical Report ICASE Report 97-47, NASA Langley Research Center, 1997.

5. I. Gustafsson. A class of first-order factorization methods. *BIT*, 18:142–156, 1978.

6. Victoria E. Howle and Stephen A. Vavasis. Preconditioning complex-symmetric layered systems arising in electrical power modeling. In *Proceedings of the Copper Mountain Conference on Iterative Methods*, Copper Mountain, Colorado, March 1998. 7 unnumbered pages.

7. Pravin M. Vaidya. Solving linear equations with symmetric diagonally dominant matrices by constructing good preconditioners. An unpublished manuscript. A talk based on the manuscript was presented at the IMA Workshop on Graph Theory and Sparse Matrix Computation, October 1991, Minneapolis.

A Threaded Sliding Window Executor for Irregular Computation on a NOW

Eric A. Schweitz and Dharma P. Agrawal

Electrical and Computer Engineering Department
North Carolina State University
Raleigh, NC 27695-7911

Abstract. In this paper, we introduce a sliding window algorithm for the parallelization of irregular coarse-grained computations on a network of workstations. This algorithm extends the well-known inspector-executor model for parallelization of coarse-grained irregular loops. The new algorithm is compared to an existing inspector-executor library as well as our Threaded Executor algorithm.

1 Introduction

With the simultaneous decline in costs and increase in performance of workstation machines, there has been a great deal of recent research activity in the area of using a network of workstations (NOW) [5, 6, 12, 15] as a low-cost hardware platform for parallel calculations. A significant area of focus for researchers has been in the development of new and better hardware and software technology to connect these powerful workstations and improve the parallel performance of the NOW-based parallel computation paradigm.

At present, networks are often composed of state-of-the-art workstations interconnected by an older network technology, such as Ethernet, which was not originally designed for network-based parallel computation. This results in a rather dramatic disparity between the speed of the processor and the speed of the network. This mismatch in relative speeds favors a computation which has significantly more computation than communication, known as a coarse-grained computation [16]. A fine-grained calculation (having a smaller computation-communication ratio) would be expected to perform less efficiently than one that is coarse-grained because existing network technology would typically become a bottleneck to overall parallel performance. One particular challenge, then, is to find better algorithms which can both more effectively utilize the hardware and realize good speed-up when a NOW is used to perform parallel computation.

Another area of active research is the automatic parallelization of existing sequential programs into a parallel form which can achieve better performance [16]. Automatic parallelization helps to relieve the programmer of tedious and possibly error-prone re-writing of sequential code into an equivalent parallel form. Much of the work in automatic parallelization has centered on the extraction of parallelism from loops in sequential code, since loops are often a source of natural parallelism which can be exploited for speed-up. Related to this research is the design of runtime systems which can be used as components in an automatic parallelization system [10, 11].

In Sec. 2, we look at the inspector-executor (IE) [7, 8] which is a runtime system that can be used in the parallelization of irregular computations, and also discuss recent work related to IE and some of the limitations of that work. We review the Threaded Executor (TE) [13] and its limitations in Sec. 3. In Sec. 4, we introduce the Sliding Window Executor (SWE) and explain how it addresses the limitations of the aforementioned methods. This is followed by the presentation of our preliminary results with SWE on a NOW. Finally, we conclude the paper by discussing future directions and applications of this work.

2 Irregular Computation, Inspector-Executor (IE)

An example of an irregular computation is shown in Fig. 1(a). Assuming there is no aliasing between the arrays A[] and B[] in the example, then it is clear that without prior knowledge of the values of the indexing array IB[], it is not statically determinable which values of B[] are required for a particular iteration of the loop. When transforming such a loop into parallel code, the arrays may be distributed across processors in the system with the intent of executing different iterations on different processors. Trivial solutions, such as keeping an entire copy of B[] local to each processor or broadcasting all the values of B[] between nodes, are not effective or efficient for performance and space reasons. To better address the issue, the IE model was developed to do a run-time analysis and allow parallel execution of such a loop.

```
                              Inspector(IB, Sched,...)
                              ...
                              Executor(Sched, B,...)
DO I = 1, N                   DO I = LL, LU
    A[I] = A[I] + B[IB[I]]        A[I] = A[I] + B[IB[I]]
END DO                        END DO

(a) sequential code          (b) IE code
```

Fig. 1. An irregular computation loop and IE transformation

The IE model [7, 8] is a two-phase, run-time approach for allowing an irregular problem to be executed in a parallel manner. This technique has been well researched and applied to many different areas [1, 3, 9, 11, 16, 17].

The first phase of the algorithm, known as the inspector, performs an analysis on the access pattern of the original loop. The goal of this analysis is to determine which values residing on other processors are needed by the current processor in order to perform its calculations correctly. This access information is termed a "schedule."

The second phase, known as the executor, uses the schedule produced by the inspector phase to gather (scatter) values of a distributed array from (to) the other processors cooperating in the calculation. Such a restructuring of an irregular loop of Fig. 1(a) is shown in Fig. 1(b). In Fig. 1(b), the loop has been restructured such that each processor is responsible for some subset of the original iteration space, here shown as the symbolic bounds LL to LU.

The schedule is dependent on the local processor's iterations and the resultant access pattern and not on a specific distributed array, so as long as the access pattern (the values of IB[] in Fig. 1(b)) doesn't change the schedule can be re-used, even to access different distributed arrays. This allows some optimization of the executor since some of the information exchanged between processors need not be re-sent and thus the costs of sending the schedule initially can be amortorized through repeated use.

The nature of the data movement between processors depends on several factors including: the nature of the computation itself, the algorithms used to produce the schedule, the data distribution, and the values of the indexing array. Because of these factors, the executor might consist of several steps which gather information from the neighboring processors, scatter results back to other processors, and possibly even perform inductive operations on partial results as they are transferred.

One well-known and feature-rich implementation of the IE is the CHAOS library [11]. Part of our work has been to alter CHAOS to use the PVM communications library [4]. PVM is itself a communications library which provides a message passing facility for parallel computation on a NOW. CHAOS attempts to concentrate inter-processor communications into a well-defined point in the code, which we have shown as the Executor() call in Fig. 1(b). While this is an excellent approach for extracting fine-grained parallelism from some irregular loops, it may not be the best solution for a coarse-grained loop with loop-carried data dependencies. In Fig. 2(a), a coarse-grained loop with potential loop-carried data dependencies is shown. Our original motivation in parallelizing such loops stemmed directly from earlier work in trying to parallelize a weather modeling program [14] for parallel computation on a NOW. To address this problem, the TE was developed.

3   Multi-Threading, Threaded Executor (TE)

Our TE utilizes multi-threading to achieve better performance in a NOW environment. Multi-threading is a programming model in which a program consists of more than one thread of control. These threads share a common address space which allows threads to share data easily.

There are many different types of threading systems, but for this work, we used Pthreads [2], a standard, portable specification of a particular model of multi-threading. One big advantage to multi-threading is that if the threads are executed on an operating system which supports thread scheduling then unblocked threads may proceed while blocked threads wait for resources. Furthermore, on hardware that can execute multiple threads concurrently (such as an Symmetric Multi-Processor (SMP) machine), the threads can achieve true parallel execution.

```
                                    DO I = LL,LU
DO I = 1,N                             VAL = rfetch(B[F(IB[I],X1)])
   B[I] = G(B[F(IB[I],X1)],X2)          B[I] = G(VAL,X2)
   ...                                  ...
END DO                              END DO
```

(a) sequential code   (b) in-loop IE

Fig. 2. Loop with unknown data dependency between iterations

As mentioned previously, a loop with loop-carried data dependencies does not necessarily fit the CHAOS library's model of a synchronous executor which distributes all required values prior to the execution of the loop. The example in Fig. 2(a) is clearly more complex than the loop in Fig. 1(a). The right-hand side introduces a non-trivial set of dependencies which could clearly be across iterations. For example, in iteration i the value of the expression F(IB[I],X1) may be $i - c$, which would give rise to a data dependency between iterations i and $i - c$ with respect to the array B[] in the original, sequential loop. Since we have a lexicographical ordering of $0 < i - c < i$, the value of B[$i - c$], if sent before the loop executes (see Fig. 1(b)), may not be the value desired by iteration i. Also, if iteration $i - c$ is executed on another processor, the new value of B[$i - c$] will not reside locally prior to executing the loop.

TE [13] addresses this issue by separating the computational and communications components of the irregular computation problem into separate threads. This separation allows more parallelism to be extracted than the IE. One motivation for the TE is that the executor functionality of the IE can be distributed throughout the loop, as shown in Fig. 2(b). The local processor calls the executor function **rfetch()** which fetches the requested value from either local memory or from a cache, maintained by the communications thread, of "non-local" values as is required to satisfy the fetch request.

The communications threads, one of which runs on each node in the NOW, form a sort of distributed agent model in that the computation thread does not need to know the origin of the values it needs, nor exactly when those values become available. In other words, the computation thread is not concerned with where the values have been computed nor how they have arrived; rather, it makes requests for desired values and expects those requests to be answered (perhaps with some delay). Our conceptual distributed communication agent is therefore responsible for the distribution of values from the node upon which the value resides to the node(s) which need that value to perform some calculation.

Communications are conducted between processors via a request and response pair of messages. While the request message adds overhead, there are several reasons that this demand-driven mechanism can achieve better performance than by a more naive broadcasting of values. By requesting only the required values, the overhead and bandwidth of transferring unneeded values between nodes is saved. Secondly, if the problem is well-partitioned (most required data resides locally during calculations) then the request/response traffic can easily reduce the overall message traffic. Furthermore, if the schedule remains constant then its transfer can be amortorized through repeated use.

In [13] we describe a scheduling technique for the bundling of response messages. Essentially, we analyze the request schedule together with knowledge of the computation schedule of results and assign priorities to the requested values. As values are computed, the communications thread can either defer sending the result to other processors or send the value (bundled with any other values which were previously deferred) as soon as possible if the value has "critical" priority. In TE, computation is given priority over communication until a "critical" result is computed, at which point the priority is reversed. This allows values to propagate to their destinations as early as possible, to avoid starving the non-local computation threads. Bundling and scheduling the responses prevents the local node from spending cycles and bandwidth sending information which will not immediately help the other processor(s) progress with its (their) calculations.

From our previous experiences with the TE, we observed that increased parallelism could be exploited if it was possible to execute iterations of the loop in an *out-of-order* fashion. While the computation thread was waiting for a value from a "slow processor" before proceeding with a calculation, the values for the next iteration in the loop may have already arrived from other "fast processors" and be ready for execution. With this observation in mind, we have developed the SWE to extract this potential parallelism.

4 Sliding Window Executor (SWE)

Like the TE, the SWE uses multi-threading to separate the communication portion from the computation portion of the code. By separating these two aspects, the functional parallelism between the communication and computation components of the code can be exploited and the overall execution time of the irregular loop can be reduced, given the hardware and/or operating environment which can take advantage of threaded code. Our SWE also uses the same priority-based messaging system as the TE and described, briefly, in the previous section.

The SWE extends the TE by allowing for the *out-of-order* execution of loop iterations. Iterations are allowed to execute in a non-deterministic manner, so long as the dependencies of the original sequential code are observed. Because a NOW may be heterogeneous, loads on processors may vary greatly over the course of the calculation, and network traffic may also be bursty and impede communications, an out-of-order execution allows the local processor to proceed with the calculations of "future" iterations while, in the case of TE, the processor would be waiting for a message before continuing.

TE enforces an in-order execution. On the other extreme, one could envision a system which allows any processor to select any iteration, and so long as the dependencies are met, perform the calculation for that iteration. Unfortunately, the bookkeeping of maintaining a completely out-of-order execution would scale with the size of the original loop and would become prohibitive in terms of space. We bound the amount of information required by using the concept of a sliding window. This window has a predetermined size which both bounds the overhead and sets a limit on how much parallelism is examined at any given time. As we will see below, while this first seems like a trade-off, it is actually beneficial to limit the amount of parallelism considered.

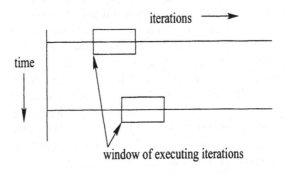

Fig. 3. The Sliding Window of Iterations

The sliding window gets its name from the way in which it passes over the iterations being considered for execution. As shown in Fig. 3, the window slides along the set of iterations as the processor makes progress on its assigned computations. The iterations of the loop which are within the window are logically in a "may be executed" state and they may be run in any order, constrained only by the original loop-carried dependencies arising in the sequential loop. Iterations to the left of the window in the figure are "committed" iterations. These iterations have been completely computed and their results have been saved to the program's data-space, including possibly overwriting the contents of certain memory locations. Iterations to the right of the window are those that still need to be calculated but are not yet under consideration for possible execution.

Perhaps surprisingly, little bookkeeping information is actually maintained for the sliding window. There must be a temporary buffer to store the results for the iterations in the window as well as two bits per iteration in the window to signal if the iteration has been run to completion. To track where the window is, we also maintain a variable holding the index of the last committed iteration. As the loop executes, the executor fills the window with iterations which "may be executed" and these are executed and committed, if possible. The communications thread monitors which iterations have been committed to schedule the transfer of values that have been computed and need to be sent to a requester.

If an iteration is blocked waiting for a value to arrive from another node, it is marked as "blocked" using one of the bits mentioned above. Once the value arrives, the `rfetch()` will succeed; the computation can be completed; and, finally, the iteration is marked "completed" (using the other bit). The iteration may then wait to be committed; because, while execution is out-of-order, we commit the results in an in-order fashion. This prevents overwriting a location with a "future" value before a logically earlier iteration has a chance to complete. The purpose of the temporary buffer then is to store these future values while completed, but not yet committed, iterations wait to be committed in-order.

In TE, we used only two threads on a node. One was responsible for the computation while the other participated in communicating the results between the nodes. However, with a window of iterations awaiting execution, the SWE can utilize a pool of computation threads to execute iterations in a truly parallel fashion. Our experiments showed that on a single CPU node, having multiple computation threads was a disadvantage because of the added context switching overhead between the competing, equivalent-priority computation threads. On the other hand, using different threads for the computation and communication can be a significant win because it allows independence between these operations (calculations can continue while the communications thread waits for the operating system to service a request) and because we can assign a priority system to the threads and thereby limit the number of context switches. Recall that we always give computation priority, unless we are waiting for a value or have computed a critical value in which case the priority is inverted. Still, a pool of computation threads can indeed improve performance given sufficient hardware resources as we will see in the next section, and this added flexibility should be considered a merit for SWE.

Assuming a single computation thread then, there are two synchronization points between the computation thread and the communication thread. The first is that values which arrive from other nodes in the system must be placed in a buffer area so that the computation thread can access them. Because the threads share the same address space, the communication thread may place the value directly into a location where the calculation can access it without copying the values. The communications thread also associates a lifetime with entries in this table. By monitoring which iterations are committed, the communications thread can retire entries from the table and thereby keep entries current.

The second synchronization point is when the computation thread calculates a critical-path value as described above. In our current algorithm, the communications thread maintains all request information from other processors. By monitoring the advance of the lower edge of the window, the communication thread can determine when such critical-path results are completed and send a message with the results to the requesting processor. Again, a more detailed discussion can be found in [13].

5   Results

Experiments have been conducted on a NOW comprised of five SPARC-10 workstations connected via an Ethernet network. The PVM library [4] has been used
for the inter-task communications interface. The programs perform a stencil
calculation using nine indirect elements as shown in Fig. 4. For the no data
dependency test, we assume X′ ≠ X whereas the tests with data dependency assume that X′ = X. This calculation was originally taken from a complex loop in a
weather modeling simulation program. That loop also performed a stencil-type
calculation and the values of IX[] vary from grid-point to grid-point. In our
results, higher workloads represent coarser granularity of the inner loop calculations, where granularity is the number of statements in the loop body. The tests
are run on ten indexing arrays (IX[]) whose values are randomly generated. The
random values of IX[] do not necessarily reflect those of the original motivation,
but they serve to eliminate bias introduced by the problem partitioning and data
distribution. The results are the averages of each of these trials.

```
FOR I=1,N
    ...
    X'[I] = FUNC(X[IX[I]-C-1], X[IX[I]-C], X[IX[I]-C+1],
                 X[IX[I]-1], X[IX[I]], X[IX[I+1]],
                 X[IX[I]+C-1], X[IX[I]+C], X[IX[I]+C+1])
ENDFOR
```

Fig. 4. The basic calculation

In Fig. 5(a), we compare the CHAOS IE with PVM as a communications
layer to both the TE and the SWE. Because CHAOS doesn't support loop-
carried dependencies, the loops for both the TE and the SWE are altered so
that they do not have loop-carried dependencies as well. In all three cases, it is
possible to send all the values that are needed in the computations prior to the
beginning the calculations (as is done in CHAOS).

From Fig. 5(a) though, we can see that TE and SWE outperform CHAOS
on larger granularity workloads. This is primarily because the communications
aspect of the calculation has been dispersed over time by moving the communications workload into the loop. This enables both the TE and SWE to begin
working on the calculations while data is still being sent over the network and
achieve a form of DOACROSS parallelism.

TE and SWE are very close in performance in this case because very little
extra parallelism can be found in the window through out-of-order execution.
Once the values have been sent out to their destinations, all three algorithms
spend their time on calculations exclusively. TE and SWE outperformed CHAOS
from 14% to 44%, depending on the granularity of the loop.

In Fig. 5(b), the relative performance of three algorithms, a brute-force broadcast (BFB), TE, and SWE, are compared on a loop with loop-carried data de-

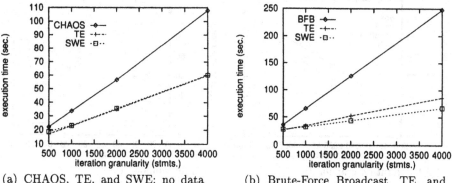

(a) CHAOS, TE, and SWE: no data dependencies

(b) Brute-Force Broadcast, TE, and SWE: with data dependencies

Fig. 5. Comparing SWE to other techniques

pendencies. The BFB is the naive approach to solving the problem, where all nodes broadcast their values to every node in the system and then proceed to perform all the calculations.[1] CHAOS does not handle arbitrary complex loop-carried data dependencies, so in this case the comparison is made with the simple brute-force approach.

The results show that both the TE and SWE perform much better than the BFB approach by taking advantage of the parallelism in the loop. SWE performs from 5% to 22% better than the TE as the granularity of the loop becomes relatively coarser. The window allows iterations to be executed out-of-order and the processor to stay busy performing useful calculations while waiting for values from "slow processors" to arrive.

At smaller granularities, however, TE outperforms SWE because there is not enough work available in a given window to keep the processor busy while it is waiting on values from other processors. Factoring in the additional bookkeeping overhead of the SWE algorithm, the TE exhibits superior performance in this case. Still, there are different strategies that might be useful in deriving benefit from out-of-order execution of the loop's iterations. For example, one might use a technique such as loop fusion [16] to increase the granularity of the loop; or, the size of the window itself can be increased and tuned so as to provide a sufficient volume of computational work for the processor to stay busy when waiting for results from other nodes.

Fig. 6 shows how window size is related to granularity size. For smaller granularities, a larger window gives better overall performance because the larger window exposes more potential parallelism. But when the granularity of the iterations increases, a smaller window can give better performance. The smaller window means that stalled iterations, iterations which are initially waiting on a

---

[1] As in the rest of this paper, we are assuming that the loop in question is part of a larger parallel code, since, obviously such a solution would perform worse than our original sequential code.

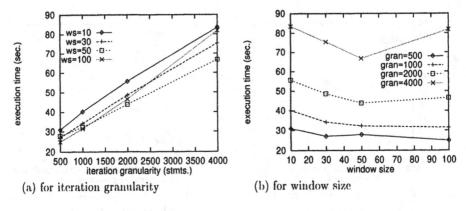

(a) for iteration granularity   (b) for window size

Fig. 6. Comparing SWE with different window sizes

value, on average will be given a chance to complete sooner and thus send their results on to other nodes which may themselves be stalled and waiting for those values to proceed. A bigger window delays the completion of stalled loops by assigning the computation thread(s) too much work to do.

Decreasing the window size too far can also negatively impact performance by restricting the amount of work too greatly and not providing enough parallelism to the system. In this situation, the processor is spending time waiting for results from other processors because it is able to complete all the available work in the window before the result arrives. By increasing and tuning the window size, we can make sure the processor spends more time doing useful work and less idle time waiting.

Another advantage of TE and SWE are that because these algorithms are threaded, they can achieve real parallelism on processing nodes which have the necessary hardware and operating system resources to exploit this characteristic. SMP workstations are becoming more prevalent and provide such hardware resources. On an SMP machine with two processors, the communication and computation threads of the TE can be run in parallel, each on its own processor, to achieve even better performance. Fig. 7 shows how this parallelism can lead to overall performance gains (in our experiments, 14% to 17%) in the TE.

The SWE algorithm, by exposing even more parallelism, allows an SMP NOW to be exploited even further. If we schedule the communications thread on one processor, worker threads can be running in parallel on any other processors in the SMP machine as well as running on the same processor as the communications thread when the communications thread is blocked or otherwise idle. In our experiments on a two-processor SMP NOW with two worker threads and a communications thread, SWE outperformed TE on a single processor by 45% to 55% and the SMP TE by 35% to 45%.

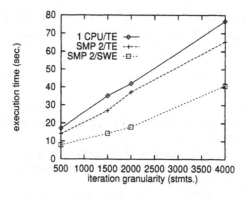

Fig. 7. TE, SWE on SMP network

6 Conclusion

We have presented an extension to the TE, called the SWE. SWE takes advantage of multi-threading to improve independence between communication and computation which can be exploited as true parallelism with hardware support such as SMP. Furthermore, multi-threading can improve performance on a single-CPU node in a NOW by allowing the computation and communication parts of the task to be scheduled more efficiently. For example, while one thread is blocked waiting for the network to become available, another thread can continue to work on the subsequent calculations in its work-list.

Also like TE, SWE allows loops with data-dependencies to be converted into a parallel form which can potentially improve overall performance by allowing (some) calculations to precede communications. This can improve performance in a coarse-grained, NOW environment over a synchronous Executor phase which runs prior to starting the calculations loop. Calculations that have all inputs available can proceed immediately, thus eliminating the barrier that no computation begins until each node has sent and received all requested values from the other participating nodes.

The big advantage of the SWE over the TE is that it allows iterations within the loop to be performed out-of-order. Iterations within a designated window are executed in any order which allows more parallelism and can improve performance of an irregular loop with loop-carried dependencies even further. Essentially, the window allows iterations which are ready to continue to be worked on by the processor while other iterations may be blocked and waiting for the communications thread to provide values from another processor. Additionally, the window serves to limit the amount of work which is considered at any given time. If too much work is available, the computation thread(s) may spend too much time working on "future" values and not commit critical values in a timely enough manner to transfer them to other processors. Thus, tuning the window size is quite important for performance gain. The other advantage of SWE is it naturally allows for a pool of worker threads to perform computation in parallel given proper hardware resources.

Acknowledgement: We thank the anonymous referees for their time and valuabled inputs which greatly helped us to improve this paper.

References

1. Agrawal, G., Sussman, A., and Saltz, J.: An integrated runtime and compile-time approach for parallelizing structured and block structured applications. Trans. Par. Dist. Sys. **6(7)** (1995) 747–754
2. Butenhof, D.: *Programming with POSIX Threads.* Addison-Wesley (1997)
3. Chong, F., et. al.: Multiprocessor Support for Fine-grained Irregular DAGs. Par. Proc. Let. **5(4)** (1995) 671–683
4. Geist, A., et. al.: *PVM 3 User's Guide and Reference Manual.* Oak Ridge National Lab. Tech. Rep. ORNL/TM-12187 (1994)
5. Haddaya, A. and Park, K: Mapping parallel iterative algorithms onto workstation networks. Int'l Symp. High Perf. Dist. Comp. (1994) 211–218
6. Horie, T. and Kurame, H.: Evaluation of parallel performance of large scale computing using workstation network. Comp. Mech. **17(4)** (1996) 234–241
7. Koelbel, C., Mehrotra, P., and Van Rosendale, J.: Supporting shared data structures on distributed memory machines. Symp. Prin. Prac. Par. Prog. (1990) 177–186
8. Mirchandaney, R., et. al.: Principles of runtime support for parallel processors. Int'l Conf. Supercom. (1988) 140–152
9. Ponnusamy, R., et. al.: An integrated runtime and compile-time approach for parallelizing structured and block structured applications. Trans. Par. Dist. Sys. **6(8)** (1995) 737–753
10. Rauchwerger, L. and Padua, D.: The privatizing doall test: a run-time technique for doall loop identification and array privatization. Int'l Conf. Supercom. (1994) 33–43
11. Saltz, J., et.al. *A Manual for the CHAOS Runtime Library* U. of Maryland: Dept. of Comp. Sci. Tech. Rep. CS-TR-3437 (1995)
12. Saulnier, E.T. and Bortscheller, B.: Data transfer bottlenecks over SPARC-based computer networks. Conf. Local Comp. Net. (1995) 289–295
13. Schweitz, E. and Agrawal, D.: A multi-threaded distributed agent for coarse-grained irregular computation. Int'l Conf. Par. Dist. Comp. Sys. (1997) 285–290
14. Semazzi F., et. al.: A global nonhydrostatic semi-Lagrangian atmospheric model without orography. Mon. Weather Rev. (1995) 2534–2544
15. Subhlok, J., et. al.: Exploiting task and data parallelism on a multicomputer. Sym. Prin. Prac. Par. Prog. (1993) 13–22
16. Wolfe, M.J.: *High Performance Compilers for Parallel Computing.* Addison-Wesley (1996)
17. Wu, J., et. al.: Distributed memory compiler design for sparse problems. Trans. Comp. **44(6)** (1995) 737–753

Parallel Profile Matching for Large Scale Webcasting

Matthias Eichstaedt[1] and Qi Lu[1] and Shang-Hua Teng[2*]

[1] IBM Almaden Research Center, 650 Harry Road, San Jose, CA 95120
[2] Department of Computer Science, University of Illinois, Urbana IL 61801, USA.

Abstract. Profile matching is a key problem for webcasting systems. In such a system, each user has a "personal" profile. Each information document is matched with profiles in a profile-database, and pushed only to those users whose profiles match the content of the document. We present an efficient and scalable parallel profile matching algorithm which can handle a large subscription volume and diversity of information content. Our algorithm automatically partitions the profile database for load balancing and minimizes the interaction among processors in parallel profile matching. It has a dynamic load balancing mechanism for handling profile updates. We describe the implementation of our algorithm in the context of the Grand Central Station (GCS) project at the IBM Almaden Research Center. The initial performance evaluation indicates that parallel profile matching can be scaled up gracefully via dynamic adaptation.

1 Introduction

Webcasting, or Internet push, has attracted growing attention because of its potential for re-shaping the Internet. Pioneered by PointCast [1] and other ensuing products [2, 3, 4], webcasting systems automatically deliver information to the users based on user profiles. Information that is frequently updated and that is of regular interest to the users becomes prime target for webcasting delivery such as headline news and stock quotes. More and more digital content is being offered in webcasting channels to Internet users or in corporate intranets for productivity enhancement.

One of the main problems in webcasting today is the lack of personalization. As most webcasting users can attest to, a subscribed channel often contains a significant amount of information that is irrelevant to their interest. For example, users cannot customize their subscription to only receive information about their favorite teams when subscribing to a sports channel. Moreover, the bandwidth wasted by delivering irrelevant content exacerbates the burden on the network infrastructure, thus preventing widespread deployment. The solution is to enable users to filter subscribed channels according to their needs via their profiles and more importantly to match profiles against available content on the server side. Thus, only information pertaining to the users personal interest needs to be displayed and delivered over the network, which significantly enhances usability while reducing network traffic.

At the heart of content personalization is the profile matching problem. Each user has a "personal" profile that is represented as a logical expression over primitive

---

* Part of the work is done while at IBM Almaden Research Center.

predicates. A profile matching algorithm matches an information document with all profiles in a profile-database and return those profiles whose boolean expression is satisfied by the content of the document. This is a computationally expensive problem because of the large volume of subscriptions and the diversity of information content [5, 6]. Furthermore, webcasting systems must perform profile matching "on-line" because usually they have to deliver information in a timely fashion. In the same vein, they have to maintain a dynamically changing profile database.

Therefore, our first objective is scalability because webcasting systems can typically expect tens of thousands, even up to millions, of subscribers as witnessed by PointCast. It is imperative that the profile engine has the ability to be gracefully scaled up. This enables the profile engine to absorb new subscribers without severely degrading performance. The second objective is high performance because delay-sensitive information such as stock quotes must be matched and delivered in a timely manner. The third objective is adaptability because the profile matching mechanisms must dynamically adjust to evolutions in digital content and user profiles to sustain high scalability and performance.

We present a fast parallel profile matching algorithm. Our algorithm has been implemented and used for matching a variety of digital content against a large collection of user profiles in the context of the Grand Central Station (GCS) [7, 8], a recent project at the IBM Almaden Research Center. It has been built upon our previously developed sequential profile matching program. It automatically partitions the profile database for load balancing and minimizes the interactions among processors. It has a dynamic load balancing mechanism for handling profile updates.

We first give a brief architectural overview of GCS to provide the context of our profile matching algorithm and its implementation. GCS combines both information discovery and webcasting-based information dissemination into a single system. Figure 1 illustrates the GCS information discovery infrastructure which contains a collection of Gatherers/Collectors to systematically gather data from any source in any format and summarize them into an XML format [14].

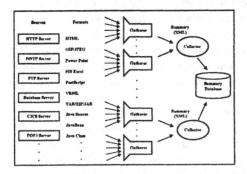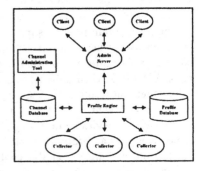

Fig. 1. IBM GCS Information Discovery and Webcasting Architecture

The heart of GCS webcasting is the *profile engine* (see Figure 1), which maintains a large *profile database* and matches it against incoming data received from GCS *Collectors*. Data satisfying certain profiles will automatically be delivered to the corresponding users by the *admin server*. Users interact with the client to subscribe to web channels by specifying filters to personalize a subscribed channel. The profile engine consults a channel database to automatically compile data into a hierarchy of channels, defined by the webcasting system as well as by individual users.

Section 2 defines the profile matching problem and the multi-way profile partitioning problem. The goal is to reduce the large scale matching problem to a collection of independent matching problems. Section 3 provides a multi-way profile partitioning algorithm and its use in our parallel profile matching program. Section 4 addresses the issue of dynamic load balancing. Section 5 discusses certain implementation issues and presents some experimental results in the context of GCS.

2 Profile Matching

We first discuss the profile language used in IBM's GCS in order to provide some background for this paper.

2.1 The Profile Language

Users express their interests using a profile language (GCSPL), which is designed after Rufus [15]. The content filters are incorporated into channels of interests which can be defined by a channel administration tool; these channel expressions are automatically converted into profile language expressions. GCSPL is a Boolean structured language employing parameterized predicates as the tool for content selection. Its extensible structure is designed to accommodate the diversity of webcasting content. If an image channel wants to allow users to filter content based on image characteristics, we simply need to extend GCSPL with a set of image selection predicates. The following list gives a subset of the primitive predicates in GCSPL. The final product will support predicates ranging from image content (e.g., shapes and color patterns) to document classification. Assuming I is the data document, then our primitive predicates have the following semantic.

- Keyword(x): I contains the keyword x.
- Adj(x, y): Keywords x and y are adjacent in I in the given order.
- Adj.$n(x, y)$: Same as Adj(x, y) except that x and y can be n words apart.
- Near(x, y): Keywords x and y are adjacent in I in any order.
- Near.$n(x, y)$: Keywords x and y are at most n word apart in I in any order.
- Channel(x): I satisfies the definition of channel x.
- MediaType(x): The media type of I is x.
- Time(x): I is last modified within x days.
- Source(x): The URL of I is x.

The keywords x and y may be qualified with parameters such as x.f (first letter of x capitalized), x.i (Ignore case in x), x.s (Stemming for x), and x.r (Respect case in x). Using GCSPL, a sample profile can be:

"Channel(AlmadenEvents) AND Adj(database.i, technology.i) OR Channel(Sports) AND Keyword(baseball)".

And the definition of channel AlmadenEvents can be:
"Source(nntp://news.almaden.ibm.com/ibm.csconf.talks) OR Source(nntp://news.almaden.ibm.com/ibm.almaden.calendar)".

2.2   The Profile Matching Problem

Let $P = \{p_1, ..., p_m\}$ be a set of primitive predicates. Each predicate p_i has a cost or a weight $w(p_i)$ measuring the average cost to evaluate the predicate. Let $E = \{e_1, ..., e_n\}$ be a set of profile expressions. Let I be a data document. The profile matching problem is to return the set of indices for all profile expressions that is satisfied by I.

Each profile expression has a Boolean tree representation, in which each leaf is labeled with a primitive predicate in P and each internal node is labeled with a Boolean operator (AND/OR/NOT). We can merge common sub-expressions to compress a Boolean tree to an equivalent Boolean DAG in which each primitive predicate can occur at most once. Figure 2 illustrates a compressed Boolean DAG for the following profile expressions.

e_1=(Channel(IBM-news) AND Keyword(database)) OR (Channel(Sports) AND Keyword(baseball.i))
e_2=(Channel(Sports) AND Keyword(baseball.i)) OR (Channel(Seminars) AND Keyword(database))

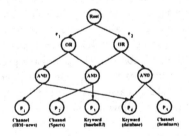

Fig. 2. A Compressed Boolean DAG

Webcasting systems typically expect millions of subscribers. It is imperative that the profile engine has the ability to be gracefully scaled up. The compressed DAG representation enables the profile engine to absorb new subscribers without severely degrading performance because users have shared interests. Given a set of profile expressions E, we can create a single compressed DAG, in which each expression of E is associated with a internal node of DAG. To evaluate expressions in E against a document, we only need to evaluate each primitive predicate once. After reaching a certain scale, most common interests expressed in predicates such as Channel(Sports) and Keyword(baseball) will have ample representations. This means that admitting new subscribers is unlikely to introduce too many new predicates or sub-expression, allowing the system to be scalable.

3   Parallel Profile Matching Algorithms

Suppose $E = \{e_1, ..., e_n\}$ is a set of profile expressions over a predicate set $P = \{p_1, ..., p_m\}$ whose weights are given by $W = \{w(p_1), ..., w(p_m)\}$. Let D be the reduced DAG for expressions in E. We have k processors and we decompose D into k DAGs $D_1, ..., D_k$ and map each sub-DAG to a processor. The problem of decomposing D into k DAGs is called the k-way profile partitioning problem.

The main cost of the calculation is the evaluation of primitive predicates. To achieve the best possible speedup we need to balance the work across multiple processors. We also need to minimize the dependency or the redundancy among the processors in order to simplify the parallel profile matching program and to reduce the overhead. Our scheme explores a trade-off between redundancy of evaluation and the simplicity of the parallel program. We present two approaches to conduct the matching via partitioned profiles.

3.1   Multi-Way Profile Partitioning

In this first approach, each profile expression is completely contained in one of the sub-DAGs, i.e. for each i, there exists a j e_i is completely contained in D_j.

> PROFILE PARTITIONING: Given a collection of profile expressions $E = \{e_1, ..., e_n\}$ over a set of primitive predicates $P = \{p_1, ..., p_m\}$ with respective costs/weights $\{w(p_1), ..., w(p_m)\}$ and an integer k, construct k DAGs, $D_1, ..., D_k$, such that: For each i, there exists j such that e_i is contained in D_j and $\max_{j=1}^{k} w(D_j)$ is minimized, where $s(D_j)$ is the total cost of the predicates in D_j.

In this approach, a primitive predicate could be mapped to and evaluated by more than one processor. The main objective is thus to minimize the number of duplicates and balance the work across the processors. With this partition, the parallel profile matching method is to (1) map D_i onto processor i; (2) broadcast a document I to all processors; and (3) execute the sequential profile matching algorithm with processor i on I and D_i. In other words, after the partitioning step, the profile matching problem is decomposed into k independent subproblems so that processors can collaboratively conduct the matching without any additional communication. This scheme simplifies the parallel program and makes it more portable over many parallel, distributed platforms.

In the second approach, processors first work together to evaluate the set of predicates that are shared by more than one sub-DAG. Then each processor evaluates the rest of each own DAG by running a sequential algorithm.

A primitive predicate p_s is a boundary predicate in $D_1, ..., D_k$ if p_s is contained in more than one sub-DAG. Let B be the collection of boundary predicates. There are several alternatives for processors to collaboratively evaluate boundary predicates.

- *Uniform Partition followed by broadcast*: (1) divide B into k subsets $B_1, B_2, ...,$ B_k such that $\max_{i=1}^{k} w(B_i)$ is minimized; (2) map B_i onto processor i, where $i = 1...k$; (3) when receiving a document I, evaluate I against predicates in B_i on processor i; (4) each processor broadcasts its results to all other processors; (5) each processor individually completes the evaluation of I based on the results received from the $k - 1$ other processors.

- *Group partition followed by pointwise communication*: Each predicate in B belongs to a set of processors. One of these processors evaluates the predicate and sends the result to the others that have this boundary predicate. (1) map D_i onto processor i; (2) Broadcast a receiving I to all processors; (3) assign each boundary predicate p_s to a processor i, and processor i evaluate p_s and send the result to all processors whose DAGs contain p_s; (4) each processor performs the sequential profile matching algorithm on I and D_i.

Comparing with the first approach to profile evaluation, the second approach could be more efficient because boundary predicates are evaluated only once, but at the expense that the profile matching programs on all processors need to be synchronized when a document is received in order to collaboratively evaluate the boundary predicates. An additional communication step for broadcasting the values of boundary predicates is also needed. The first approach, in addition to its simplicity, is more suitable to the *lazy evaluation* rule that is used in sequential profile matching. The lazy evaluation rule works as follows. If one operand of an AND operator is false (of an OR operator is true), then we do not need to evaluate its second operand. With lazy evaluation, we might not need to evaluate all predicates in order to evaluate a Boolean DAG. The lazy evaluation based sequential program can be incorporated into the first approach directly. However, in the second approach, we have to evaluate all boundary predicates before a lazy evaluation based sequential program is applied.

3.2   A Set Decomposition Problem

The profile partitioning problem can be modeled by an elegant combinatorial optimization problem, which we call SET DECOMPOSITION.

> SET DECOMPOSITION: Given a set $\Gamma = \{g_1, ..., g_m\}$ with weights $\{w_1, ..., w_m\}$ and subsets $A_1, ..., A_n$ of Γ and an integer k, construct k subsets of Γ, $G_1, ..., G_k$ such that for each i there exists j such that $A_i \subseteq G_j$. We would like to minimize $\max_{j=1}^{k}(\sum_{s \in G_j} w(s))$.

There is a natural mapping from PROFILE PARTITIONING to SET DECOMPOSITION. Given the Boolean expressions $E = \{e_1, ..., e_n\}$ over primitive predicates $P = \{p_1, ..., p_m\}$, we divide E into k subsets $E_1, ..., E_k$. We then form the DAG D_i for each subset E_i. Each expression $e \in E$ is defined as a subset of P. The primitive predicates are mapped to the set Γ, the cost for evaluating a predicate is mapped to its, and the subset of predicates used by expression e_i is mapped to A_i.

SET DECOMPOSITION is NP-hard. We present several approximation/heuristic algorithms to this problem. In Section 5, we show that these algorithms perform well on our experimental data.

Greedy Clustering Our starting point is a greedy scheme, motivated by list scheduling [13]. For $B \subseteq \Gamma$, let $w(B) = \sum_{g_j \in B} w_j$ be the total weight of elements in B. For $B, C \in \Gamma$, let $overlap(B, C) = w(B \cap C)$. We have $w(B \cup C) = w(B) + w(C) - overlap(B, C)$. Let π be a permutation of $\{1, \ldots, n\}$. π can be given by sorting $w(A_1), \ldots, w(A_n)$ in descending ordering. We permutate the set A with π and now assume A is given in the order of the permutation. Below, initially, $G_1, \ldots, G_k$ are empty sets.

Algorithm Greedy-Mapping
1. For $i = 1$ to n,
    (a) For each $j = 1, 2, \ldots, k$, let
        $fitness(j) = w(A_i \cup G_j) = w(A_i) + w(G_j) - overlap(A_i, G_j)$.
    (b) Let j be the index with the smallest fitness.
    (c) $G_j = G_j \cup A_i$.
2. Return $G_1, \ldots, G_k$.

We first show that this greedy approach performs well in the case where sets do not have "too much" overlap. Define the overlap rate, denoted by ply, as

$$ply(A) = (\sum_{j=1}^{n} w(A_i)) / w(\cup_{j=1}^{n} A_j).$$

Let W_{opt} be the size of the largest set in the optimal solution of SET DECOMPOSITION. Let W_{greedy} be the size of the largest set produced by the greedy algorithm above.

Lemma 1.

$$W_{greedy} \le (ply(A) + 1)W_{opt}.$$

Proof: There are two lower bounds for the weight of all sets in SET DECOMPOSITION:

1. Each set has at least a weight of $W_{max} = \max_i w(A_i)$.
2. Each set has at least a weight of $W_{avg} = w(\cup_{i=1}^{n} A_i)/k$.

We can show that $W_{greedy} \le ply(A)W_{avg} + W_{max} \le (ply(A) + 1)W_{opt}$. □

In practice, as shown in Section 5, the greedy method performs much better than the theoretical result above. First, its performance is not very sensitive to W_{max} if π is the permutation for descending ordering of cost values. Secondly, the algorithm exploits a large amount of overlaps within individual partitions, making its performance much closer to the optimal.

The simplicity and efficiency of *Greedy-Mapping* makes it very attractive for practical implementations. If we use a hash function to compute the overlap between two sets, then we can implement the algorithm in $O(k \sum_{i=1}^{n} |A_i|)$ time. We now present several improved algorithms for profile partitioning. Because the number of user profiles is of a large scale (e.g., 1,000,000 or more), we will ensure all our refined algorithms take no more than $O(k \sum_{i=1}^{n} |A_i|)$ time.

Giving overlaps more credit Greedy-Mapping may not produce a solution close to the optimal solution in the worst case. We give one such "bad" example to motivate our improved methods.

Example: Let $A_{1,1}, A_{1,2}..., A_{k,k}$ be $n = k^2$ sets, where $A_{i,j} = \{g_{i,1}, g_{i,2}, ..., g_{i,k}, b_j\}$. Notice that A_{i,j_1} and A_{i,j_2} differ by only one entry, b_{j_1} versus b_{j_2}. The optimal decomposition is to merge all $A_{i,*}$ to give G_i of size $2k$. Greedy-Mapping, however, first assigns $A_{1,j}$ to G_j, for all $1 \leq j \leq k$. Then it iteratively assigns $A_{i,j}$ for all $2 \leq i \leq k$ and $1 \leq j \leq k$. The cost of its decomposition is more than $k^2 + 1$ and its ratio to the optimal solution is $k/2$.

Looking closely, Greedy-Mapping assigns $A_{2,2}$ to G_2 instead of G_1, because the merge of $A_{2,2}$ with G_2 (which is $A_{1,2}$) has size $2k+1$ and overlap 1, while the union of $A_{2,2}$ with G_1 has size $2k+2$ but overlap k. The following algorithm shows a better trade-off between the overlap and the size, by giving the overlap more credit. Let $\beta \geq 1$ be a constant in the following algorithm.

Algorithm (β-Mapping)
1. For $i = 1$ to n,
    (a) for each $j = 1, 2, \ldots, k$, let
    $$fitness(j) = w(A_i \cup G_j) = w(A_i) + w(G_j) - \beta \cdot overlap(A_i, G_j).$$
    (b) Let j be the index with the smallest fitness.
    (c) $G_j = G_j \cup A_i$.

When $\beta = 1$, β-Mapping is Greedy-Mapping. If we choose $\beta = 1.1$ and apply β-Mapping to the example above, we find a decomposition of size $3k$, because it assigns $A_{i,j}$, for all $1 \leq j \leq k$ together with $A_{1,i}$ to G_i. Interestingly, $A_{1,i}$ is assigned to G_i because it was assigned as the first k sets. However, if the first k sets in the permutation are $A_{1,1}, A_{2,1}, ..., A_{p,1}$ then the 1.1-mapping finds the optimal decomposition. We need to have a better strategy to choose the first k sets!

β-Mapping enjoys the same simplicity and efficiency. Proper choice of β gives it much better performance over Greedy-Mapping. *How big should we choose β?* Our experiments seem to suggest that there are a two constants α_1 and α_2 such that the performance depends on β according to $\alpha_1\beta + \alpha_2/\beta$. If β is too large, say ∞, then β-Mapping tends to assign all sets into one component. The example above shows that a too tiny β may not give enough credit to overlaps. Such observation is very important since we do not know the best β in advance. If we model the cost function as $\alpha_1\beta + \alpha_2/\beta$, we can determine the parameter α_1 and α_2 after two choices of β, we then apply β-Mapping with $\beta = \sqrt{\alpha_2/\alpha_1}$. Our scheme can find the better β according to the structure of the profiles.

Clustering: plant the first k seeds The performance of Greedy-Mapping and β-Mapping are sensitive to the assignment of the first k sets, called the seeds of the components. Here our method is motivated by multi-way graph partitioning via clustering. To partition a graph into k subgraphs, the clustering method starts with k nodes as seeds and grows all seeds simultaneously to obtain k-equally sized subgraphs. At each step, each seed tries to include its neighboring nodes into its components; conflicts are resolved by making edges local to each component. For better partitions, the k seeds should be pairwisely "far" away from each other.

Define the overlap ratio $\phi(A_i, A_j)$ of A_i and A_j as $overlap(A_i, A_j)/\sqrt{w(A_i)w(A_j)}$. Two subsets are "close" if their overlap ratio is large. We can choose k mutually far-away seeds as follows: Let S_1 be the set with the largest weight. For $i = 2$ to k, let S_i be the set with the largest weight among those whose overlap ratio with $G_1, \ldots, G_{i-1}$ is the smallest. Several related heuristics for finding mutually far away seeds have been shown to have good mathematical guarantee on their performance [11]. After selecting the first k seeds, we then apply β-Mapping to the rest of the sets.

Incremental Clustering We can extend the algorithm for finding k mutually far away seeds one step further. We use the two natural lower bounds on SET DECOMPOSITION, $W_{max} = \max_i w(A_i)$ and $W_{avg} = (\sum_{i=1}^n c(A_i))/k$.

Let $W = \max(W_{max}, W_{avg})$. Choose a constant $1 < \alpha < 2$.

> **Algorithm (Incremental-Clustering)**
> 1. For $i = 1$ to k,
>   (a) If $i = 1$ then let j be the index of the set in A with the largest cost; otherwise, let j be the index of the set in A with the largest cost among those whose overlap ratio with $G_1, \ldots, G_{i-1}$ is the smallest. Let $G_i = A_j$. Remove A_j from A.
>   (b) While $(w(G_i) < \alpha W)$ and A is not empty:
>   Let j be the index of the set in A whose overlap ratio with G_i is the largest. If $w(G_i \cup A_j) \leq \alpha W$, then let $G_i = G_i \cup A_j$. Remove A_j from A; otherwise exit the while loop.
> 2. If A is not empty, apply β-Mapping to the rest of sets in A.

Incremental Clustering grows one cluster at a time until a certain size is reached. We then choose a seed that is farthest away from the current clusters to grow a new cluster. We then apply β-Mapping to map the rest of the sets.

Other Heuristics We can apply other optimization techniques such as simulated annealing and the multilevel method [12]. We can also apply local optimization techniques such as swap to improve the quality of the decomposition.

4 Dynamic Nature of Profile Matching: Updates and Load Re-balancing

The profile matching problem is a dynamic problem: New users continuously subscribe to the webcasting system, introducing new profiles and predicates. The current users might modify their profiles and the average cost for evaluating each primitive predicate might change, introducing load imbalance among the processors. In addition, new processor could be available for profile matching or some existing processors may quit, all introducing load imbalance.

In our system, conceptually, we have a front-end processor and a collection of k processors for profile matching. The front-end processor is responsible for receiving data documents and broadcast them to the k processors. It also maintains the interface for profile addition and updates and the profile partition. More details about the system design will be discussed in Section 5.

When the front-end processor receives a new profile, it assigns the profile to a matching processor. When users update their profile, the front-end processor needs to inform the matching processors of the updates. When the load becomes very unbalanced either because of the change of cost or because of the addition of new matching processors, the front-end processor needs to generate a new partition or rebalance the existing partitions. The following properties are desirable for dynamic update and load re-balancing algorithms: (1) **Low impact updates**: The addition of a new profile and the update of an existing profile should have a minimum impact to most of the matching processors. (2) **Parallel Load Re-Balancing**: The load re-balancing algorithm itself should be efficient and parallel itself.

When a new user subscribes to the system with a profile expression e, the front-end processor can apply either Greedy-Mapping or β-Mapping to assign e to one of the matching processors. In algorithms, the addition of new profiles only has impact on one processor. This is one of the reasons why this set of algorithms is attractive for practical profile partitioning. Moreover, the quality of the new partition is consistent with the partition generated by repartitioning the profiles.

When a user modifies her profile expression from e to e', we need to delete e from the processor whose partition contains e and reassigns e' using Greedy-Mapping or β-Mapping. Therefore, not more than two processors are involved.

When the load of the current partition is very unbalanced due to profile updates and cost function updates, we could repartition the profiles from scratch, or apply local optimizations. Local optimization refers to a technique where profiles are moved from an overloaded processor to one with less workload. Our strategy is to hybridize these two approaches. Collaboratively, all processors compute the current $W = \max(W_{max}, W_{avg})$. Let $S_1 \ldots, S_k$ be the current partition. For each i, processor i applies the first step of Incremental-Clustering to the profiles in S_i to create a new S_i whose cost is no more than αC. Then each processor i returns the unassigned profile in S_i back to the front-end processor. Notice that these unassigned sets are very small in general. The front-end processor then applies Greedy-Mapping or β-Mapping to assign the rest of the profiles, which are not assigned in the previous step. Note that processors can perform the first step in parallel. Then second step can be treated as adding new profiles to the system.

5 Implementation and Experiments

Our profile matching algorithm is implemented and used in IBM's GCS project. The GCS profile matching algorithms, along with the rest of the GCS system, have been implemented using the Java programming language. This section presents an initial performance evaluation of our method.

The performance evaluation focuses on verifying whether the main design objectives of our profile matching algorithms have been achieved. Specifically, we would like to address the following issues.

- What is the ratio of the largest partition to the total size of the profile database as a function of k? This ratio determines the performance improvement of our parallel profile-matching program over the sequential one.
- What is the impact of the ordering of profiles on the quality of the partition generated by Greedy-Mapping and β-Mapping. We would like to show that our program is not too sensitive to the ordering because for dynamic profile updating and partitioning, we cannot control the nature of the ordering.
- What is the dependency of the partitioning quality on the choice of β in β-Mapping.

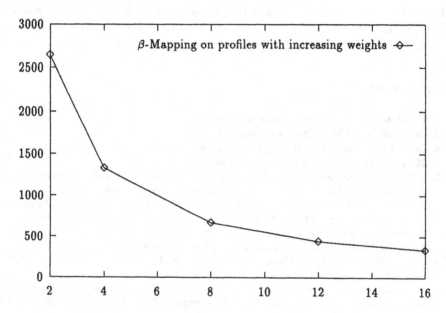

Fig. 3. The size of the largest partition as a function of β.

Figure 3 shows an experiment with β-Mapping ($\beta = 1.1$); the x-axis is for the number of processors k while the y-axis is for the weight of the largest partition generated by β-Mapping. Due to the lack of current deployment size, we use a random procedure to generate profiles of varying costs and sizes. Our experiment uses a database of 5000 profiles generated by this random profile generation program over the basic predicates defined in Section 2.1, in which we assume the following costs: Keyword(): 20, Adj(,): 300, Adj.n(,): 100, Near(,): 200, Near.n(,): 100, Channel(): 5, MediaType(), Time(), Source(): 1. The unit for the y-coordinate of Figure 5 is 1000. Among 5000 random profiles, 30% are large profiles in the sense that they contain 30 to 60 text predicates. The other 70% are small profiles which contain 0 to 2 text predicates. The word parameters in text predicates are randomly selected from a dictionary containing about 3500 words collected from articles in the several popular newsgroups such as "comp.lang.java.programmer". The random selection is weighted with word occurrence frequency. Notice that the same predicate symbol on different keywords defines different basic predicates. In addition, each profile contains 1 `Channel()`, 0 to 2 `MediaType()`, 0 or 1 `Time()`, and 0 to 2 `Source()` predicates. The *ply* of the profile database is about 3.17. The experiment shows a close to optimal speed-up even if we apply β-Mapping on profiles with increasing weights, which is considered to be one of the worst kinds of orderings. When we order the profiles with decreasing weights, we got a little better result. In our final version, we hope to include experiments from a real deployment.

6 Acknowledgment

The authors wish to thank the entire GCS team, particularly John Thomas and Joe Gebis, for their support in GCS webcasting development. We would like to especially thank Professor Gunter Schlageter for his encouragement and support.

References

1. PointCast Inc., 1997, http://www.pointcast.com/
2. BackWeb Inc., 1997, http://www.backweb.com/
3. DataChannel Inc., 1997, http://www.datachannel.com/
4. Marimba Inc., 1997, http://www.marimba.com/
5. T.W. Yan, H. Garcia-Molina, A tool for wide-area information dissemination, *Proceedings of the 1995 USENIX Technical Conference*, pages 177-86, 1995.
6. T.W. Yan, H. Garcia-Molina, Structures for selective dissemination of information under the boolean model. *ACM Transactions on Database Systems*, 19(2):332-64, 1994.
7. Information on the Fast Track, *IBM Research Magazine*, Vol. 35, No.3, 1997: 18-21.
8. IBM: All searches start at Grand Central, *Network World*, Nov. 11, 1997, 1-2.
9. The Harvest Information Discovery and Access System, 1996, http://harvest.transarc.com/
10. C.M. Bowman, P.B. Danzig, D.R. Hardy, U. Manber and M.F. Schwartz, The Harvest information discovery and access system, *Computer Networks and ISDN Systems 28* (1995) pp. 119-125.
11. S.-H. Teng. Greedy algorithms for low energy and mutually distant sampling. *J. Algorithms*, 1998.
12. S.-H. Teng. Coarsening, sampling, and smoothing: elements of the multilevel method. The IMA Volumes in Mathematics and Its Applications, R. Schreiber ed. Springer-Verlag, 1998.
13. R.L. Graham. Bounds for certain multiprocessor anomalies, *Bell System Technical Journal*(1966) 1563-1581.
14. Extensible Markup Language (XML), 1997, http://www.w3.org/XML/
15. K. Shoens, A. Luniewski, P. Schwarz, J. Stamos, J. Thomas, The Rufus system: information organization for semi-structured data, *Proceedings of 19th International Conference on Very Large Data Bases*, Dublin, Ireland, Aug. 1993.
16. S. Wu, U. Manber, A fast algorithm for multi-pattern searching, Technical Report TR94-17, Department of Computer Science, University of Arizona, Tucson, May 1994.

Large-Scale SVD and Subspace-Based Methods for Information Retrieval*

Hongyuan Zha, Osni Marques, and Horst D. Simon

Lawrence Berkeley National Laboratory/NERSC, Berkeley, CA 94720
{zha,osni,simon}@nersc.gov

Abstract. A theoretical foundation for *latent semantic indexing* (LSI) is proposed by adapting a model first used in array signal processing to the context of information retrieval using the concept of subspaces. It is shown that this subspace-based model coupled with *minimal description length* (MDL) principle leads to a statistical test to determine the dimensions of the latent-concept subspaces in LSI. The effect of weighting on the choice of the optimal dimensions of latent-concept subspaces is illustrated. It is also shown that the model imposes a so-called *low-rank-plus-shift* structure that is approximately satisfied by the cross-product of the term-document matrices. This structure can be exploited to give a more accurate updating scheme for LSI and to correct some of the misconception about the achievable retrieval accuracy in LSI updating. Variants of Lanczos algorithms are illustrated with numerical test results on Cray T3E using document collections generated from World Wide Web.

1 Introduction

Latent semantic indexing (LSI) is a concept-based automatic indexing method that aims at overcoming the fundamental synonymy and polysemy problems which plague traditional lexical-matching indexing schemes [4]. LSI is an extension of the vector-space model for information retrieval [15]. In the vector-space model, the collection of text documents is represented by a *term-document* matrix $A = [a_{ij}] \in \mathcal{R}^{m \times n}$, where a_{ij} is the number of times term i appears in document j, and m is the number of terms and n is the number of documents in the collection. Consequently, a document becomes a column vector, and a user's query can also be represented as a vector of the same dimension. The similarity between a query vector and a document vector is usually measured by the cosine of the angle between them, and for each query a list of documents ranked in decreasing order of similarity is returned to the user. LSI extends this

---

* This work was supported by the Director, Office of Energy Research, Office of Laboratory Policy and Infrastructure Management, of the U.S. Department of Energy under Contract No. DE-AC03-76SF00098 and NSF grant CCR-9619452. Hongyuan Zha's permanent address: 307 Pond Laboratory, Department of Computer Science and Engineering, The Pennsylvania State University, University Park, PA 16802-6103. zha@cse.psu.edu.

vector-space model by modeling the term-document relationship using a *reduced-dimension representation* computed by the singular value decomposition (SVD) of the term-document matrix A. Specifically let

$$A = P\Sigma Q^T, \quad \Sigma = \text{diag}(\sigma_1, \ldots, \sigma_{\min(m,n)}), \quad \sigma_1 \geq \cdots \geq \sigma_{\min(m,n)}, \quad (1)$$

be the SVD of A. Then the representation is given by the best rank-k approximation $A_k \equiv P_k \Sigma_k Q_k^T$, where P_k and Q_k are formed by the first k columns of P and Q, respectively, and Σ_k is the k-th leading principal submatrix of Σ. Corresponding to each of the k reduced dimensions is associated a latent-concept which may not have any explicit semantic content yet helps to discriminate documents [1, 4].

The effectiveness of LSI measured by, for example, increased average precision, has been demonstrated for several text collections [4]. There are also text collections for which LSI is not significantly better than the much simpler original vector-space model. However, very little theoretical results have been derived for LSI except some explanation of its effectiveness based on multidimensional scaling and multiple regression. The first research issue of this paper is to place LSI on a firm theoretical foundation by adapting a subspace-based model first used in array signal processing to the context of IR [18, 19]. Our model explicitly exhibits the *linear* relations between terms and the so-called latent concepts. It also leads to a statistical test that can be used to determine the optimal dimension k used in Equation (1). The model we developed serves as a connection between IR and the field of numerical linear algebra and gives novel computational methods for LSI. The model may be extended in several directions in order to provide improved retrieval accuracy. Here we also want to emphasize the importance of considering LSI in the broader picture of dimension reduction and feature extraction whereby a suitable transformation from the instance space to the feature space is performed. Many of the techniques developed in this paper can also be applied to a wider areas of applications some of which are listed below:

- Text classification using naive Bayesian method.
- Correspondence analysis for collaborative filtering to match users with similar interests.
- Statistical methods for NLP based on co-occurrence data based on syntactic structures.
- Post-processing of result set returned by search engines to extract to so-called hub and authoritative web pages.

The rest of the paper is organized as follows: In Section 2, we introduce the subspace-based model and discuss how to use MDL to determine the optimal dimension of the latent-concept subspace. We also evaluate the performance of our method using the Medline text collection [3]; In Section 3 we explore the low-rank-plus-shift structure of the term-document matrix and show how the structure can be used to develop novel algorithms for various computational problems in LSI. We also illustrate our algorithms using LSI updating problem as an example; Section 4 is devoted to Lanczos algorithms and numerical test results. Section 5 concludes the paper and point out some topics for future studies.

2   A subspace-based model and MDL principle

The essence of LSI is to describe terms and documents as linear combinations of so-called latent concepts or variables [4]. However, these linear relations are often used as an *ad hoc* interpretation of the effects of SVD that are used to compute a reduced-dimension representation in Equation (1). In particular no applicable method has been devised to determine the dimension k: a range of k from 100 to 500 or more have been suggested based on empirical evidences [4]. However, the optimal k is collection-dependent and there is a need for a systematic way for its determination. This certainly has important ramification for the computation of LSI. Our approach here is to postulate this type of linear relations between the terms and the latent concepts at the very beginning to build a subspace-based model, and to infer all the important structural information associated with LSI from this model.

   In the vector-space model each term in a document collection is represented by a row vector of the *weighted* term-document matrix A. To facilitate the presentation, we will consider each term as a column vector $t_i \in \mathcal{R}^n$ by a simple transposition, where n is the number of documents in the text collection. We postulate the following model for the i-th term t_i,

$$t_i = C w_i + \epsilon_i,$$

where $C \in \mathcal{R}^{n \times k}$ and $w_i \in \mathcal{R}^k$. The columns of C can be interpreted as representing the underlying latent concepts associated with the particular text collection, and w_i is the weight vector when combined with C gives the i-th term t_i. The vector ϵ_i represents noise and we assume that the noise components are uncorrelated, i.e., $\mathcal{E}\{\epsilon_i\epsilon_i^T\} = \sigma^2 I_n$ with σ unknown, where $\mathcal{E}\{\cdot\}$ is the expectation operator. Assuming furthermore that w_i and ϵ_i are also uncorrelated with each other, we obtain

$$T \equiv \mathcal{E}\{t_i t_i^T\} = CWC^T + \sigma^2 I_n, \tag{2}$$

where $W = \mathcal{E}\{w_i w_i^T\}$. The linear subspace span$\{C\}$ spanned by the columns of C is what we call *latent-concept subspace* and its orthogonal complement is the noise subspace. Let W be nonsingular and C be of full rank. Then the rank of CWC^T is equal to k, where for all practical purposes $k \ll n$. Let the eigenvalues of CWC^T be

$$\lambda(CWC^T) = \{\mu_1, \ldots, \mu_k, 0, \ldots, 0\}.$$

Then it can be deduced that the eigenvalues of T are

$$\lambda(T) = \{\mu_1 + \sigma^2, \ldots, \mu_k + \sigma^2, \sigma^2, \ldots, \sigma^2\}, \tag{3}$$

i.e., T has $n - k$ smallest eigenvalues that are equal to σ^2. In practice we will not have access to the ideal correlation matrix T. However, T can be estimated by the term-document matrix A as follows

$$T \approx A^T A / m,$$

where m is the number of terms in the text collection. Our purpose is therefore to estimate the dimension k as well as the latent-concept subspace span$\{C\}$ from the weighted term-document matrix A.

It seems that from Equation (3) we can determine k simply by counting how many smallest eigenvalues of T are equal. Unfortunately in practice the smallest eigenvalues of $A^T A$ are never equal to each other and therefore more sophisticated methods are needed. One of these tests that have been used in the context of array signal processing is based on the minimum description length (MDL) principle which roughly says that we should select the most parsimonious model that has the minimum code length that is used to encode the model [18, 19]. Specifically, let the eigenvalues of $A^T A$ be $\lambda(A^T A) = \{\lambda_1, \ldots, \lambda_n\}$. Then the latent-concept subspace dimension k is obtained as the index that achieves the minimum of MDL(i), i.e.,

$$k = \arg \min_{0 \leq i \leq n-1} \text{MDL}(i),$$

where the function MDL(i) is defined as

$$\text{MDL}(i) = m(n - i) \log \left(\frac{\sum_{j=i+1}^{n} \lambda_j / (n - j)}{\left(\Pi_{j=i+1}^{n} \lambda_j \right)^{1/(n-i)}} \right) + i(2n - i + 1) \log(m)/2, \quad (4)$$

where again m is the number of terms and n is the number of documents in the text collection. Once k is determined, the latent-concept subspace can be determined by computing a partial SVD of the term-document matrix A as is done in [4], i.e., let the SVD of A be

$$A = U_k \Sigma_k V_k^T + U_k^\perp \Sigma_k^\perp (V_k^\perp)^T.$$

Then span$\{C\}$ = span$\{V_k\}$. We should notice that only the latent-concept subspace is uniquely determined from the model but not each of the individual latent concepts. Actually the orthogonal representation for these latent concepts is just one of many possibilities.

Now if q represents the query vector, the score s used in ranking documents can be computed as

$$\begin{aligned} s &= q^T A_k = q^T U_k \Sigma_k V_k^T \\ &= (U_k^T q)^T (U_k^T A_k) = (U_k^T q)^T (U_k^T A) \\ &= \underbrace{(\Sigma_k^\alpha U_k^T q)^T}_{\text{renormalization}} \underbrace{(\Sigma_k^{1-\alpha} U_k^T A)}_{\text{renormalization}} \end{aligned} \quad (5)$$

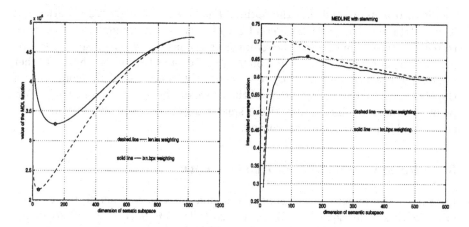

Fig. 1. MDL function values (left) and average precision(right) for the MEDLINE collection

where α is a parameter that can be adjusted for the purpose of renormalization.

EXAMPLE. We now use the MEDLINE text collection [3] to illustrate the accuracy of using our model and MLD for estimating the dimensions of the latent-concept subspaces. For MEDLINE the term-document matrix is 3681×1033 and the number of queries is 30, where we used the stemming option supplied by the SMART system [3]. Since we are also interested in effects of weighting on the latent-concept subspace dimension, we used two different weighting schemes len.lex and lxn.bpx for the MEDLINE text collection to generate the weighted term-document matrices [15]. As a measure of retrieval accuracy, we used 11-point average precision, a standard measure for comparing information retrieval systems [7]. We used $A_k = U_k \Sigma_k V_k^T$ from the SVD of A to compute the score of a query as indicated in Equation (5) with no renormalization. To compensate for stemming, we multiply the number of terms by two to obtain the m used in the MDL function (4). On the left of Figure 1 we plotted the MDL functions values against k while on the right we plotted the average precisions. Small circles indicate where maximum or minimum is achieved. Notice that for the average precisions only results for dimensions up to 550 are given. From the plot we can observe the well-known fact that for the MEDLINE text collection precisions go up with the increase of k up to a point, and then precisions actually go down [4]. The following table lists where maximum or minimum is achieved for both lxn.bpx and len.lex.

	lxn.bpx	len.lex
MDL (min)	141	38
Precision (max)	148	56

The dimension estimate for lxn.bpx are certainly better than that for len.lex, but both of them seem to be within reasonable range of the optimal k. We should also keep in mind that the optimal k is based on the given 30 queries and it can certainly change for a different set of queries. For example, we have also tested the case of measuring the average precisions based on subsets of the 30 queries, the optimal k's are around 50 for len.lex and 150 for lxn.bpx. Another interesting observation from Figure 1 is that the sharpness of the concavity of the MLD curves matches well that of the convexity of the precision curves. This point is important since in practice a small range of k is equally suitable as far as achievable precisions are concerned.

3   The low-rank-plus-shift structure

In practice the latent-concept subspace dimension $k \ll n$. Therefore the correlation matrix T in Equation (3) has the so-called *low-rank-plus-shift structure*. This structure was implicitly used (via MDL) to determine k in the previous section. Now we show that this structure also has important ramifications in several computational issues associated with LSI. First we show that if $A^T A$ has the low-rank-plus-shift structure, then the optimal low-rank approximation of A can be computed via a divide-and-conquer approach. To proceed we introduce some notation: for any matrix $A \in \mathcal{R}^{m \times n}$, we will use $\text{best}_k(A)$ to denote its best rank-k approximation, and its singular values are assumed to be arranged in nonincreasing order,

$$\sigma_1(A) \geq \sigma_2(A) \geq \cdots \geq \sigma_m(A).$$

Theorem 1. *Let $A = [A_1, A_2] \in \mathcal{R}^{m \times n}$ with $m \geq n$. Furthermore assume that*

$$A^T A = X + \sigma^2 I, \quad \sigma > 0, \tag{6}$$

where X is symmetric and positive semi-definite with $\text{rank}(X) = k$. Then there are integers $k_1 \leq k$ and $k_2 \leq k$ with $k_1 + k_2 \geq k$ such that

$$\text{best}_k([\text{best}_{k_1}(A_1), \text{best}_{k_2}(A_2)]) = \text{best}_k([A_1, A_2]).$$

REMARK. In essence the above result states that if $A^T A$ has the low-rank-plus-shift structure, then an optimal low-rank approximation of A can be computed by merging the optimal low-rank approximations of its two submatrices A_1 and A_2. The result can be generalized to the case where A is partitioned into several blocks $A = [A_1, A_2, \ldots, A_s]$.

Theorem 1 naturally leads to a divide-and-conquer approach for computing an optimal low-rank approximation for a large matrix A: divide A into several blocks, and compute an optimal low-rank approximation for each of the block *in parallel*, then merge the low-rank approximations into an optimal low-rank approximation for A. Due to space limitation, the high performance computing issues related to the divide-and-conquer approach will not be further discussed here.

We now turn to the discussion of updating problems in LSI: In rapidly changing environments such as the World Wide Web, the document collection is frequently updated with new documents and terms constantly being added, and there is a need to find the latent-concept subspaces for the updated document collections [1]. Let $A \in \mathcal{R}^{m \times n}$ be the original term-document matrix, and $A_k = P_k \Sigma_k Q_k^T$ be the best rank-k approximation of A. There are three types of updating problems in LSI [1]. Here we concentrate only on updating documents:[17]

Let $D \in \mathcal{R}^{m \times p}$ be the p new documents. Compute the best rank-k approximation of $B \equiv [A_k, D]$.

Notice that in updating we use the matrix $[A_k, D]$ instead of using the true new term-document matrix $[A, D]$ as would have been the case in traditional SVD updating problems. So it is a critical issue whether the replacement of A by its best rank-k approximation is justified for there is always the possibility that this process may introduce unacceptable error in the updated latent-concept subspace. It can be deduced from Theorem 1 that

$$\text{best}_k[\text{best}_k(A), D] = \text{best}_k[A, D],$$

if $[A, D]$ has the low-plus-shift structure, therefore justifying the use of $[A_k, D]$. In the following contrary to what has been claimed in the literature, we show that no retrieval accuracy degradation will occur if updating is done correctly. We first present our new updating algorithm [17]:

UPDATING DOCUMENTS. Let the QR decomposition of $(I - P_k P_k^T)D$ be

$$(I - P_k P_k^T)D = \hat{P}_k R,$$

where $\hat{P}_k$ is orthonormal, and R is upper triangular. For simplicity we assume R is nonsingular.[1] It can be verified that

$$B \equiv [A_k, D] = [P_k, \hat{P}_k] \begin{bmatrix} \Sigma_k & P_k^T D \\ 0 & R \end{bmatrix} \begin{bmatrix} Q_k^T & 0 \\ 0 & I_p \end{bmatrix}.$$

Notice that $[P_k, \hat{P}_k]$ is orthonormal. Now let the SVD of

$$\hat{B} \equiv \begin{bmatrix} \Sigma_k & P_k^T D \\ 0 & R \end{bmatrix} = [U_k, U_k^\perp] \begin{bmatrix} \hat{\Sigma}_k & \\ 0 & \hat{\Sigma}_p \end{bmatrix} [V_k, V_k^\perp]^T, \tag{7}$$

where U_k and V_k are of column dimension k, and $\hat{\Sigma}_k \in \mathcal{R}^{k \times k}$. Then the best rank-$k$ approximation of B is given by

$$B_k \equiv ([P_k, \hat{P}_k]U_k)\hat{\Sigma}_k \left(\begin{bmatrix} Q_k & 0 \\ 0 & I_p \end{bmatrix} V_k \right)^T.$$

In [1, 11], only $[\Sigma_k, P_k^T D]$ instead of $\hat{B}$ in (7) is used to construct the SVD of B. The R matrix in $\hat{B}$ is completely discarded. The SVD thus constructed is

---

[1] If $(I - P_k P_k^T)D$ is not of full column rank, R can be upper trapezoidal.

certainly not the *exact* SVD of B, and can not even be a good approximation of it if the norm of R is not small. This situation can happen when the added new documents alter the original low-dimension representation significantly. Our numerical experiments bear this out.

EXAMPLE. We use the MEDLINE text collection again. The weighting scheme used is `lxn.bpx`. The latent-concept subspaces are computed using a two-step method based on updating: for a given s we compute a rank-k approximation of the first s columns of the term-document matrix using the Lanczos SVD process, and then we add the remaining documents to produce a new rank-k approximation using updating algorithms. In Table 1, $k = 100$, p is the number of new documents added, Meth_1 is the updating algorithm discussed above and Meth_2 is that used in [1, 11]. Row 3 and row 4 of the table gives the average precisions in percentage. As is expected Meth_1 performs much better than Meth_2 for those seven combinations of p and s. What is surprising is that Meth_1 performs even better than rank-k approximation using the whole term-document matrix for which the average precision is 65.50%.

Table 1. Comparison of average precisions for MEDLINE collection

p	100	200	300	400	500	600	700
s	933	833	733	633	533	433	333
Meth_1	65.36	65.52	66.58	67.16	66.98	66.56	66.48
Meth_2	64.26	64.40	58.48	50.78	46.90	44.04	44.97
Increm	65.36	65.61	65.61	66.33	66.65	66.58	66.75

Instead of updating a group of p new documents all at once, we also carry out a test by breaking these p new documents into subgroups of 100 documents each, and use the updating algorithms to update one subgroup at a time. Row 5 of Table 1 gives the computed average precisions for $k = 100$ for our updating algorithm. Since the algorithms in [1, 11] always discard the R matrix in (7) therefore it makes no difference to the updated low-rank approximation whether it is computed with all the new documents all at once or incrementally with each subgroup at a time. More examples are presented in [17].

4 Lanczos Methods

In this section we present variants of Lanczos algorithms for solving singular value problems associated with large sparse matrices. These methods use the matrices in the singular value problems only in the form of matrix-vector multiplication. Our approach is to consider the associated symmetric eigenproblem instead, and the matrix-vector multiplication will be of the form

$$Ax = GG^Tx \text{ or } G^TGx,$$

where G is the term-document matrix. Sparsity is not lost by (further) transformation of the matrices.

The Krylov subspace containing j vectors associated with an $n \times n$ matrix A and a starting vector q of unitary length is defined as

$$\mathcal{K}(A, q, j) = \text{span}(q, Aq, \ldots A^{j-1}q). \tag{8}$$

It turns out that, as j increases, the vector $A^{j-1}q$ in the Krylov subspace converges toward the eigenvector associated with the dominant eigenvalue of A. Also, only the multiplication of A by a vector is required for the generation of the Krylov subspace. The Lanczos algorithm builds a basis for a Krylov subspace. The projection of the eigenproblem into such a basis corresponds to a smaller problem involving a tridiagonal matrix (which is symmetric). One of the advantages of generating a basis for (8) is that many approximations for eigensolutions of can be obtained as j increases. Such approximations are computed through a Rayleigh-Ritz procedure. The Gram-Schmidt orthonormalization process could be applied to all vectors of subspace (8), in the natural order $q, Aq, A^2q, \ldots$, so as to construct a basis. Nevertheless, one can show that for the orthonormalization of the i-th Krylov vector, it suffices to take into account only the two previous orthonormalized vectors [5, 13]. Moreover, the basis can be built vector by vector, without explicitly generating the Krylov subspace itself. That is what the Lanczos method computes, through the sequence of operations summarized in Algorithm 1. Note that only the multiplication of A by a vector is needed. That operation, as well as the subsequent operations involving only vectors, can be performed in parallel. In other words, the matrix A and the vectors involved can be split among the available processors on a parallel computer [2].

After j steps, the vectors generated by Algorithm 1 can be arranged as $Q = [\, q_1 \; q_2 \ldots q_j \,]$, satisfying $Q_j^T Q_j = I$. Also, $Q_j^T A Q_j = T_j$, where T_j is a tridiagonal matrix whose entries are $t_{i,i} = \alpha_i$, $i = 1, 2 \ldots j$, and $t_{i+1,i} = t_{i,i+1} = \beta_i$, $i = 1, 2 \ldots j - 1$. In other words, the projection of the eigenproblem into Q_j is given by T_j. An approximate solution $(\tilde{\lambda}_k, \tilde{x}_k)$ is then given by the *Ritz value* $\tilde{\lambda}_k = \theta_k$ and by the *Ritz vector* $\tilde{x}_k = Q_j s_k$, where (θ_k, s_k) is the solution of the reduced problem $T_j s_k = \theta_k s_k$. In many cases, extreme eigenvalues (and corresponding eigenvectors) of T_j lead to good approximations of extreme eigenvalues (and corresponding eigenvectors) with $j \ll n$. Since T_j is a tridiagonal matrix of small dimension, its associated eigenproblem is inexpensive to compute. Convergence checks can also be carried out in inexpensive ways [6].

Several implementations of the Lanczos algorithm have been developed [10, 6, 12]. A block Lanczos algorithm, for instance, is useful when many eigenvalues are sought or the eigenvalue distribution is clustered. It also allows better data management on some computer architectures. Associated with the block strategy there is a block Krylov subspace defined as

$$\mathcal{K}(A, \hat{Q}, j) = span(\hat{Q}, A\hat{Q}, \ldots A^{j-1}\hat{Q}), \tag{9}$$

where $\hat{Q} = [\, \hat{q}_1 \; \hat{q}_2 \ldots \hat{q}_p \,]$ is a full rank $n \times p$ matrix, that is to say, $\hat{Q}^T \hat{Q} = I$, and p is the *block size*. A block counterpart for Algorithm 1, for instance, can be easily

Algorithm 1. Basic Lanczos	**Algorithm 2.** Bi-diagonal Lanczos

Algorithm 1. Basic Lanczos

set $q_0 := 0$, $r_0 \neq 0$

set $\beta_0 := \sqrt{r_0^T r_0}$

set $q_1 := \dfrac{r_0}{\beta_0}$

for j=1,2...

   a) $r_j := A q_j$

   b) $r_j := r_j - q_{j-1}\beta_{j-1}$

   c) $\alpha_j := q_j^T r_j$

   d) $r_j := r_j - q_j \alpha_j$

   e) $\beta_j := \sqrt{r_j^T r_j}$

   f) $q_{j+1} := \dfrac{r_j}{\beta_j}$

end for

Algorithm 2. Bi-diagonal Lanczos

set $u_0 := 0$, $s_0 \neq 0$

set $\beta_0 := \sqrt{s_0^T s_0}$

set $v_1 := \dfrac{s_0}{\beta_0}$

for j=1,2...

   a) $r_j := G v_j$

   b) $r_j := r_j - u_{j-1}\beta_{j-1}$

   e) $\gamma_j := \sqrt{r_j^T r_j}$

   f) $u_j := \dfrac{r_j}{\gamma_j}$

   a') $s_j := G^T u_j$

   b') $s_j := s_j - v_j \gamma_j$

   e') $\beta_j := \sqrt{s_j^T s_j}$

   f') $v_{j+1} := \dfrac{s_j}{\beta_j}$

end for

defined. In that case, the vectors involved are replaced by rectangular matrices of appropriate dimensions and the projection of the original eigenproblem into the basis generated is a block symmetric tridiagonal matrix.

In finite precision arithmetic, a loss of orthogonality among the vectors of the basis generated by the Lanczos algorithm is generally observed after some steps. It is related to roundoff errors and also, to the convergence of a pair $(\tilde{\lambda}_k, \tilde{x}_k)$ and, therefore, to the eigenvalue distribution of the associated problem [13]. Once orthogonality is lost, $Q_j^T Q_j \neq I$, and redundant copies of eigenpairs emerge. As an immediate option to avoid such problem, one could apply a full reorthogonalization strategy. That is, at each step orthogonalize r_j (in Algorithm 1) against all $j-1$ previous vectors q computed. However, such a scheme would strongly increase the number of operations. On the other hand, some preventive measures based on potentially dangerous vectors can be used to keep the basis orthogonality within a certain level, such as selective orthogonalization and partial reorthogonalization [14, 16]. In the first case, whenever necessary, r_j is orthogonalized against converged Ritz vectors $\tilde{x}$. In the second case, whenever necessary, r_j is orthogonalized against q_i, $i = 1, 2 \ldots j$. Simple recurrence formulas can be used to indicate when those strategies should be applied. The selective orthogonalization and partial reorthogonalization can be also easily extended to block versions of the Lanczos algorithm.

For the computation of the SVD we concentrate on the strategy of suing $A = G^T G$ or $A = GG^T$ and applying Algorithm 1. In that case, one can use whichever of these eigenvalue problems is of smaller order to compute the shorter set of singular vectors, say V, assuming $m > n$. In step a) of Algorithm 1, one can compute $t = Gq_j$ and then $r_j = G^T t$. Given V, U can be obtained as $U = AV\Sigma^{-1}$. Similarly, when $m < n$, U can be computed. Then, $t = G^T q_j$ and $r_j = Gt$ in step a) of Algorithm 1, $V = A^T U \Sigma^{-1}$.

EXAMPLE. In the LSI applications, only the largest singular values and corresponding singular vectors are considered relevant. We now show an example of a large sparse matrix for which we have computed a few singular values and singular vectors. The matrix corresponds to a data set containing 100,000 documents (the number of rows of the matrix), 2,559,430 documents (the number of columns of the matrix), 421,057,104 entries (the number of nonzero entries in the matrix), and occupying 5 GBytes of disk space. Actually, this matrix is a fairly small example of the applications we have in mind. For instance, the documents in the data set could be associated with an important fraction of all URL's available on the Web, which are roughly 150,000,000 at the time of this writing. The number of singular values and singular vectors required in such cases are likely to be as large as 1,000.

Our computations were performed on a 512 parallel-processor T3E-900, with 256 Mbytes of memory per processor, installed at NERSC. For the matrix described in the previous paragraph, 40 processors were needed to partition the matrix into pieces and run an MPI based implementation of the block Lanczos algorithm. We apply the Lanczos algorithm to the operator AA^T, obtaining approximations for eigenvalues and eigenvectors of AA^T, that is to say, pairs (λ, x). Therefore, the singular values can be computed by taking the square root of λ, and each x corresponds to a left singular vector. Figure 2.a shows the eigenvalues approximated by the Lanczos algorithm (with block size set to one) after 33 steps, and Figure 2.b their square roots. By examining the residuals of $(AA^T)x - \lambda x$ (estimated at low cost by the Lanczos algorithm), we can infer how many pairs have converged. Those residuals are shown in Figure 2.c, where we can see that 5 singular values and values have converged. Finally, Figure 2.d shows the execution time (in seconds) for the algorithm. Although the variation is almost linear for 33 steps, the execution time is supposed to grow quadratically after a certain number of steps. That is due to additional operations needed to preserve semi-orthogonality of the basis of vectors generated by the Lanczos algorithm.

5 Concluding remarks

The subspace-based model we developed provides a theoretical foundation for LSI. In particular, the model can be used with MDL to determine the dimension of the latent-concept subspaces. We showed that the model also imposes a low-rank-plus-shift structure which can be used to handle several computational problems arising from LSI. Here we want to point out that the subspace-based

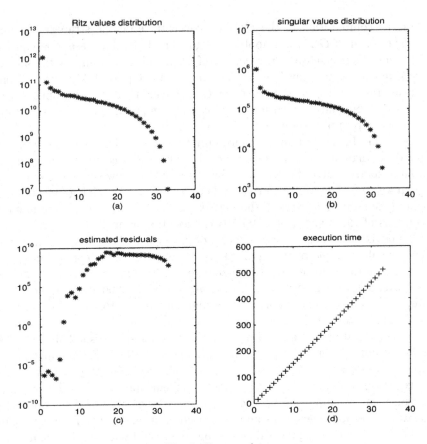

Fig. 2. Test results

model model can be extended in several ways and the extensions may lead to improved retrieval accuracy:

- We certainly can consider the colored noise case where the covariance matrix of ϵ_i is assumed to be a known matrix up to a constant, i.e., $\mathcal{E}\{\epsilon_i\epsilon_i^T\} = \sigma^2\Sigma$.
- A more ambitious approach is to consider the term frequencies in the *un-weighted* term-document matrices as small counts and use the idea of like-lyhood function from generalized linear model. The starting point is to look at marginal distributions of the term-vector and see if they match certain probability distributions such as Poisson distribution.
- We can also replace the linear relations between the terms and latent concepts with nonlinear ones such as those used in generalized additive models.

These directions certainly need to be pursued in future research.

In terms of computational issues, we notice that only the largest singular values and corresponding singular vectors are considered relevant. Therefore, the approaches outlined previously (bidiagonal reduction or eigenvalues of AA^T) are likely to have desirable convergence properties. However, size favors taking the approach of finding the eigenvalues of AA^T. However, a large subset, say 1000, of singular values and singular vectors may be required. It is likely that the Lanczos algorithm will require at least twice that number of vectors (steps) to compute the required subset. In other words, the computation of a large subset of singular values and singular vectors needs a large subspace. Therefore, preserving orthogonality in an efficient way becomes a very important issue. Also, deflation techniques can be used to prevent converged solutions to reappear. In that respect, one of the next phases of our research will be to experiment with combinations of orthogonalization and deflation strategies, implemented in parallel, for the computation of a large subset of singular values and singular vectors.

References

[1] M.W. Berry, S.T. Dumais and G.W. O'Brien. Using linear algebra for intelligent information retrieval. *SIAM Review*, 37:573-595, 1995.

[2] L. S. Blackford, J. Choi, A. Cleary, E. D'Azevedo, J. W. Demmel, I. Dhillon, J. J. Dongarra, S. Hammarling, G. Henry, A. Petitet, K. Stanley, D. Walker, and R. C. Whaley. *ScaLAPACK User's Guide*. SIAM, Philadelphia, USA, 1997.

[3] Cornell SMART System, `ftp://ftp.cs.cornell.edu/pub/smart`.

[4] S. Deerwester, S.T. Dumais, T.K. Landauer, G.W. Furnas and R.A. Harshman. Indexing by latent semantic analysis. *Journal of the Society for Information Science*, 41:391–407, 1990.

[5] G. H. Golub and C. F. Van Loan. *Matrix Computations*. The Johns Hopkins University Press, Baltimore, USA, third edition, 1996.

[6] R. G. Grimes, J. G. Lewis, and H. D. Simon. A Shifted Block Lanczos Algorithm for Solving Sparse Symmetric Eigenvalue Problems. *SIAM J. Matrix Anal. Appl.*, 15:228-272, 1994.

[7] D. Harman. TREC-3 conference report. NIST Special Publication 500-225, 1995.

[8] G. Kowalski. Information Retrieval System: Theory and Implementation. Kluwer Academic Publishers, Boston, 1997.

[9] R. Krovetz and W.B. Croft. Lexical ambiguity and information retrieval. *ACM Transactions on Information Systems*, 10:115–141, 1992.

[10] B. Nour-Omid, B. N. Parlett, T. Ericsson, and P. S. Jensen. How to Implement the Spectral Transformation. *Mathematics of Computation*, 48:663–673, 1987.

[11] G.W. O'Brien. Information Management Tools for Updating an SVD-Encoded Indexing Scheme. M.S. Thesis, Department of Computer Science, Univ. of Tennessee, 1994.

[12] O.A. Marques. BLZPACK: Description and User's Guide. CERFACS, TR/PA/95/30, 1995.

[13] B. N. Parlett. *The Symmetric Eigenvalue Problem*. Prentice Hall, Englewood Cliffs, USA, 1980.

[14] B. N. Parlett and D. S. Scott. The Lanczos Algorithm with Selective Orthogonalization. *Mathematics of Computation*, 33:217–238, 1979.

[15] G. Salton. Automatic Text Processing. Addison-Wesley, New York, 1989.

[16] H. D. Simon. The Lanczos Algorithm with Partial Reorthogonalization. *Mathematics of Computation*, 42:115–142, 1984.

[17] H.D. Simon and H. Zha. Low rank matrix approximation using the Lanczos bidiagonalization process with applications. Technical Report CSE-97-008, Department of Computer Science and Engineering, The Pennsylvania State University, 1997.

[18] G. Xu and T. Kailath. Fast subspace decomposotion. *IEEE Transactions on Signal Processing*, 42:539–551, 1994.

[19] G. Xu, H. Zha, G. Golub, and T. Kailath. Fast algorithms for updating signal subspaces. *IEEE Transactions on Circuits and Systems*, 41:537–549, 1994.

Thick-Restart Lanczos Method
for Symmetric Eigenvalue Problems*

Kesheng Wu and Horst D. Simon

Lawrence Berkeley National Laboratory/NERSC, Berkeley, CA 94720.
{kwu, hdsimon}@lbl.gov.

This paper describes a restarted Lanczos algorithm that particularly suitable for implementation on distributed machines. The only communication operation is requires outside of the matrix-vector multiplication is a global sum. For most large eigenvalue problems, the global sum operation takes a small fraction of the total execution time. The majority of the computer is spent in the matrix-vector multiplication. Efficient parallel matrix-vector multiplication routines can be found in many parallel sparse matrix packages such as AZTEC [9], BLOCK-SOLVE [10], PETSc [3], P_SPARSLIB[1]. For this reason, our main emphasis in this paper is to demonstrate the correctness and the effectiveness of the new algorithm.

1 Introduction

Given an $n \times n$ matrix A, its eigenvalue λ and the corresponding eigenvector x are defined by the following relation,

$$Ax = \lambda x, \qquad (x \neq 0).$$

In this paper, we will only consider the symmetric eigenvalue problems where A is symmetric. The algorithm given here can be easily extended to hermitian eigenvalue problems or symmetric generalized eigenvalue problems.

The symmetric Lanczos algorithm is an effective method for solving symmetric eigenvalue problems [8, Section 9.1.2] [13] [16, Section 6.6] [15]. It solves the eigenvalue problem by first building an orthonormal basis, see Algorithm 1, then forming approximate solutions using the Rayleigh-Ritz projection [8, 13, 15]. Because it only accesses the matrix through matrix-vector multiplications, it is commonly used when the matrices are too large to store or not explicitly available. It is also a popular eigenvalue algorithm for parallel environments since the parallel matrix-vector multiplication routines are easy to construct and widely available.

There are a number of algorithms that have similar characteristics as the Lanczos algorithm, for example, the Arnoldi method, see Algorithm 2 [1, 17],

---

* This work was supported by the Director, Office of Energy Research, Office of Laboratory Policy and Infrastructure Management, of the U.S. Department of Energy under Contract No. DE-AC03-76SF00098.

[1] P_SPARSLIB source code is available at http://www.cs.umn.edu/~saad/.

and the Davidson method [5, 6]. In fact, for symmetric problems both the Arnoldi method and the unpreconditioned Davidson method are mathematically equivalent to the Lanczos method. Because the Lanczos method explicitly takes advantage of the symmetry of the matrix and avoids computing quantities that are zero, it uses fewer arithmetic operations per step than the others. In exact arithmetic, the values of $h_{i,j}, j = 1, \ldots, i-2$, in step *(c)* of Algorithm 2 are zero. Because of this, the Lanczos algorithm only computes $\alpha_i (\equiv h_{i,i})$. Because $h_{i-1,i} = h_{i,i-1}$, the Lanczos algorithm uses $\beta_{i-1} (\equiv h_{i-1,i})$ in the place of $h_{i,i-1}$. Steps *(c)* and *(d)* of Algorithms 1 and 2 are known as the orthogonalization steps. Clearly, the orthogonalization step of the Lanczos algorithm is much cheaper. We do not show the details of the Davidson method because it uses even more arithmetic operation per step than the Arnoldi method and it is commonly used with preconditioning.

When the algorithms are implemented using floating-point arithmetic, the expected zero dot-products are no longer zero. This phenomenon is related to loss of orthogonality among the Lanczos vectors. It causes the Ritz values and Ritz vectors computed by the Lanczos method to be different from what is expected in the exact arithmetic. We will explore this issue by some numerical examples later.

ALGORITHM 1 *The Lanczos iterations starting with r_0. Let $\beta_0 = \|r_0\|$, and $q_0 = 0$.*
For $i = 1, 2, \ldots,$

(a) $q_i = r_{i-1}/\|r_{i-1}\|$,
(b) $p = Aq_i$,
(c) $\alpha_i = q_i^T p$,
(d) $r_i = p - \alpha_i q_i - \beta_{i-1} q_{i-1}$,
(e) $\beta_i = \|r_i\|$.

ALGORITHM 2 *The Arnoldi iterations starting with r_0.*
For $i = 1, 2, \ldots$

(a) $q_i = r_{i-1}/\|r_{i-1}\|$,
(b) $p = Aq_i$,
(c) $h_{j,i} = q_j^T p, j = 1, \ldots, i$,
(d) $r_i = p - \sum_{j=1}^{i} h_{j,i} q_j$,
(e) $h_{i,i+1} = \|r_i\|$.

An effective way of recovering the orthogonality is re-orthogonalization which needs to access all computed Lanczos vectors. Usually, the minimum number of Lanczos iterations required to compute a given eigenvalue is unknown until the eigenvalue is actually found. Since one Lanczos vector is generated at each step, the memory space required to store the Lanczos vectors may be exceedingly large. To control the maximum memory needed to store the Lanczos vectors, the Lanczos algorithm is restarted. Recently, a number of successful restarting schemes have been reported, for example, the implicit restarting scheme of the implicitly restarted Arnoldi method [19] and the thick-restart scheme of the dynamic thick-restart Davidson method [21, 22]. The implicitly restarted Arnoldi method has also been implemented in ARPACK to solve many different kinds of eigenvalue problems [18]. The implicitly restarted Arnoldi method for the symmetric eigenvalue problem is mathematically equivalent to the *implicitly restarted Lanczos method.* However, in the current implementation, ARPACK does not explicitly take advantage of the symmetry of the matrix. In this paper we will study a restarted Lanczos method for symmetric eigenvalue problems that takes advantage of the symmetry.

The main goal of this paper is to develop a restarted version of the Lanczos algorithm. The details of the thick-restart Lanczos method is given in Section 2. A few numerical example are shown in Section 3 to demonstrate the potential of the new method. A brief summary is given in section 4.

2 Thick-restart Lanczos iteration

In this section, we first review the restarting schemes used with the Arnoldi method and the Davidson method and then describe a restarted Lanczos algorithm in exact arithmetic. The issues related to loss of orthogonality and implementation in finite precision arithmetic are discussed elsewhere [23]. We will show a number of numerical examples in the next section to illustrate how well it works.

Our interest in restarted Lanczos algorithm is in large part sparked by the success of the implicitly restarted Arnoldi method [11, 18, 19]. One of the most straightforward schemes to restart the Lanczos method is to save one Ritz vector. If more than one eigenvalue is wanted, this scheme can be extended in many ways, for example, add all wanted Ritz vectors together to form one starting vector, or use a block version of the Lanczos algorithm that has the same block size as the number of eigenvalues wanted, or work on one eigenvalue at a time and lock the converged ones [15]. These options are simple to implement but not nearly as effective as the implicit restarting scheme [19]. What makes the implicit restarting scheme more efficient are the following.

i. It not only reuses the basis vectors, but also the projection matrix in the Rayleigh-Ritz projection, therefore reduces the number of matrix-vector multiplications.

ii. It can restart with an arbitrary number of starting vectors.

iii. It can use arbitrary shifts to enhance the quality of the starting vectors.

The implicit restarting scheme with exact shifts, i.e., using the unwanted Ritz values as the shifts, is equivalent to restarting with Ritz vectors for symmetric eigenvalue problems [12, 20, 22]. There are a number of restarting schemes for the Davidson method that use Ritz vectors [7]. Researches show that the ability to reuse a large portion of the previous basis greatly enhances their overall efficiencies [12, 20, 22]. Using the exact shifts is generally a good choice. A number of researchers have shown that the *optimal* shifts are Leja points [2]. Significant reduction in the number of matrix-vector multiplications can be achieved by using Leja points as shifts when the maximum basis size is very small, say less than 10. However, when the maximum basis size is larger than 10, using the Leja shifts or the exact shifts in the implicitly restarted Arnoldi method does not change the number of matrix-vector multiplications used. When moderate basis size is used, we can use the Ritz vectors to restart as in the thick-restart scheme. The thick-restart scheme has the first two advantages of the implicit restarting scheme. It is simple to implement and just as effective as the implicit restarting scheme in most cases [20]. Based on this observation, we will design a restarted Lanczos method that restarts with Ritz vectors, i.e., a thick-restart Lanczos method.

Next, we can proceed to derive the restarted Lanczos method. Assume that the maximum number of Lanczos step can be taken before restart is m. After m steps of Algorithm 1, the Lanczos vectors satisfy the following Lanczos recurrence,

$$AQ_m = Q_m T_m + \beta_m q_{m+1} e_m^T, \tag{1}$$

where $Q_m = [q_1, \ldots, q_m]$, e_m is last column of the identity matrix I_m, and $T_m = Q_m^T A Q_m$ is an $m \times m$ symmetric tridiagonal matrix constructed from α_i and β_i as follows,

$$T_m = \begin{pmatrix} \alpha_1 & \beta_1 & & & & \\ \beta_1 & \alpha_2 & \beta_2 & & & \\ & \ddots & \ddots & \ddots & & \\ & & \ddots & \ddots & \ddots & \\ & & & \beta_{m-2} & \alpha_{m-1} & \beta_{m-1} \\ & & & & \beta_{m-1} & \alpha_m \end{pmatrix} .$$

Using the Rayleigh-Ritz projection, we can produce approximate solutions to the eigenvalue problem. Let (λ, y) be an eigenpair of T_m, then λ is an approximate eigenvalue of A, and $x = Q_m y$ is the corresponding approximate eigenvector. They are also known as the Ritz value and the Ritz vector.

When restarting, we first determine an appropriate number of Ritz vectors to save, say k, then choose k eigenvectors of T_m, say Y, and compute k Ritz vectors, $\hat{Q}_k = Q_m Y$. The following derivation can be carried out by assuming Y to be any orthonormal basis of a k-dimensional invariant subspace of T_m. Since the matrix T_m is symmetric and there is no apparent advantage of using a different basis set, we will only use Ritz vectors in the restarted Lanczos algorithm. To distinguish the quantities before and after restart, we denote the quantities after restart with a hat (ˆ). For example, the projected matrix T_m after restart is $\hat{T}_k \equiv Y^T T_m Y$. Since we have chosen to restart with Ritz vectors, the matrix $\hat{T}_k$ is diagonal and the diagonal elements are the Ritz values. Immediately after restart, the new basis vectors satisfy the following relation,

$$A\hat{Q}_k = \hat{Q}_k \hat{T}_k + \beta_m \hat{q}_{k+1} s^T, \tag{2}$$

where $\hat{q}_{k+1} = q_{m+1}$ and $s = Y^T e_m$. We recognize that this equation is an extension of Equation 1. One crucial feature of the Lanczos recurrence is maintained here, i.e., the residual vectors of the basis $\hat{Q}_k$ are in one direction. In Algorithm 1, the Lanczos recurrence is extended one column at a time by augmenting the current basis with q_{m+1}. In the same spirit, we can augment the basis $\hat{Q}_k$ with

$\hat{q}_{k+1}$. If there is no relation between $\hat{Q}_k$ and $\hat{q}_{k+1}$, we can build an augmented Krylov subspace from $\hat{q}_{k+1}$ [4, 14]. The thick-restart Arnoldi method and the thick-restart Davidson method generate basis of the same subspace when no preconditioning is used. They are shown to be equivalent to the implicitly restarted Arnoldi method [20, 22]. Because the Arnoldi method and the Lanczos method are mathematically equivalent to each other, this restarted Lanczos method is also equivalent to the implicitly restarted Arnoldi method with the exact shifts. Based on this equivalence, the basis generated after restarting is another Krylov subspace even though we do not know the starting vector.

To compute the new Lanczos vectors after restart we can use the Gram-Schmidt procedure as in the Arnoldi algorithm, see Algorithm 2. Fortunately a cheaper alternative exists because of the symmetry of the matrix. Let's first look at how to compute $\hat{q}_{k+2}$. Based on the Gram-Schmidt procedure, the expression for $\hat{q}_{k+2}$ is as follows,

$$
\begin{aligned}
\hat{\beta}_{k+1}\hat{q}_{k+2} &= (I - \hat{Q}_{k+1}\hat{Q}_{k+1}^T)A\hat{q}_{k+1} \\
&= (I - \hat{q}_{k+1}\hat{q}_{k+1}^T - \hat{Q}_k\hat{Q}_k^T)A\hat{q}_{k+1} \\
&= (I - \hat{q}_{k+1}\hat{q}_{k+1}^T)A\hat{q}_{k+1} - \hat{Q}_k\beta_m s.
\end{aligned}
\tag{3}
$$

The above equation uses the fact that $\hat{Q}_k^T A\hat{q}_{k+1} = \beta_m s$, which is due to Equation 2 and the orthogonality of the Lanczos vectors. The scalar $\hat{\beta}_{k+1}$ in the above equation is equal to the norm of the right-hand side so that $\hat{q}_{k+2}$ is normalized. This equation shows that $\hat{q}_{k+2}$ can be computed more efficiently than a typical step of the Arnoldi method. Since the vector $\hat{Q}_k^T A\hat{q}_{k+1}$ is known, we only need to compute $\hat{\alpha}_{k+1}$ as in step (c) of Algorithm 1 and replace step (d) with the following, $\hat{r}_{k+1} = \hat{p} - \hat{\alpha}_{k+1}\hat{q}_{k+1} - \sum_{j=1}^{k}\beta_m s_j\hat{q}_j$, where $\hat{p} = A\hat{q}_{k+1}$. While computing $\hat{q}_{k+2}$, we also extended the matrix $\hat{T}_k$ by one column and one row,

$$
\hat{T}_{k+1} = \begin{pmatrix} \hat{T}_k & \beta_m s \\ \beta_m s^T & \hat{\alpha}_{k+1} \end{pmatrix},
$$

where $\hat{\alpha}_{k+1} = \hat{q}_{k+1}^T A\hat{q}_{k+1}$. Obviously, the Lanczos recurrence relation, Equation 1, is maintained after restart, more specifically, $A\hat{Q}_{k+1} = \hat{Q}_{k+1}\hat{T}_{k+1} + \hat{\beta}_{k+1}\hat{q}_{k+2}e_{k+1}^T$, where $\hat{\beta}_{k+1} = \|\hat{r}_{k+1}\|$. Even though $\hat{T}_{k+1}$ is not tridiagonal as in the original Lanczos algorithm, it does not affect the restarted Lanczos recurrence as we will show next.

After we have computed $\hat{q}_{k+i}$ ($i > 1$), to compute the next basis vector $\hat{q}_{k+i+1}$, we again go back to the Gram-Schmidt procedure, see Equation 3.

$$
\begin{aligned}
\hat{\beta}_{k+i}\hat{q}_{k+i+1} &= (I - \hat{Q}_{k+i}\hat{Q}_{k+i}^T)A\hat{q}_{k+i} \\
&= (I - \hat{q}_{k+i-1}\hat{q}_{k+i-1}^T - \hat{q}_{k+i}\hat{q}_{k+i}^T - \hat{Q}_{k+i-2}\hat{Q}_{k+i-2}^T)A\hat{q}_{k+i} \\
&= (I - \hat{q}_{k+i}\hat{q}_{k+i}^T - \hat{q}_{k+i-1}\hat{q}_{k+i-1}^T)A\hat{q}_{k+i} - \hat{Q}_{k+i-2}(A\hat{Q}_{k+i-2})^T\hat{q}_{k+i} \\
&= A\hat{q}_{k+i} - \hat{\alpha}_{k+i}\hat{q}_{k+i} - \hat{\beta}_{k+i-1}\hat{q}_{k+i-1},
\end{aligned}
$$

where $\hat{\alpha}_{k+i}$ is $\hat{q}_{k+i}^T A\hat{q}_{k+i}$ by definition and $\hat{\beta}_{k+i}$ is the norm of the right-hand side. The above equation is true for any i grater than 2. From this equation we see that computing $\hat{q}_{k+i}$ $(i > 2)$ requires the same amount of work as in the original Lanczos algorithm, see Algorithm 1. The matrix $\hat{T}_{k+i} \equiv \hat{Q}_{k+i}^T A \hat{Q}_{k+i}$ can be written as follows,

$$\hat{T}_{k+i} = \begin{pmatrix} \hat{T}_k & \beta_m s & & & & \\ \beta_m s^T & \hat{\alpha}_{k+1} & \hat{\beta}_{k+1} & & & \\ & \hat{\beta}_{k+1} & \hat{\alpha}_{k+2} & \hat{\beta}_{k+2} & & \\ & & \ddots & \ddots & \ddots & \\ & & & \ddots & \ddots & \ddots \\ & & & & \hat{\beta}_{k+i-1} & \hat{\alpha}_{k+i} \end{pmatrix}.$$

This restarted Lanczos iteration also maintains the recurrence relation described in Equation 1. Since the Lanczos recurrence relation is satisfied by the restarted Lanczos algorithm, the above equations for computing $\hat{q}_{k+i}$ are not only true after restarting the initial Lanczos iterations, they are true after every restart. The recurrence relation is a three-term recurrence except the first step after restart. Therefore most of the Lanczos vectors can be computed as efficiently as in the original Lanczos algorithm. In addition, using the Lanczos recurrence relation we can estimate the residual norms of the approximate eigenpairs cheaply.

The thick-restart Lanczos algorithm is basically specified by the above equations, next we will discuss one detail concerning the storage of T_m. As mentioned before, if Y is a collection of eigenvectors of T_m, the matrix $\hat{T}_k$ is diagonal, and the diagonal elements can be stored as first k elements of $\hat{\alpha}_i$, $i = 1, ..., k$. The array $(\beta_m s)$ is of size k, it can be stored in the first k elements of $\hat{\beta}_i$, $i = 1, ..., k$. After restart, the arrays $\hat{\alpha}_i$ and $\hat{\beta}_i$ are as follows,

$$\hat{\alpha}_i = \lambda_i, \qquad \hat{\beta}_i = \beta_m y_{m,i}, \qquad i = 1, ..., k, \tag{4}$$

where λ_i is the ith saved eigenvalue of T_m, the corresponding eigenvector is the ith column of Y, and $y_{m,i}$ is the mth element of the ith column. At restart the first k basis vectors satisfy the following relation,

$$A\hat{q}_i = \hat{\alpha}_i \hat{q}_i + \hat{\beta}_i \hat{q}_{k+1}.$$

It is easy to arrange the algorithm so that $\hat{q}_i$ and q_i are stored in the same memory in a computer. The hat is dropped in the following algorithm.

ALGORITHM 3 *Restarted Lanczos iterations starting with k Ritz vectors and corresponding residual vector r_k satisfying $Aq_i = \alpha_i q_i + \beta_i q_{k+1}$, $i = 1, ..., k$, and $q_{k+1} = r_k/\|r_k\|$. The value k may be zero, in which case, α_i and β_i are uninitialized, and r_0 is the initial guess.*

1. **Initialization**.
   (a) $q_{k+1} = r_k/\|r_k\|$,
   (b) $p = Aq_{k+1}$,

(c) $\alpha_{k+1} = q_{k+1}^T p,$

(d) $r_{k+1} = p - \alpha_{k+1} q_{k+1} - \sum_{i=1}^{k} \beta_i q_i,$

(e) $\beta_{k+1} = \|r_{k+1}\|,$

2. **Iterate.** For $i = k+2, k+3, \ldots,$

(a) $q_i = r_{i-1}/\beta_{i-1},$

(b) $p = Aq_i,$

(c) $\alpha_i = q_i^T p,$

(d) $r_i = p - \alpha_i q_i - \beta_{i-1} q_{i-1},$

(e) $\beta_i = \|r_i\|.$

The difference between Algorithm 1 and 3 is in the initialization step. In Algorithm 1, at the first iteration, the step (d) is modified to be $r_1 = p - \alpha_1 q_1$. In Algorithm 3, step (d) in the initialization stage performs more SAXPY than during normal iterations. It should be fairly easy to modify an existing Lanczos program based on Algorithm 1 into a restarted version. To convert a complete eigenvalue program to use the above restarted Lanczos algorithm, the Rayleigh-Ritz projection step needs to be modified as well because the matrix T_m is not tridiagonal in the restarted Lanczos algorithm. Some of the options to deal with T_m include treating it as an full matrix, treating it as a banded matrix, and using Givens rotations to reduce it to a tridiagonal matrix. After deciding what to do, we can use an appropriate routine from LAPACK or EISPACK to find all eigenvalues and eigenvectors of T_m. At this point, the restarted Lanczos eigenvalue program performs convergence tests as in non-restarted versions.

After the convergence test, we will know whether all eigenvalues are computed to requested accuracies. If we have not found all wanted eigenvalues, we will restart the Lanczos algorithm. The main decision here is what Ritz pairs to save and how many. Many simple choices are described in literatures [11, 18, 20, 21, 22]. Most of them save a number of Ritz values near the wanted ones. For example, if we want to compute n_d largest eigenvalue of a matrix, we may choose to save $k = n_c + \min(n_d + n_d, m/2)$ largest Ritz values, where n_c are the number of converged eigenvalues. Usually, we also require that $k < m - 3$. Once k is decided, we can prepare all the necessary data to restarted Algorithm 3, $Q_m Y \rightarrow Q_k$, $q_{m+1} \rightarrow q_{k+1}$ and $\alpha_i, \beta_i, i = 1, \ldots, k$ can be computed from from Equation 4.

3 Numerical tests

In this section we will demonstrate the usefulness of the restarted Lanczos method by applying it on two eigenvalue problems. The restarted Lanczos method in this experiment maintain full orthogonality among the basis vectors. This scheme might use more time than a version that maintains semi-orthogonality, however it is easier to implement. we perform re-orthogonalization if $\alpha_{k+1}^2 + \sum_{i=1}^{k} \beta_i^2 > r_{k+1}^T r_{k+1}$ at step (1.d), or $\alpha_i^2 + \beta_{i-1}^2 > r_i^T r_i$ at step (2.d). These conditions are equivalent to what is used in ARPACK. The thick-restart Lanczos method saves $k = n_c + \min(n_d + n_d, m/2)$ Ritz vectors when restart. It is possible enhance the performance of the eigenvalue method by adopting a clever restarting strategy [21]. However, as we will see later, this simply scheme is enough to demonstrate the potential of the thick-restart Lanczos method.

The test matrix used here is call NASASRB. It is a symmetric matrix generated from finite element model of a shuttle rocket booster from NASA. The matrix has 54,870 rows and 2,677,324 nonzero elements. It is positive definite with the smallest eigenvalue of 4.74 and the largest eigenvalue of 2.65×10^9. The two examples attempt to compute the five extreme eigenvalues and their corresponding eigenvectors, where the first one computes the five largest eigenvalues and the second one computes the five smallest eigenvalues. The convergence tolerance is set to $\|r\| \leq 10^{-8}|\lambda|$. To validate the correctness of the eigenvalues, we compare the results from the thick-restart Lanczos method against the results from PARPACK. The version of PARPACK used in these tests is based on ARPACK version 2.1 dated 3/19/97. The above convergence test is also used in PARPACK for symmetric eigenvalue problems[2]. The number of vectors saved at restart in PARPACK is different from what used in the thick-restart Lanczos method. This will cause the two methods to use different numbers of matrix-vector multiplications. When the two methods use similar number of matrix-vector multiplications, the most significant difference between the two methods is the amount of work spent in orthogonalization.

The experiment is conducted on the Cray T3E 900 located at National Energy Research Scientific Computing (NERSC) center[3]. In the experiments reported here, only a small number of processors are used.

Through this experiment, we would like to show that the thick-restart Lanczos method with full re-orthogonalization can compute accurate solutions. Since the Lanczos method is an iterative scheme, the Ritz pairs can be improved by applying more iterations. What causes the solutions to be inaccurate is the loss of orthogonality among the Lanczos vectors. Ideally, we would like to compare the quantities computed by the Lanczos program against those computed in exact arithmetic. Since we can not produce the exact solutions, we resort to use the following two quantities to measure the errors. When there is loss of orthogonality, the actual residuals of the computed Ritz pairs would be different from the predictions. This difference between the predicted residual norms and the actual residual norms is the first error measure. The loss of orthogonality affects the accuracies of α_i and β_i, which in turn affects the Ritz values computed. In exact arithmetic, the Ritz values would be same as the Rayleigh quotients of their corresponding Ritz vectors. Our second measure of error is the difference between the Ritz values and their corresponding Rayleigh quotients. In the experiments, the actual residual norms and the Rayleigh quotients are computed using floating-point arithmetic as well. We consider the differences as zero if they are less than the expected errors of computing them. The error in multiplying A with an unit vector is bounded by $\epsilon_u \|A\|$. The upper bound of error in a

---

[2] Additional information on PARPACK and ARPACK can be found at http://www.caam.rice.edu/software/ARPACK.

[3] NERSC can be accessed from the world wide web at http://www.nersc.gov.

thick-restart Lanczos method
MATVEC: 185, restarts: 46, time: 12.0 sec (2 PE)

i	λ_i	$\frac{x_i^T A x_i}{x_i^T x_i} - \lambda_i$	$\lvert \beta_m e_m^T y \rvert$	$\lVert A x_i - \lambda_i x_i \rVert$
1	2648056755.2108812	8.58E-06	7.69E-05	7.67E-05
2	2647979344.2127852	-4.29E-06	1.19E-04	1.19E-04
3	2634048614.9911947	1.00E-05	3.68E-03	3.68E-03
4	2633679289.2081351	1.43E-06	6.56E-03	6.56E-03
5	2606151408.4051809	1.72E-05	1.06E+01	1.06E+01

PARPACK(10)
MATVEC: 184, restarts: 39, time: 14.6 sec (2PE)

i	λ_i	$\frac{x_i^T A x_i}{x_i^T x_i} - \lambda_i$	$\lvert \beta_m e_m^T y \rvert$	$\lVert A x_i - \lambda_i x_i \rVert$
1	2648056755.2108836	4.29E-06	5.37E-06	1.18E-05
2	2647979344.2127848	-1.91E-06	6.08E-07	3.33E-06
3	2634048614.9912062	-1.91E-06	7.10E-05	7.10E-05
4	2633679289.2081342	4.77E-07	1.32E-04	1.32E-04
5	2606151408.4051933	5.72E-06	1.63E+01	1.63E+01

Table 1. The largest eigenvalues of NASASRB and their errors (basis size 10).

thick-restart Lanczos method
MATVEC: 90, restarts: 8, time: 6.1 sec (2 PE)

i	λ_i	$\frac{x_i^T A x_i}{x_i^T x_i} - \lambda_i$	$\lvert \beta_m e_m^T y \rvert$	$\lVert A x_i - \lambda_i x_i \rVert$
1	2648056755.2108898	-9.54E-07	4.92E-01	4.92E-01
2	2647979344.2127857	-1.91E-06	7.48E-01	7.48E-01
3	2634048614.9912052	4.77E-07	1.41E+00	1.41E+00
4	2633679289.2081337	0.00E+00	2.32E+00	2.32E+00
5	2606151408.4051962	1.91E-06	7.22E+00	7.22E+00

PARPACK(20)
MATVEC: 157, restarts: 11, time: 13.4 sec (2 PE)

i	λ_i	$\frac{x_i^T A x_i}{x_i^T x_i} - \lambda_i$	$\lvert \beta_m e_m^T y \rvert$	$\lVert A x_i - \lambda_i x_i \rVert$
1	2648056755.2108912	-3.34E-06	7.11E-08	4.56E-06
2	2647979344.2127795	3.34E-06	8.02E-09	6.88E-06
3	2634048614.9912057	0.00E+00	5.39E-07	3.21E-06
4	2633679289.2081347	-9.54E-07	9.91E-07	3.35E-06
5	2606151408.4052033	-2.86E-06	8.46E-01	8.46E-01

Table 2. The largest eigenvalues of NASASRB and their errors (basis size 20).

residual norm explicitly computed using $\|r\| = \|Ax - \lambda x\|$ can be as large as $\epsilon_u \|A\|$. Therefore, if the differences between the predicted residual norms and the actual residual norms are smaller than $\epsilon_u \|A\|$ we consider them zero. Similarly, the error in computing a Rayleigh quotient $x_i^T A x_i / x_i^T x_i$ is also $\epsilon_u \|A\|$, if the differences between the Ritz values and the Rayleigh quotients $\left(\frac{x_i^T A x_i}{x_i^T x_i} - \lambda_i\right)$ are smaller than $\epsilon_u \|A\|$ we consider them zero.

The tables 1 and 2 show the results of computing the five largest eigenvalues and their corresponding errors. The tables list the computed Ritz values (λ_i), the differences between the Ritz values and the Rayleigh quotients of the computed Ritz vectors $\left(\frac{x_i^T A x_i}{x_i^T x_i} - \lambda_i\right)$, the estimated residual norms ($|\beta_m e_m^T y|$), and the actual residual norms ($\|A x_i - \lambda_i x_i\|$). The tables also show the number of matrix-vector multiplications (MATVEC), the number of restarts, and the time used by the two different methods.

Let's first compare the predicted residual norms and the actual residual norms. Table 1 shows results from finding the largest eigenvalues of NASASRB with basis size of 10. This example only needs about 40 restarts. The two methods solve the eigenvalue problem to about the same accuracy using roughly the same number of matrix-vector multiplications. The discrepancies between predicted residual norms and computed residual norms are on the order of 10^{-5} for PARPACK, which is 14 orders of magnitude smaller than the norm of the matrix. The same discrepancies for the thick-restart Lanczos method is slightly smaller, 10^{-7}. When we change the basis size to 20, see Table 2, the thick-restart Lanczos method uses much less matrix-vector multiplications to compute the five largest eigenvalues. The difference between the predicted and the actual residual norms are too small to be displayed for the thick-restart Lanczos method. The difference between predicted residual norms ($|\beta_m e_m^T y|$) and the actual computed residual norms for PARPACK are about 10^{-6}, which are close to $\epsilon_u \|A\|$.

The second indicator of error is the differences between the computed Ritz values and the Rayleigh quotients, $\left(\frac{x_i^T A x_i}{x_i^T x_i} - \lambda_i\right)$. In Table 1, these differences are on the order of 10^{-6} for most eigenvalues computed by the two methods. The same is true in Table 2. There are two entries for the thick-restart Lanczos method where the differences are about 10^{-5}.

Overall, the quality of the solutions found by the thick-restart Lanczos method is about the same as those computed by PARPACK. Because the Lanczos algorithm uses fewer arithmetic operations, the thick-restart Lanczos method uses less time to compute solutions in both above examples.

In many practical applications, the user wants to know the smallest eigenvalues of a matrix and the smallest eigenvalues are often much harder to compute than the largest ones. For this reason, we will show an example of computing the smallest eigenvalues of NASASRB. Similar to Tables 1 and 2, Table 3 shows the eigenvalues and the corresponding errors of this test problem. Using the non-restarted Lanczos method, 11,600 Lanczos steps are needed to compute the smallest eigenvalue to reasonable accuracy. Using basis size of 100, the smallest Ritz values computed by the restarted Lanczos method are larger thick-restart Lanczos method

MATVEC: 50467, restarts: 51, time: 8546 sec (8 PE)

i	λ_i	$\frac{x_i^T A x_i}{x_i^T x_i} - \lambda_i$	$|\beta_m e_m^T y|$	$\|A x_i - \lambda_i x_i\|$
1	4.7434139714076551	-2.40E-07	3.87E-15	1.83E-06
2	5.0304978087926671	-1.66E-07	3.34E-15	1.85E-06
3	57.269882851333911	3.04E-07	3.98E-07	1.09E-06
4	59.325143838635135	2.04E-06	1.30E-09	3.02E-06
5	114.48805524461864	-1.65E-07	2.27E-04	2.27E-04

PARPACK

MATVEC: 46761, restarts: 61, time: 8547 sec (16 PE)

i	λ_i	$\frac{x_i^T A x_i}{x_i^T x_i} - \lambda_i$	$|\beta_m e_m^T y|$	$\|A x_i - \lambda_i x_i\|$
1	388.43347600844902	-5.14E-07	6.99E+03	6.99E+03
2	1961.8238927546274	-1.94E-07	1.69E+04	1.69E+04
3	9890.9668426024327	-7.96E-09	5.92E+04	5.92E+04
4	18693.850673534958	4.62E-08	8.49E+04	8.49E+04
5	34854.798171100148	-2.88E-07	5.84E+04	5.84E+04

Table 3. The five smallest eigenvalues of NASASRB and their errors (basis size 1,000).

than ten after 50,000 matrix-vector multiplications. Using basis size of 1,000, the restarted Lanczos method declare the four smallest eigenvalues converged after about 50,000 matrix-vector multiplications. We did not attempt to find the fifth one since computing it will not change our observations. The eigenvalues found by the restarted Lanczos method agree with previously computed results using non-restarted Lanczos method. On the Cray T3E at NERSC, the longest jobs are limited to four hours long, different numbers of processors are used to allow the two methods to finish. The time and the number of processors used are also shown in Table 3. The PARPACK routine is stopped after 60 restarts. No Ritz values are converged in this case. This example probably can be better handled if a larger basis is used or a shift-and-invert operator is used instead. The main point of this test is to show that the restarted Lanczos method can find solutions to *difficult* eigenvalue problems. Even in this difficult case, the restarted Lanczos method computes accurate solutions.

In Table 3, the differences between Ritz values λ_i and the Rayleigh quotients of their corresponding Ritz vectors are on the order of $\epsilon_u \|A\|$. In fact the differences are roughly 16 orders of magnitudes smaller than the matrix norm which are smaller than the same differences when computing the largest eigenvalues, see Tables 1 and 2. The differences are about the same size for both the Lanczos method and PARPACK.

The discrepancies between the estimated residual norms and the computed residual norms are as large as 10^{-6} for the restarted Lanczos method with is again close to $\epsilon_u \|A\|$ for this test matrix. The discrepancies for PARPACK is too small to be noticed because the residual norms are relatively large.

4 Summary

In this paper we described a restarted version of Lanczos method that uses almost the same amount arithmetic per step as the original Lanczos method. Theoretically, the thick-restarted Lanczos method is equivalent to the implicitly restart Arnoldi method with the exact shifts are used as in ARPACK. However, because the thick-restart Lanczos method uses fewer arithmetic, we expect it to use less time than the implicitly restarted Arnoldi method. This is verified in a small number of experiments.

The thick-restart Lanczos method is equivalent to the implicitly restarted Arnoldi method implemented in ARPACK. However, because the difference in the number of vectors saved at restart, they use different number of matrix-vector multiplications to computed the largest eigenvalues of NASASRB, see Tables 2 and 3. Another important factor in the overall performance of the two restarted methods is the basis size. Comparing Tables 1 and 2, we see that significant difference in the number of matrix-vector multiplications used when the basis size change. Without restarting, the Lanczos method uses 75 steps to computed the five largest eigenvalues. The restarted Lanczos method uses only 90 matrix-vector multiplications when basis size is 20. In this case, a basis size of 20 is almost optimal choice. However, a basis size of 10 is not nearly as good.

Overall, the thick-restart Lanczos method is a efficient alternative to the implicitly restarted Arnoldi method on symmetric eigenvalue problems. To effective use a restarted method, the user has to choose an appropriate basis size and restarting strategy. These two issues are important research topics for the future.

References

[1] W. E. Arnoldi. The principle of minimized iteration in the solution of the matrix eigenvalue problem. *Quarterly of Applied Mathematics*, 9:17–29, 1951.

[2] J. Baglama, D. Calvetti, and L. Reichel. Iterative methods for the computation of a few eigenvalues of a large symmetric matrix. *BIT*, 36:400–421, 1996.

[3] S. Balay, W. Gropp, L. C. McInnes, and B. Smith. PETSc 2.0 users manual. Technical Report ANL-95/11, Mathematics and Computer Science Division, Argonne National Laboratory, 1995. Latest source code available at URL http://www.mcs.anl.gov/petsc.

[4] A. Chapman and Y. Saad. Deflated and augmented Krylov subspace techniques. Technical Report UMSI 95/181, Minnesota Supercomputing Institute, University of Minnesota, 1995.

[5] M. Crouzeix, B. Philippe, and M. Sadkane. The Davidson method. *SIAM J. Sci. Comput.*, 15:62–76, 1994.

[6] Ernest R. Davidson. The iterative calculation of a few of the lowest eigenvalues and corresponding eigenvectors of large real-symmetric matrices. *J. Comput. Phys.*, 17:87–94, 1975.

[7] Ernest R. Davidson. Super-matrix methods. *Computer Physics Communications*, 53:49–60, 1989.

[8] G. H. Golub and C. F. van Loan. *Matrix Computations*. The Johns Hopkins University Press, Baltimore, MD 21211, third edition, 1996.

[9] S. A. Hutchinson, J. N. Shadid, and R. S. Tuminaro. AZTEC user's guide. Technical Report SAND95-1559, Massively parallel computing research laboratory, Sandia National Laboratories, Albuquerque, NM, 1995.

[10] M. T. Jones and P. E. Plassmann. Blocksolve95 users manual: scalable library software for parallel solution of sparse linear systems. Technical Report ANL-95/48, Mathematics and Computer Science Division, Argonne national laboratory, Argonne, IL, 1995.

[11] Richard B. Lehoucq. *Analysis and implementation of an implicitly restarted Arnoldi iteration*. PhD thesis, Rice University, 1995.

[12] Ronald B. Morgan. On restarting the Arnoldi method for large nonsymmetric eigenvalue problems. *Mathematics of Computation*, 65(215):1213–1230, July 1996.

[13] Beresford N. Parlett. *The symmetric eigenvalue problem*. Prentice-Hall, Englewood Cliffs, NJ, 1980.

[14] Y. Saad. Analysis of augmented Krylov subspace techniques. Technical Report UMSI 95/175, Minnesota Supercomputing Institute, University of Minnesota, 1995.

[15] Yousef Saad. *Numerical Methods for Large Eigenvalue Problems*. Manchester University Press, 1993.

[16] Yousef Saad. *Iterative Methods for Sparse Linear Systems*. PWS publishing, Boston, MA, 1996.

[17] Miloud Sadkane. A block Arnoldi-Chebyshev method for computing the leading eigenpairs of large sparse unsymmetric matrices. *Numer. Math.*, 64(2):181–193, 1993.

[18] D. Sorensen, R. Lehoucq, P. Vu, and C. Yang. *ARPACK: an implementation of the Implicitly Restarted Arnoldi iteration that computes some of the eigenvalues and eigenvectors of a large sparse matrix*, 1995.

[19] D. S. Sorensen. Implicit application of polynomial filters in a K-step Arnoldi method. *SIAM J. Matrix Anal. Appl.*, 13(1):357–385, 1992.

[20] A. Stathopoulos, Y. Saad, and K. Wu. Thick restarting of the Davidson method: an extension to implicit restarting. In T. Manteuffel, S. McCormick, L. Adams, S. Ashby, H. Elman, R. Freund, A. Greenbaum, S. Parter, P. Saylor, N. Trefethen, H. van der Vorst, H. Walker, and O. Wildlund, editors, *Proceedings of Copper Mountain Conference on Iterative Methods*, Copper Mountain, Colorado, 1996.

[21] A. Stathopoulos, Y. Saad, and K. Wu. Dynamic thick restarting of the Davidson and the implicitly restarted Arnoldi methods. *SIAM J. Sci. Comput.*, 19(1):227–245, 1998.

[22] Kesheng Wu. *Preconditioned Techniques for Large Eigenvalue Problems*. PhD thesis, University of Minnesota, 1997. An updated version also appears as Technical Report TR97-038 at the Computer Science Department.

[23] Kesheng Wu and Horst Simon. Thick-restart Lanczos method for symmetric eigenvalue problems. Technical Report 41412, Lawrence Berkeley National Laboratory, 1998.

Portable Parallel Adaptation of Unstructured 3D Meshes

P.M. Selwood, M. Berzins, J.M. Nash and P.M. Dew

School of Computer Studies
The University of Leeds
Leeds LS2 9JT, West Yorkshire
United Kingdom

Abstract. The need to solve ever-larger transient CFD problems more efficiently and reliably has led to the use of mesh adaptation on distributed memory parallel computers. PTETRAD is a portable parallelisation of a general-purpose, unstructured, tetrahedral adaptation code. The variation of the tetrahedral mesh density both in space and time gives rise to dynamic load balancing problems that are time-varying in an unpredictable manner. The performance of a C/MPI version of PTETRAD will be demonstrated and the implementation of complex parallel hierarchical data-structures discussed. The need to make coding of such applications easier is addressed through the design of a novel abstract interface. The relationship of this interface to existing software and hardware systems will be described and the performance benefits illustrated by means of an example. The portable implementation of this interface by means of shared abstract data types will be considered.

1 Introduction

The use of parallel computers for the solution of large, complex computational partial differential equations problems has great potential for both significant increases in the number of mesh elements and the significant reduction of solution times. For transient problems accuracy and efficiency constraints may also require the use of mesh adaptation since solution features on different length scales are likely to evolve.

The use of mesh adaptation for such problems raises a number of issues. The irregular nature of the unstructured meshes and complex geometry requires the use of sophisticated load balancing techniques to ensure that each processor has the same number of mesh elements. The constantly changing nature of the mesh means that periodic re-balancing and data movement is required to maintain this load balance. The programming of complex tree-based data structures in parallel can be a difficult and error-prone task when working at the low-level required by current message passing systems. Given the long standing use of unstructured meshes in industrial applications it is perhaps surprising that there have been relatively few attempts to address these issues and to use tetrahedral-based transient solvers on parallel machines. Jimack [6] provides a recent survey. In this paper we will summarise and briefly extend our previous work in this area on load balancing and performance issues [12, 15] and then consider how to make make programming these important applications easier in greater detail. This will be addressed in the context of PTETRAD, a parallel implementation of a general purpose serial code, TETRAD (TETRahedral ADaptation), for the adaptation of unstructured tetrahedral meshes [14]. The technique used in this code is that of local refinements/de-refinements of the mesh to ensure sufficient density of the approximation space throughout the spatial domain at all times. Although TETRAD has been used with both finite element and finite volume solvers (cell-centred and cell-vertex), in this paper a cell-centred finite volume scheme is applied to systems of hyperbolic conservation laws in three space dimensions of the form

$$\frac{\partial \underline{u}}{\partial t} + \frac{\partial \underline{F}(\underline{u})}{\partial x} + \frac{\partial \underline{G}(\underline{u})}{\partial y} + \frac{\partial \underline{H}(\underline{u})}{\partial z} = 0 \, , \tag{1}$$

such as the Euler equations for example, and is a parallel version of the algorithm described in detail [14]. This is a conservative cell-centred scheme which is a second-order extension of Godunov's Riemann problem-based scheme using piecewise linear reconstructions of the primitive variables within each element and explicit two-stage-per step time-stepping algorithm, [14]. The mesh refinement is based upon the adaptive refinement of a coarse root mesh of tetrahedra which covers the spatial domain. The flexibility of the data structures held within the adaptation code means that the exact nature of the solver may vary (e.g. finite element or finite volume) provided it uses a tetrahedral mesh and is able to work with a partition of the elements of this mesh.

An overview of the PTETRAD solver and of the parallel adaptive and load balancing algorithms is given in the next section of the paper. A numerical example illustrates the performance and scalability of the code. Section 3 then discusses the developments in both languages and cache coherency algorithms and describes the new proposed software abstraction to support unstructured mesh calculations. A simple experiment is used to illustrate the effectiveness of the approach. Finally in Section 4, we consider the efficient and portable implementation of the abstraction using shared abstract data types. The paper concludes by discussing the viability of this approach and outlining future work.

2 The Parallel Adaptation of Unstructured 3D Meshes

The parallel computation of adaptive unstructured 3d meshes is typically based around distinct phases of execution, marked by the timesteps of the updated solution values, the mesh adaption points, and the possible redistribution of the mesh elements. Once the data is partitioned, the parallel version of the solver is straightforward to code due to the face data structure that exists within the adaptation software (see Figure 1 for example). To avoid any conflicts at the boundary between two sub-domains a standard "owner computes" rule is used for each of the faces when solving the approximate Riemann problems to determine fluxes. The use of halo elements ensures that the owner of each face has a copy of all of the data required to complete these flux calculations provided the halo data is updated before each of the two stages of the time-step calculation. In contrast, the parallel data structures are less straightforward. TETRAD utilises a tree-based hierarchical mesh structure, with a rich interconnection between mesh objects. Figure 1 indicates the TETRAD mesh object structures in which the main connectivity information used is 'element to edge to node to element' and a complete mesh hierarchy is maintained by both element and edge trees.

For parallelisation of TETRAD, there are two main options for partitioning a hierarchical mesh. The first is to partition the grid at the root or coarsest level while the second is to partition the leaf-level mesh, i.e. the actual computational grid. The pros and cons of these two approaches are discussed in [13] and the approach used here is that of partitioning the coarse mesh. The main disadvantage of this, that of possible suboptimal partition quality, can be avoided if the initial, coarse mesh is scaled as one adds more processors.

Given a partitioned mesh, need new data-structures are needed in order to support inter-processor communication and to ensure data consistency. Data consistency is handled by assigning ownership of mesh objects (elements, faces, edges and nodes). As is common in many solvers such as those used by [1] halo elements, a copy of inter-processor boundary elements (with their associated data) are used to reduce communication overheads. In order to have complete data-structures (e.g. elements have locally held nodes) on each processor, halo copies of relevant edge, node and face objects are stored. If a mesh object shares a boundary with many processors, it may have a halo copy on each of these. All halos have the same owner as the original mesh object. In situations where halos may have different data than the original, the original is used to overwrite the halo copies and thus is definitive. This is used to help prevent inconsistency between the various copies of data held.

2.1 Adaptation Algorithms

Both TETRAD ([14]) and its parallel implementation, PTETRAD ([13]), use a similar strategy to that outlined in [9] to perform adaptation. Edges are first marked for refinement/de-refinement (or neither) according to some estimate or indicator. Elements with all edges marked for refinement are refined regularly into eight children. The remaining elements which have one or more edge to be refined use so-called "green" refinement. This places an extra node at the centroid of each element and is used to provide a link between regular elements of differing levels of refinement. The types of refinement are illustrated in Figure 2. Green elements are not refined further as this may adversely affect mesh quality, but are first removed and then uniform refinement is applied to the parent element.

Mesh de-refinement takes place immediately before the refinement of a mesh and only when all edges of all children of an element are marked for de-refinement and when none of the neighbours of an element to be deleted are green elements or have edges which have been marked for refinement. This restriction is to

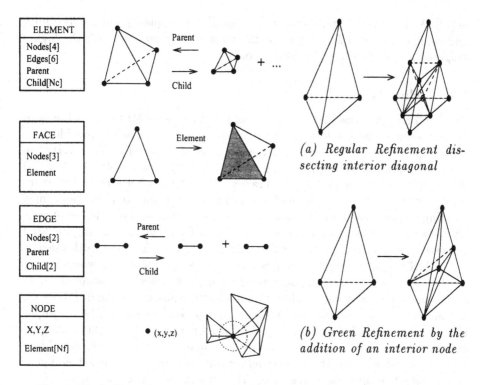

Fig. 1. Mesh Data-Structures in TETRAD

Fig. 2. Refinement methods

prevent the deleted elements immediately being generated again at the refinement stage which follows. Important implementation issues of these adaptive algorithms such as performing parallel searches in order to allow refinement of edges of green elements (which requires coarsening followed by regular refinement), maintaining mesh consistency and dealing with halo data in parallel are considered by [13].

2.2 Load Balancing

In order for the parallel solver to perform efficiently at each stage of the solution process, the work load of each processor should be about equal. If this equality of load is initially achieved through appropriately partitioning the original finite element/volume mesh across the processors then it is clear that the use of parallel adaptation for transient problems will cause the quality of the partition to deteriorate as the solution develops. Hence a parallel load-balancing technique is required which is capable of modifying an existing partition in a distributed manner so as to improve the quality of the partition whilst keeping the amount of data relocation as small as possible.

Parallel versions of Metis and Jostle are used to compute these new partitions; a comparison between these and a more recent algorithm for PTETRAD is given in [15]. All the algorithms produce partitions of similar quality, but the flexibility and configurability of Jostle gives it the advantage when used with the highly weighted graphs produced by adaptation.

2.3 Coding Issues

Parallel TETRAD was implemented using ANSI C with MPI due to the need for portability. The low-level of the message passing approach is not ideal for the complex data-structures and large amounts of communication involved in parallel adaptation. This type of application (with irregular, unstructured data) is currently poorly supported by libraries and compilers however and message passing is the only real option. As far as possible, communications are performed by using nonblocking MPI functions to avoid deadlock and allow a degree of overlap between computations and communications. Communications are coalesced wherever possible to minimise total latencies and maximise bandwidth usage.

A 'BSP' style of programming is used [10] as each processor works in 'supersteps', e.g. de-refine mesh and regular refine mesh, where no communications are done, separated by data update communications phases. Unlike BSP however not all the data is held consistently at each superstep boundary. For example, in de-refinement, halo edges may be removed from adaptation lists as the processor does not have enough information to make a correct decision. This is later corrected after all the regular refinement and subsequent construction of communications links has taken place. The construction of communication links involves much message-passing and so is a superstep boundary. The synchronisation of the edge refinement lists takes place after this and thus does not fully conform to BSP style.

The main difficulty of working at the message passing level is that it is very difficult to maintain consistency between mesh objects and their copies. Debugging situations where inconsistency occurs is particularly awkward as the inconsistency may not cause problems until the end of the superstep where communication occurs. Moreover, this problem tends to manifest itself in that send and receive buffers will not match up in size (as communication is coalesced) and discovering exactly which mesh object causes the problem can involve many hours work. While debugging tools such as SGI's Workshop Debugger are of some use, they do not yet have the same ease of use as their serial counterparts. Combined with the lack of high level language support for irregular parallelism, this creates one of the major obstacles for the more widespread use of irregular parallelism.

The portability of the ANSI C/MPI based approach of PTETRAD involved consideration of a variety of issues including the avoidance of deadlock and issues of buffering and the resulting code has been tested on a variety of platforms including a Cray T3D, an SGI PowerChallenge, an SGI Origin 2000 and on a workstation network of SGI O2s. Only one line is changed for the differing machines and this is purely for memory efficiency. The 64 bit Cray T3D gives the same accuracy for **double** and **int** variables as for a 32 bit machine, one need use only **float** and **short**. This can result in a significant memory saving as all the physical solution variables, and many of the flags, ids, etcetera used in the data structures may be halved.

2.4 Numerical Gas Jet Example

There are many factors that affect the scalability of the complete adaptive solution of a PDE. These include the number of remeshes required, the effectiveness of the remesh, the amount of work per adaptation step and the total depth of refinement. An example is given here that illustrates how the number of time-steps (or equivalently the amount of solver work) per remesh affects the scalability of the entire process. The other factors are discussed in [13]. This problem involves the solution of the gas jet problem discussed in Speares and Berzins [14]. For this problem remeshing takes place at fixed intervals rather than being driven by an error estimate. An initial mesh of 34,560 element is used with two levels of refinement giving an equivalent fine mesh resolution of 1.2 million elements. Jostle was used for repartitioning with a variety of imbalance thresholds; the best are shown in each case.

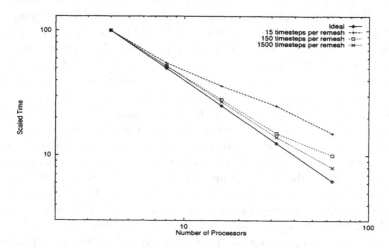

Fig. 3. Scalability with varying numbers of timesteps per remesh

Figure 3 shows the times (scaled for comparison purposes) of the whole solution process using varying amounts of work per remesh. The solver utilises explicit time integration and is computationally cheap. Varying the number of timesteps per remesh is thus a reasonable way of varying the work done as different solvers may require tighter CFL conditions (with smaller timesteps) or be computationally more demanding by using implicit methods with linear solves (e.g for multiple reacting chemical species). Remeshing itself does not scale particularly well, and the costs associated with redistribution of the mesh can be quite high. It is therefore not surprising that scalability improves as the amount of work per remesh is increased. This illustrates that this approach to the parallel adaptive solution of PDEs is best suited to those problems (such as fast flows or chemical reacting flows) that have more work inbetween remeshes, see [13].

3   Supporting Portable Parallel Adaptation

One of the main problems to overcome is to be able to efficiently support the communication of the required data within these phases, given that the mesh elements have been partitioned among the processors. Since the partitioning of the mesh and the redistribution of mesh elements are both carried out at run-time, a static compiler analysis is inappropriate. However, although a phase of execution has the characteristic that the data communicated is unpredictable, the communication patterns are repetitive. There have been many attempts to take advantage of this for irregular parallel problems at the operating system, language and applications level.

3.1   Relevant Work in Supporting Data Locality

Support at the operating system level comes from the use of adaptive paging protocols, which can take some advantage of the repetitive communications patterns to reduce the overall network traffic. The Reactive NUMA (R-NUMA) system [5] is a cache coherency protocol which combines the advantages of a conventional Cache Coherent NUMA (CC-NUMA) protocol with the Simple COMA (S-COMA). CC-NUMA improve data locality by caching shared data accesses (invoking the appropriate coherency protocol). S-COMA additionally allows the local memory of the processor to be used to store pages of shared data (using the same coherency protocol). This can potentially improve data locality by utilising the larger size of the main memory, but operating system overheads make this a more expensive option. An R-NUMA system combines the two approaches. *Reuse* pages contain data which is frequently accessed locally and *communication* pages are mainly used to exchange data between processors. The former can use the S-COMA protocol to reduce the network traffic caused due to cache capacity misses, and the latter can use the CC-NUMA approach to allow the sharing of sparse data at the cache-line level. The system can dynamically decide when a page switches between reuse and communication by noting the number of capacity and conflict misses. Applying the system to a partitioned mesh allows internal mesh elements to be located on reuse pages and the shared (halo) elements to use communication pages. This is under the assumption that these two distinct types of mesh elements can be arranged to lie on distinct pages, which would imply some form of domain-specific knowledge about the application being executed.

3.2   Related Language Developments

Language support is characterised by the provision of a framework for expressing irregular data structures. This is typically through the use of both static compiler analysis, and cache coherency protocols which can take advantage of unpredictable but repetitive communications patterns. For example, the C** language [8] uses data-parallelism to perform parallel operations on a data collection within a global namespace. The support for data-parallelism allows a static compiler analysis to identify distinct communication phases during execution, the communications patterns in each phase being predicted by a run-time cache protocol. The C** coherency protocol first incrementally builds a schedule to support the given communications, (by noting the incoming requests) while the second stage uses this schedule in order to prefetch the required data. A serious limitation is that an initially empty schedule for each phase, results in no cache blocks being requested at first. The protocol approach is similar to the CHAOS runtime system [2]. However, CHAOS requires the protocol stages to be explicitly defined within the application, and does not provide the necessary facility for the incremental update of a schedule.

3.3 Related Applications Developments

Support at the applications level is typified by the use of the skeleton/template approach, where domain-specific knowledge of the type of application to be executed can be used to support high performance in a portable manner. An example is the M-Tree abstract data type [16], which aims to capture the data structure and computational structure of adaptive numerical problems, using a regional mesh tree (in which each node represents a region of the domain and children specify sub-domains). Example applications are in the area of the adaptive-mesh heat flow problem, and adaptive multigrid. Related studies have used the terms *Distributed Shared Abstractions* (DSAs) [3] and *Information Sharing Mechanisms* [7], and *Shared Abstract Data Types*, as described in Section 4. The common characteristic is the representation of shared abstractions as structures that may be internally distributed across the nodes of a machine, so that their implementations can be altered while maintaining software portability.

3.4 The SOPHIA Interface

In this section an abstraction to support unstructured mesh computations is introduced. The **SO**ftware **P**refetch **H**alo **I**nterface **A**bstraction formalises the halo concept in such a way as to allow data to be prefetched when it is needed. The central idea is to specify explicitly the data that needs to be prefetched in order to optimise a calculation with a complex data structure. This specification thus allows best use to be made of the underlying system while also making use of application specific knowledge without incurring the penalties of general language support for irregular data structures. In particular, shared data is prefetched in the form of halos.

The SOPHIA interface takes the form of the following primitives:

1. `SOPHIA_Fetch(local_data, shared_data)`
   This establishes a halo and its related communication patterns based on the distributed data and the required sharing. A full local copy of the required remote data is made to enable local computations to be made exactly as they would be in serial. In order for the fetch to be made, it is required that data on interprocess boundaries, together with off-processor connectivity,

is specified at the initial partition stage. This connectivity is then stored either as a processor-pointer pair (for distributed memory machines) or just as a pointer (for shared memory) in order that the data structures under consideration may be traversed in order to complete the halo prefetch.

2. `SOPHIA_Update(shared_data, data_field)`
This updates the given shared data with the current values of the specified data fields. By using knowledge of the application, only the necessary specified data fields are updated rather than the whole halo and thus communications can be minimised.

3. `SOPHIA_Invalidate(shared_data)`
This removes a given halo from local memory. Careful use of invalidation followed by a new fetch enables e.g. changing the order of a solver partway through a CFD simulation.

This interface allows us to lift the abstraction above that of explicit messages passing, but with careful implementation (see Section 4) the performance benefits of message passing should not be lost. It is particularly suited to applications, such as mesh adaptation, with irregular, complex data that varies significantly over time due to the ability to change the halos held by use of invalidation. Moreover the users knowledge of the application can be harnessed to ensure that halo updates are efficient.

This approach has some similarities with BSP, in particular `SOPHIA_Fetch` plays a similar role to `bsp_push_reg` in that shared data is established for later communications. A BSP programming style with strict supersteps is not assumed however. Similarities also exist with CHAOS++ `Gobjects` [2], although SOPHIA is aimed more at dynamic calculations with the use of halo invalidation.

In the case of an irregular mesh calculation on tetrahedra, SOPHIA can be used in the following manner. Prior to a calculation, the distributed mesh consists of a local mesh with some processor-pointer pairs completing connectivity with remote subdomains. `SOPHIA_Fetch` creates local storage for remote elements which will be required as halos, copies the remote data into this storage and replaces the processor-pointer pair with a pointer to the new *local* copy. A communications schedule linking elements and these new halos is also established. During the computation, calls to `SOPHIA_Update` ensure that halos have the most recent values for appropriate solution values. Should a different halo be required (either due to a mesh adaptation or a change of numerical scheme), `SOPHIA_Invalidate` is used to remove existing halo elements and reset pointer-processor links with remote subdomains.

3.5 Experiment

A simple experiment demonstrates quite clearly how knowledge of the problem being solved and its associated sharing patterns can be utilised to improve the performance of a parallel application. The application considered is the parallel finite volume solver described in Section 2. In this solver two data updates per time-step are performed, in the SOPHIA manner, in order to ensure that halo elements have current data values for subsequent computations. This is currently achieved by using MPI packing to copy data $en - bloc$ to reduce the number of messages (and thus the total latency) used to a minimum. As there is prior knowledge of both the halo structure and the solver, this is a simple and quick operation.

This is contrasted with computation performed in a reactive cache style framework. Halo updates are done by request. That is, when data is required for the computation that is owned by a remote processor and a local copy is no longer valid, a request is sent for the correct data. The relevant component of the mesh is then updated with the most recent data. This has again been implemented with MPI. However, unlike a true reactive cache program, updates are done in a single phase rather than interleaved with computation. On most current hardware this approach should be faster as computations need not be interrupted in order to fetch remote data, although this may have less impact on systems where it is possible to truly overlap computations and communications.

Processors	SOPHIA	Reactive
4	104	125
8	55	81
16	32	48
32	15	26
64	9	18

Table 1. Timings (in secs) for Cray T3D. 15 time-steps on 97,481 element mesh

Processors	SOPHIA	Reactive
4	97	126
8	50	83
16	22	37

Table 2. Timings (in secs) for SGI Origin 2000. 15 time-steps on 387,220 element mesh

Table 1 gives timings for 15 time-steps using the reactive and SOPHIA style halo updates on a mesh of 97,481 elements on a Cray T3D. Similarly, Table 2 gives timings for 15 time-steps on a mesh of 387,220 elements on an SGI Origin 2000. It is clear that for both cases the SOPHIA style communications are faster than with reactive communications. Moreover as the number of processors increases so does the relative difference between the two communication styles. This difference is caused mainly by the increase in the number of messages (and hence the increase in total incurred latency) as the halo to compute element ratio increases with the number of processors.

4 Shared Abstract Data Types for Portable Performance

The previous section has shown the potential benefits of using the SOPHIA approach in that the low-level efficiency of message passing is combined with the specification of communications at a high level. Although SOPHIA can be implemented using MPI and thus readily ported between parallel platforms, this does not imply that this is an optimal solution on all platforms or that using some other form of communications mechanism might give significant performance improvements. The application of *Shared Abstract Data Types* (SADTs)

[4, 11] makes it possible to support such performance requirements. SADTs are an extension of ADTs to include concurrency. An SADT instance may be made visible to multiple processors, which may then concurrently invoke operations. The abstraction barrier enables implementations to take advantage of parallelism where appropriate, while shielding the user from details such as communication and synchronisation. The SADTs are used to hide the often complex concurrency issues involved in the use of fine-grain parallelism to support irregular forms of parallelism [11], as well as making use of coarse-grain parallelism. The application is written using a combination of coarse-grain parallelism, to structure the main phases of the code, and the SADTs to support more dynamic forms of parallelism. Weakened forms of data consistency can be used to maximise performance, where this will not affect the correctness of the application using it [4, 3].

An SADT can be used to support the updating of halo information, generated by the **SOPHIA_Update** interface call. The SADT strategy allows an optimised implementation to be developed for a given platform, while providing a portable interface. The key characteristics of such an SADT are:

Data placement: The segmentation of the PTETRAD elements across the processors allows for the concurrent access of the mesh elements and thus scalable performance.

Data consistency: Halo elements need only be fetched from their home location at specific points in the execution of the code, and then subsequently accessed locally. This weak consistency of the PTETRAD elements reduces the overheads of communication and synchronisation, supporting good practical performance.

- In the case of MPI, communications for each neighbouring mesh partition are coalesced by packing the required data into a contiguous buffer area before sending a message. This minimises the substantial startup latencies which MPI can incur.
- Using the SHMEM library, an alternative approach may be to directly write each element update to the halos. The pipelining of these updates can then be used to tolerate the network latency.

The adoption of data coalescing in MPI and the SHMEM approach of data pipelining both depend on the use of weak data consistency in order to provide high performance. This is in contrast to supporting a sequentially consistent shared data, which would require each halo update to complete before the subsequent update could then begin.

5   Concluding Remarks

In this paper the parallel PTETRAD solver has been outlined and issues concerning its parallel performance have been discussed and illustrated by numerical examples. The difficulty of programming such codes using low-level message passing has been considered and a suitable high-level abstraction developed,

SOPHIA. The effectiveness of this approach has been demonstrated using a simple example and a technique described for the portable parallel implementation of SOPHIA using shared abstract data types (SADTs). Future work in this area will be concerned with the development and coding of these SADTs in the context of SOPHIA.

References

1. J. Cabello, *"Parallel Explicit Unstructured Grid Solvers on Distributed Memory Computers"*, Advances in Eng. Software, 23, 189–200, 1996.
2. C. Chang and J. Saltz, *Object-Oriented Runtime Support for Complex Distributed Data Structures*, University of Maryland: Department of Computer Science and UMIACS Tech. Reports CS-TR-3438 and UMIACS-TR-95-35, 1995.
3. C. Clemencon, B. Mukherjee and K. Schwan, *Distributed Shared Abstractions (DSA) on Multiprocessors*, IEEE Trans. on Soft. Eng., vol 22(2), pp 132-152, 1996.
4. D. M. Goodeve, S. A. Dobson and J. R. Davy, *Programming with Shared Data Abstractions*, Irregular'97, Paderborn, Germany, 1997.
5. B. Falsafi and D. A. Wood, *Reactive NUMA: A Design for Unifying S-COMA with CC-NUMA*, ACM/IEEE Int. Symp. on Computer Architecture (ISCA), 1997.
6. P.K. Jimack *"Techniques for Parallel Adaptivity"*, Parallel and Distributed Processing for Computational Mechanics II (ed. B.H.V. Topping), Saxe-Coburg, 1998.
7. L. V. Kale and A. B. Sinha, *Information sharing mechanisms in parallel programs*, Proceedings of the 8th Int.Parallel Processing Symp., pp 461-468, 1994.
8. J. R. Larus, R. Richards and G. Viswanathan, *Parallel Programming in C** A Large-Grain Data-Parallel Programming Language*, In G. V. Wilson and P. Lu, editors, Parallel Programming Using C++, MIT Press, 1996.
9. R. Löhner, R. Camberos and M. Merriam, *"Parallel Unstructured Grid Generation"*, Comp. Meth. in Apl. Mech. Eng., 95, 343–357, 1992.
10. W. F. McColl, *An Architecture Independent Programming Model For Scalable Parallel Computing*, Portability and Performance for Parallel Processing, J. Ferrante and A. J. G. Hey eds, John Wiley and Sons, 1993.
11. J. M. Nash, *Scalable and Portable Performance for Irregular Problems Using the WPRAM Computational Model*, To appear in Information Processing Letters: Special Issue on Models for Parallel Computation.
12. P.M. Selwood, M. Berzins and P.M. Dew, *"3D Parallel Mesh Adaptivity: Data-Structures and Algorithms"*, Proc. of 8th SIAM Conf. on Parallel Proc. for Scientific Computing, SIAM, 1997.
13. P.M. Selwood, M. Berzins and P.M. Dew, *Parallel Unstructured Mesh Adaptation Algorithms; Implementation, Experiences and Scalability*, in prep. for Concurrency.
14. W. Speares and M. Berzins, *"A 3-D Unstructured Mesh Adaptation Algorithm for Time-Dependent Shock Dominated Problems"*, Int. J. Num. Meth. in Fluids, 25, 81–104, 1997.
15. N.Touheed, P.M. Selwood, M. Berzins and P.K. Jimack, *"A Comparison of Some Dynamics Load Balancing Algorithms for a Parallel Adaptive Solver "*, Parallel and Distributed Processing for Computational Mechanics II (ed. B.H.V. Topping), Saxe-Coburg, 1998.
16. Q. Wu, A. J. Field and P. H. J. Kelly, *Data Abstraction for Parallel Adaptive Computation*, in M. Kara, *et al* (eds)., Abstract Machine Models for Parallel and Distributed Computing, IOS Press, pp 105-118, 1996.

Partitioning Sparse Rectangular Matrices for Parallel Processing*

Tamara G. Kolda

Computer Science and Mathematics Division, Oak Ridge National Laboratory, Oak Ridge, TN 37831-6367. kolda@msr.epm.ornl.gov.

Abstract. We are interested in partitioning sparse rectangular matrices for parallel processing. The partitioning problem has been well-studied in the square symmetric case, but the rectangular problem has received very little attention. We will formalize the rectangular matrix partitioning problem and discuss several methods for solving it. We will extend the spectral partitioning method for symmetric matrices to the rectangular case and compare this method to three new methods — the alternating partitioning method and two hybrid methods. The hybrid methods will be shown to be best.

1 Introduction

Organizing the nonzero elements of a sparse matrix into a desirable pattern is a key problem in many scientific computing applications, particularly load balancing for parallel computation. In this paper we are interested in ordering the nonzeros of a given matrix into *approximate block diagonal form* via permutations. This problem corresponds directly to the partitioning problem in graph theory, and so is often referred to as matrix partitioning.

The partitioning problem has been well-studied in the symmetric case [1, 2, 4, 11, 12, 13, 14, 15, 16, 21, 22, 23]. The rectangular partitioning problem, however, has received very little attention; the primary reference in this area is Berry, Hendrickson, and Raghavan [3] on envelope reduction for hypertext matrices.

Let A denote a sparse rectangular $m \times n$ matrix. We will assume throughout that we are working with pattern (0-1) matrices, but the results and methods can easily be extended to nonnegative weighted matrices. Our goal is to partition A into a block 2×2 matrix so that most of the nonzeros are on the block diagonal and so that each block diagonal has about the same number of nonzeros. In other words, we wish to find permutation matrices P and Q such that

$$B \equiv PAQ = \begin{bmatrix} B_{11} & B_{12} \\ B_{21} & B_{22} \end{bmatrix} ,$$

* This work was supported by the Applied Mathematical Sciences Research Program, Office of Energy Research, U.S. Department of Energy, under contract DE-AC05-96OR22464 with Lockheed Martin Energy Research Corporation.

where B_{12} and B_{21} are as sparse as possible and the block rows or block columns each have about the same number of nonzeros. In order to avoid a trivial solution (e.g., $B_{11} = A$), we require that B_{11} have p rows and q columns where p is some integer between 1 and $m-1$ and q is some integer between 1 and $n-1$. The values p and q may or may not have been chosen in advance; typically we will want $p \approx m/2$ and $q \approx n/2$ to maintain load balance. If there exists P and Q such that B_{12} and B_{21} are identically zero, then we say that A is *block diagonalizable*. If we wish to partition A in a block $2^k \times 2^k$ matrix, we can recursively partition the block diagonals.

The matrix partitioning problem is equivalent to the *edge-weighted graph partitioning problem*: Given an undirected edge-weighted graph, partition the nodes into two sets of given sizes such that the sum of the weights of the edges that pass between the two sets is minimized. The graph partitioning problem is a well-known NP-complete problem (see problem ND14 on p. 209 in Garey and Johnson [6]). A rectangular $m \times n$ matrix corresponds to a bipartite graph [3] with m left nodes and n right nodes. There is an edge between left node i and right node j if a_{ij} is nonzero, and the weight of the edge is one. See Fig. 1 for an illustration. Suppose that we partition A so that the union of the first p rows and the first q columns form one partition and the remaining rows and columns form the other partition. The edges passing between the two partitions correspond to the nonzeros in the off-diagonal blocks of the partitioned matrix.

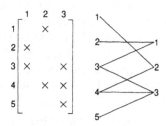

Fig. 1. The bipartite graph of a rectangular matrix.

Observe that the graph of A is disconnected if and only if the matrix A is block diagonalizable. Throughout we will assume that the graph of A is connected. If it is not, we will re-order the matrix so that it is block diagonalized with the blocks in decreasing order of size. We will only work with components that cross the boundary of the desired partition. In the discussion of the theory, we will assume that the graph of A is connected.

Many iterative methods, e.g., LSQR[20] require matrix-vector and matrix-vector-transpose multiplies with rectangular matrices. In Sect. 2 we will describe how to implement these kernels to take advantage of the partitioned matrix.

In Sect. 3, we will present several algorithms for the rectangular partitioning problem. We will discuss the well-known spectral partitioning method and show

how it can be applied to this problem. We will also introduce a new alternating partitioning method as well as two hybrid strategies.

In Sect. 4, we will compare the various partitioning methods, and show that the hybrid methods are the best.

2 Parallel Matrix-Vector Multiplication

We propose the following parallel implementations for the matrix-vector and matrix-transpose-vector multiplications. Suppose that we have $r = 2^k$ processors. We partition A into a block $r \times r$ matrix,

$$
A = \begin{bmatrix} A_{11} & A_{12} & \cdots & A_{1r} \\ A_{21} & A_{22} & \cdots & A_{2r} \\ \vdots & \vdots & \ddots & \vdots \\ A_{r1} & A_{r2} & \cdots & A_{rr} \end{bmatrix},
$$

so that most of the nonzeros are in the diagonal blocks. Here block (i, j) is of size $m_i \times n_j$ where $\sum_i m_i = m$ and $\sum_j n_j = n$.

Matrix-Vector Multiply (Block Row). We do the following on each processor to compute $y = Ax$:

1. Let i denote the processor id. This processor owns the ith block row of A, that is, $\begin{bmatrix} A_{i1} & A_{i2} & \cdots & A_{ip} \end{bmatrix}$, and x_i, the ith block of x of length n_i.
2. Send a message to each processor $j \neq i$ for which $A_{ji} \neq 0$. This message contains only those elements of x_i corresponding to nonzero *columns* in A_{ji}.
3. While waiting to receive messages, the processor computes the contribution from the diagonal matrix block, $y_i^{(i)} = A_{ii}x_i$. The block A_{ii}, while still sparse, may be dense enough to improve data locality.
4. Then, for each $j \neq i$ such that A_{ij} is nonzero, a message is received containing a sparse vector $\bar{x}_j$ that only has the elements of x_j corresponding to nonzero columns in A_{ij}, and $y_i^{(j)} = A_{ij}\bar{x}_i$, is computed. (We assume that processor i already knows which elements to expect from processor j.)
5. Finally, the ith block of the product y is computed via the sum $y_i = \sum_j y_i^{(j)}$. Block y_i is of size m_i.

Matrix-Transpose-Vector Multiply (Block Row). To compute $z = A^T v$, each processor does the following:

1. Let i denote the processor id. This processor owns v_i, the ith block of v of size m_i, and the ith block row of A.
2. Compute $z_j^{(i)} = A_{ij}^T v_i$, for each $j \neq i$ for which $A_{ij} \neq 0$. Observe that the number of nonzeros in $z_j^{(i)}$ is equal to the number of nonzero rows in A_{ij}^T, i.e., the number of nonzero columns in A_{ij}. Send the nonzero[1] elements of $z_j^{(i)}$ to processor j.

---

[1] Here we mean any elements that are guaranteed to be zero by the structure of A_{ij}. Elements that are zero by cancellation are still communicated.

3. While waiting to receive messages from the other processors, compute the diagonal block contribution $z_i^{(i)} = A_{ii}^T v_i$.

4. From each processor j such that $A_{ji} \neq 0$, receive $\bar{z}_i^{(j)}$ which contains only the nonzero elements of $z_i^{(j)}$. (Again, we assume that processor i already knows which elements to expect from processor j.)

5. Compute the ith component of the product, $z_i = z_i^{(i)} + \sum_{j \neq i} \bar{z}_i^{(j)}$. Block z_i is of size n_i.

Block column algorithms are analogous to those given for the block row layout. Observe that sparse off-diagonal blocks result in less message volume. See Hendrickson and Kolda [9] for more detail on the algorithm and for more details on potential applications.

3 Algorithms for the Rectangular Partitioning Problem

Here we will discuss how the well-known spectral method can be applied to the rectangular problem and introduce a new method that can be used on its own or in combination with other methods. The spectral method will be used as a basis for comparison to the new methods in Sect. 4.

3.1 Spectral Partitioning

In the symmetric problem, spectral partitioning based on the Fiedler vector is a well-known technique; see, for example, Pothen, Simon, and Liou [21]. Many people have studied the effectiveness of spectral graph partitioning; for example, Guattery and Miller [8] show that spectral partitioning can be bad, while Spielman and Teng [23] show that it can be good.

One natural way to approach the rectangular problem is to *symmetrize* the matrix A, yielding the $(m + n) \times (m + n)$ matrix

$$\tilde{A} = \begin{bmatrix} 0 & A \\ A^T & 0 \end{bmatrix} ,$$

and apply spectral partitioning to the symmetrized matrix. This approach is used by Berry et al. [3]. Note that the graphs of $\tilde{A}$ and A are the same.

In order to apply spectral partitioning, we compute the Laplacian of $\tilde{A}$,

$$L = D - \tilde{A} ,$$

where $D = \text{diag}\{d_1, d_2, \ldots, d_{m+n}\}$ and $d_i = \sum_j \tilde{a}_{ij}$. The matrix L is symmetric and semi-positive definite; furthermore, the multiplicity of the zero eigenvalue must be one since we are assuming that the graph of A, and hence of $\tilde{A}$, is connected [5]. Let w denote the Fiedler vector of L, that is, the eigenvector corresponding to the smallest positive eigenvalue of L. Let u denote the first m and v the last n elements of w, and sort the elements of u and v so that

$$u_{i_1} \geq u_{i_2} \geq \cdots \geq u_{i_m} ,$$
$$v_{j_1} \geq v_{j_2} \geq \cdots \geq v_{j_n} .$$

Then $\{i_1, i_2, \ldots, i_m\}$ and $\{j_1, j_2, \ldots, j_n\}$ define the row and column partitions respectively. In other words, assign rows $i_1, i_2, \ldots, i_p$ and columns $j_1, j_2, \ldots, j_q$ to the first partition and the remaining rows and columns to the second partition. Note that the ordering is independent of p and q. This means that the values of p and q may be fixed in advance as something like $\lceil m/2 \rceil$ and $\lceil n/2 \rceil$ respectively, or they may be chosen after the ordering has been computed to ensure good load balancing.

In Sect. 4 we will use this method as a basis for comparison for our new methods.

3.2 The Alternating Partitioning Method

Rather than trying to compute both the row and column partitions simultaneously as is done in the spectral method, the new method proposed in this section focuses on one partition at a time, switching back and forth. This method is derived from the Semi-Discrete Matrix Decomposition, a decomposition that has been used for image compression [19] and information retrieval [17, 18].

Before we describe the method, we will re-examine the problem. If we let $\mathcal{I}$ denote the set of row indices that are permuted to a value less than or equal to p and correspondingly let $\mathcal{I}^c$ denote the set of row indices permuted to a value greater than p and define the set $\mathcal{J}$ in an analogous way for the columns, then we can write the rectangular partitioning problem as a maximization problem,

$$\max_{\mathcal{I}, \mathcal{J}} \sum_{\substack{i \in \mathcal{I} \\ j \in \mathcal{J}}} a_{ij} + \sum_{\substack{i \in \mathcal{I}^c \\ j \in \mathcal{J}^c}} a_{ij} - \sum_{\substack{i \in \mathcal{I} \\ j \in \mathcal{J}^c}} a_{ij} - \sum_{\substack{i \in \mathcal{I}^c \\ j \in \mathcal{J}}} a_{ij} \ ,$$

$$\text{(1)}$$

$$\text{s.t. } \mathcal{I} \subseteq \{1, 2, \cdots, m\} \ , \ \mathcal{J} \subseteq \{1, 2, \cdots, n\} \ ,$$
$$|\mathcal{I}| = p, \ |\mathcal{J}| = q \ .$$

Here the objective function is the sum of the nonzeros on the block diagonal minus the sum of the elements off the block diagonal.

We can then rewrite this problem as an integer programming problem. Let x be a vector that defines the set membership for each row index; that is, $x_i = 1$ if row i is in $\mathcal{I}$, and $x_i = -1$ if row i is in $\mathcal{I}^c$, and let the vector y be defined in an analogous way for the columns. Then we can rewrite problem (1) as

$$\max x^T A y \ ,$$
$$\text{s.t. } x_i = \pm 1 \ , \ y_j = \pm 1 \ ,$$
$$x^T e = 2p - m \ , \ y^T e = 2q - n \ ,$$

$$\text{(2)}$$

where e denotes the ones vector whose length is implied by the context.

Although we cannot solve (2) exactly, we can use an *alternating method* to get an approximate solution. We fix the partition for, say, the right nodes (y), and then compute the best possible partition for the left nodes (x). Conversely, we then fix the partition for the left nodes, and compute the best possible partition for the right nodes, and so on.

Suppose that we have fixed the partition for the right nodes. To determine the best partition of the left nodes, we need to solve

$$\max x^T s \ ,$$
$$\text{s.t. } x_i = \pm 1 \ , \tag{3}$$
$$x^T e = 2p - m \ ,$$

where $s = Ay$ is fixed. The solution to this problem can be computed exactly. If we sort the entries of s so that

$$s_{i_1} \geq s_{i_2} \geq \cdots \geq s_{i_m} \ ,$$

then x defined by $x_{i_1} = x_{i_2} = \cdots = x_{i_p} = +1$ and $x_{i_{p+1}} = x_{i_{p+2}} = \cdots = x_{i_m} = -1$ is the exact solution to (3). Observe that the ordering of the elements of s does not depend on the value for p. If p has not been specified ahead of time, we would choose p to ensure load balancing. However, note that then p may be changing every iteration. An analogous procedure would be employed to find y when x is fixed.

Assuming p and q are fixed, each time we fix one side's partition and then compute the other, we are guaranteed that the value of the objective will never decrease. In other words, let $x^{(k)}$ and $y^{(k)}$ denote the partitions at the kth iteration of the method, and let f_k denote the objective value, $x^{(k)^T} A y^{(k)}$; then $f_{k+1} \geq f_k$ for all k. In the experiments presented in this paper, the method terminates when the objective value stops increasing. Alternatively, the method could terminate after at most some fixed number of iterations.

This method is called the alternating partitioning (AP) method and is specific to the rectangular problem since we are dealing with both row and column partitions and so we can alternate between working with one and then the other. In the symmetric case, we are only dealing with one partition.

We have not yet specified how to choose the first partition when we start the iterations, but that choice is important. In the standard method, we simply use the identity partition; however the next subsection will present two *hybrid* methods that use other techniques to generate a starting partition.

3.3 Hybrid Alternating Partitioning Methods

Since the alternating partitioning method is a greedy method, its key to success is having a good starting partition. Here we propose two possibilities.

Hybrid Spectral – Alternating Partitioning Method. This method uses the partition generated by the spectral method described in Sect. 3.1 as the starting partition for the alternating partitioning method.

Hybrid RCM – Alternating Partitioning Method. The Reverse Cuthill-McKee (RCM) method is not generally used as a partitioning method, but it generates a good inexpensive starting partition for the alternating partitioning method.

RCM is typically used for envelope reduction on symmetric matrices and is based on the graph of the matrix. Essentially, the method chooses a starting node and labels it 1. It then consecutively labels the nodes adjacent to it, then labels the nodes adjacent to them, and so forth. Once all the nodes are labelled, the ordering is reversed (hence the name). See George and Liu [7] for further discussion. In the nonsquare or nonsymmetric case, we apply RCM to the symmetrized matrix as was done by Berry et al. [3].

4 Experimental Results

In this section we compare the various partitioning methods presented in Sect. 3 on a collection of matrices listed in Table 1. These matrices were obtained from Matrix Market [2] with the exception of ccealink, man1, man2, and nhse400 which were provided by Michael Berry and are those used in Berry et al. [3]. These matrices range in size from 100×100 to 4000×400. All of the square matrices are structurally nonsymmetric.

Table 1. Rectangular test matrices.

Matrix	Rows	Columns	Nonzeros
bfw782a	782	782	7514
ccealink	1778	850	2388
gre__115	115	115	421
illc1033	1033	320	4732
impcol_a	207	207	572
impcol_c	137	137	411
impcol_d	425	425	1339
impcol_e	225	225	1308
man1	1853	625	3706
man2	1426	850	2388
nhse400	4233	400	5119
nnc261	261	261	1500
watson3	124	124	780
west0132	132	132	414
west0156	156	156	371
west0479	479	479	1910
utm300	300	300	3155

Tables 2, 3, and 4 show the results of partitioning the rectangular matrices into block 2×2, 8×8, and 16×16 matrices. In these experiments, all the matrices were converted to pattern (0-1) matrices, but we could have converted them to nonnegatively weighted matrices instead. When partitioning into a block 2×2

[2] http://math.nist.gov/MatrixMarket/

matrix, we choose $p = \lceil m/2 \rceil$ and $q = \lceil n/2 \rceil$. (Alternatively, we could compute p and q on the fly.) When partitioning into more blocks, we first partition into a block 2×2 matrix and the recursively partition the diagonal blocks into block 2×2 matrices until the desired number of blocks is reached.

The tables are formatted as follows. Each row corresponds to a given matrix whose name is specified in the first column. The second column lists the number of nonzeros outside of the block diagonal in the original matrix. The number of nonzeros outside of the block diagonal is equivalent to the number of edge cuts in the graph partitioning problem. Columns 3 – 6 list the number of nonzeros outside the block diagonal after applying the method listed in the column header to the original matrix. The number in parentheses is the time in seconds it took to compute the ordering. All timings were done in MATLAB. Spectral partitioning requires the second eigenvector corresponding to the second smallest eigenvalue, and this was computed using the MATLAB EIGS routine. (Note that EIGS has a random element to it, so the results for the spectral and hybrid spectral – alternating partitioning method cannot be repeated exactly.) The alternating partitioning method was implemented using our own code. The MATLAB SYMRCM routine was used to compute the RCM ordering.

Table 2. Approximate block diagonalization of rectangular matrices into block 2×2 matrices.

Matrix	Orig	Nonzeros outside of Block 2×2 Diagonal			
		Spectral	AP	RCM-AP	Spec-AP
bfw782a	2438	229 (27.52)	232 (2.04)	383 (2.38)	**147** (27.74)
ccealink	324	384 (14.92)	349 (1.99)	144 (2.50)	**106** (15.33)
gre__115	104	45 (1.14)	77 (0.12)	48 (0.18)	**40** (1.44)
illc1033	2336	588 (16.71)	**345** (1.28)	524 (1.57)	385 (16.61)
impcol_a	59	9 (1.90)	25 (0.19)	24 (0.26)	**8** (1.94)
impcol_c	46	25 (1.25)	26 (0.14)	31 (0.16)	**22** (1.24)
impcol_d	48	35 (6.74)	38 (0.45)	**20** (0.57)	25 (6.79)
impcol_e	152	26 (2.53)	46 (0.26)	49 (0.32)	**24** (2.58)
man1	803	202 (46.68)	703 (1.99)	211 (2.75)	**169** (46.13)
man2	759	189 (33.12)	382 (1.62)	204 (2.04)	**150** (33.31)
nhse400	213	186 (86.05)	169 (3.07)	74 (4.35)	**60** (83.67)
nnc261	75	73 (3.66)	**71** (0.31)	73 (0.38)	**71** (4.70)
watson3	157	87 (1.24)	119 (0.17)	**77** (0.20)	**77** (1.26)
west0132	87	33 (1.16)	36 (0.14)	30 (0.19)	**29** (1.17)
west0156	231	**7** (1.25)	37 (0.16)	18 (0.18)	**7** (1.27)
west0479	755	117 (8.64)	191 (0.58)	142 (0.73)	**112** (8.70)
utm300	266	**215** (5.38)	266 (0.61)	240 (0.75)	**215** (5.31)

We are using the spectral method as a basis for comparison for our new methods. Let us first consider the alternating partitioning method. In the 2×2 and 8×8 tests, the spectral method does better than the alternating partitioning

Table 3. Approximate block diagonalization of rectangular matrices into block 8 × 8 matrices.

Matrix	Orig	Nonzeros outside of Block 8 × 8 Diagonal			
		Spectral	AP	RCM-AP	Spec-AP
bfw782a	3872	832 (55.35)	1236 (5.02)	963 (5.72)	**804** (55.03)
ccealink	684	991 (36.90)	813 (5.61)	286 (6.37)	**216** (32.05)
gre__115	265	131 (3.92)	146 (0.53)	128 (0.64)	**118** (4.26)
illc1033	4066	2474 (31.35)	1640 (3.17)	**1398** (3.83)	1399 (32.72)
impcol_a	269	56 (5.66)	67 (0.73)	83 (0.88)	**48** (5.70)
impcol_c	170	108 (4.19)	106 (0.57)	94 (0.66)	**85** (4.14)
impcol_d	260	171 (15.02)	167 (1.37)	163 (1.82)	**128** (15.43)
impcol_e	724	303 (6.70)	173 (0.91)	153 (1.10)	**137** (6.81)
man1	1492	518 (82.38)	1249 (4.91)	578 (6.49)	**433** (83.09)
man2	2207	449 (59.49)	904 (3.90)	524 (4.95)	**326** (60.79)
nhse400	1001	1493 (143.98)	1142 (7.99)	**227** (10.25)	234 (148.47)
nnc261	596	**339** (9.23)	398 (1.04)	353 (1.27)	341 (9.99)
watson3	350	285 (4.02)	236 (0.56)	234 (0.71)	**232** (4.40)
west0132	308	97 (4.01)	111 (0.57)	104 (0.67)	**82** (4.14)
west0156	337	63 (4.13)	73 (0.59)	53 (0.76)	**42** (4.17)
west0479	1309	427 (18.58)	420 (1.65)	413 (2.08)	**274** (18.75)
utm300	1359	991 (12.19)	1211 (1.59)	**830** (1.98)	888 (12.96)

Table 4. Approximate block diagonalization of rectangular matrices into block 16 × 16 matrices.

Matrix	Orig	Nonzeros outside of Block 16 × 16 Diagonal			
		Spectral	AP	RCM-AP	Spec-AP
bfw782a	4320	1387 (61.83)	1951 (6.28)	1429 (7.17)	**1361** (62.18)
ccealink	904	1195 (38.56)	1053 (7.27)	358 (7.96)	**266** (32.63)
gre__115	291	171 (6.45)	195 (0.92)	162 (1.06)	**153** (6.35)
illc1033	4378	2776 (33.45)	2194 (4.12)	1985 (4.92)	**1920** (37.83)
impcol_a	413	125 (8.49)	115 (1.18)	127 (1.46)	**100** (8.45)
impcol_c	252	160 (8.15)	157 (1.04)	143 (1.16)	**135** (7.26)
impcol_d	444	289 (18.86)	278 (1.96)	305 (2.50)	**243** (19.83)
impcol_e	976	615 (9.95)	474 (1.55)	471 (1.86)	**450** (9.84)
man1	1837	796 (91.35)	1479 (6.21)	699 (7.98)	**602** (91.92)
man2	2519	596 (66.93)	1086 (5.06)	671 (6.27)	**456** (69.61)
nhse400	1594	1751 (161.72)	1795 (10.33)	**331** (12.64)	428 (156.84)
nnc261	890	556 (12.69)	558 (1.53)	538 (1.87)	**528** (13.08)
watson3	415	367 (6.39)	**289** (0.99)	297 (1.18)	300 (6.90)
west0132	376	162 (6.34)	152 (0.97)	141 (1.24)	**132** (6.88)
west0156	351	92 (8.11)	87 (1.04)	74 (1.23)	**70** (6.43)
west0479	1618	640 (22.55)	541 (2.33)	477 (2.84)	**414** (22.62)
utm300	1584	1504 (15.40)	1401 (2.21)	**1268** (2.70)	1268 (15.95)

method on the majority of matrices, but on the 16 × 16 tests, the alternating partitioning method does better than spectral partitioning 11 out of 17 times. In terms of time, the alternating partitioning method is approximately 10 times faster than the spectral method.

The hybrid RCM-AP method is overall better than the alternating partitioning method. It outperforms the spectral method in the majority of matrices in the 8 × 8 and 16 × 16 tests. This method is the best overall an average of 2.3 times for each block size. The time for the hybrid RCM-AP method is only slightly more than that for the alternating partitioning method, and still about 10 times less than that for the spectral method.

For all three block sizes, we see that the hybrid spectral-AP method is the best overall in terms of reducing the number of nonzeros outside the block diagonal. This method is only slightly more expensive than the spectral method in terms of time and yields consistently better results. Sometimes the improvement is remarkable; see, for example, ccealink in all four tables. Unfortunately, the spectral and hybrid spectral-AP methods are very expensive in terms of time since they require the computation of some eigenpairs.

Figure 2 shows the effect of different partitioning strategies on the west0156 matrix in the block 8 × 8 case.

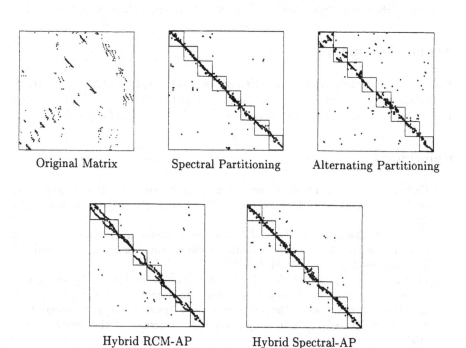

Original Matrix	Spectral Partitioning	Alternating Partitioning

Hybrid RCM-AP	Hybrid Spectral-AP

Fig. 2. Comparison of block 8 × 8 partitions on west0156.

Given these results, we recommend the hybrid spectral-AP method when quality is the top concern and the hybrid RCM-AP method when both time and quality matter.

5 Conclusions

In this work we introduced the rectangular matrix partitioning problem which is an extension of the symmetric matrix partitioning problem. The rectangular partitioning problem has a number of potential uses in iterative methods such as LSQR.

We introduced the alternating partitioning method as well as two hybrid methods and compared them with a rectangular version of the spectral partitioning method. The hybrid methods compared extremely favorably, and we recommend these as the methods of choice. The hybrid spectral-AP method is the overall best when partition quality is more important than the time to compute the partition, and the hybrid RCM-AP method is recommended when time is more important.

In other work [9, 10], this author and Bruce Hendrickson explore the multilevel method which is known to work very well for the symmetric partitioning problem [2, 11, 12, 13, 15]. We develop a multilevel method specific for bipartite graphs with various refinement strategies including the alternating partitioning method and a version of Kernighan-Lin for bipartite graphs. Eventually, we would also like to examine handling four or eight diagonal blocks directly [1, 11].

Acknowledgments

The author is indebted to Bruce Hendrickson and Dianne O'Leary for many helpful discussions. The author also thanks Eduardo D'Azevedo, Chuck Romine, and the anonymous referees for their reviews, and Mike Berry for providing data.

References

[1] Charles J. Alpert and So-Zen Yao. Spectral partitioning: The more eigenvectors, the better. In *32nd ACM/IEEE Design Automation Conference*, pages 195–200, 1995.

[2] Stephen T. Barnard and Horst D. Simon. A fast multilevel implementation of recursive spectral bisection for partitioning unstructured problems. *Concurrency: Practice and Experience*, 6:101–117, 1994.

[3] Michael W. Berry, Bruce Hendrickson, and Padma Raghavan. Sparse matrix reordering schemes for browsing hypertext. In James Renegar, Michael Shub, and Steve Smale, editors, *The Mathematics of Numerical Analysis*, volume 32 of *Lectures in Applied Mathematics*, pages 99–122. American Mathematical Society, 1996.

[4] Julie Falkner, Franz Rendl, and Henry Wolkowicz. A computational study of graph partitioning. *Math. Prog.*, 66:211–239, 1994.

[5] Miroslav Fiedler. Algebraic connectivity of graphs. *Czechoslovak Mathematical J.*, 23:298–305, 1973.

[6] Michael R. Garey and David S. Johnson. *Computers and Intractability: A Guide to the Theory of NP-Completeness.* W. H. Freeman and Company, New York, 1979.

[7] Alan George and Joseph W. Liu. *Computer Solution of Large Sparse Positive Definite Systems.* Prentice-Hall Series in Computational Mathematics. Prentice-Hall, Englewood Cliffs, 1981.

[8] Stephen Guattery and Gary L. Miller. On the quality of spectral seperators. Accepted for publication in *SIAM J. Matrix Anal. Appl.*, 1997.

[9] Bruce Hendrickson and Tamara G. Kolda. Partitioning nonsquare and nonsymmetric matrices for parallel processing. In preparation, 1998.

[10] Bruce Hendrickson and Tamara G. Kolda. Partitioning sparse rectangular matrices for parallel computations of Ax and $A^T v$. Accepted for publication in *Proc. PARA98: Workshop on Applied Parallel Computing in Large Scale Scientific and Industrial Problems*, 1998.

[11] Bruce Hendrickson and Robert Leland. Multidimensional spectral load balancing. Technical Report 93-0074, Sandia Natl. Lab., Albuquerque, NM, 87185, 1993.

[12] Bruce Hendrickson and Robert Leland. An improved spectral graph partitioning algorithm for mapping parallel computations. *SIAM J. Sci. Stat. Comput.*, 16:452–469, 1995.

[13] Bruce Hendrickson and Robert Leland. A multilevel algorithm for partitioning graphs. In *Proc. Supercomputing '95*. ACM, 1995.

[14] Bruce Hendrickson, Robert Leland, and Rafael Van Driessche. Skewed graph partitioning. In *Proc. Eighth SIAM Conf. on Parallel Processing for Scientific Computing*. SIAM, 1997.

[15] George Karypis and Vipin Kumar. A fast and high quality multilevel scheme for paritioning irregular graphs. Technical Report 95-035, Dept. Computer Science, Univ. Minnesota, Minneapolis, MN 55455, 1995.

[16] George Karypis and Vipin Kumar. Parallel multilevel graph partitioning. Technical Report 95-036, Dept. Computer Science, Univ. Minnesota, Minneapolis, MN 55455, 1995.

[17] Tamara G. Kolda. *Limited-Memory Matrix Methods with Applications.* PhD thesis, Applied Mathematics Program, Univ. Maryland, College Park, MD 20742, 1997.

[18] Tamara G. Kolda and Dianne P. O'Leary. A semi-discrete matrix decomposition for latent semantic indexing in information retrieval. Accepted for publication in *ACM Trans. Information Systems*, 1997.

[19] Dianne P. O'Leary and Shmuel Peleg. Digital image compression by outer product expansion. *IEEE Trans. Comm.*, 31:441–444, 1983.

[20] Christopher C. Paige and Michael A. Saunders. LSQR: An algorithm for sparse linear equations and sparse least squares. *ACM Trans. Mathematical Software*, 8:43–71, 1982.

[21] Alex Pothen, Horst D. Simon, and Kang-Pu Liou. Partitioning sparse matrices with eigenvectors of graphs. *SIAM J. Matrix Anal. Appl.*, 11:430–452, 1990.

[22] Horst D. Simon. Partitioning of unstructured problems for parallel processing. In *Computing Systems in Engineering*, number 2/3, pages 135–148. Pergammon Press, 1991.

[23] Daniel A. Speilman and Shang-Hua Teng. Spectral partitioning works: Planar graphs and finite element meshes. Unpublished manuscript, 1996.

Locality Preserving Load Balancing
with Provably Small Overhead

Robert Garmann*

Universität Dortmund, FB Informatik VII

Abstract. Parallelizing dynamic scientific applications involves solving the dynamic load balancing problem. The balancing should take the communication requirements of the application into account. Many problems are dealing with objects in k-dimensional space with very special communication patterns. We describe a kind of an orthogonal recursive bisection clustering and show that its dynamic adaption involves only small overhead. As a spatial clustering it is well suited to applications with local communication.

1 Introduction

Scientific computations often consist of a set of interdepending tasks that can be associated with geometric locations such that each task communicates directly only with those tasks that are mapped to locations close to the own location. The computation can be parallelized by dividing the space into spatial clusters (Fig. 1). Each processor gets a different cluster and executes the contained tasks. Sometimes this approach is called the *"owner computes" rule*. The intention of spatially clustering the tasks is to reduce the communication between the processors thereby avoiding the problems of message contention and hot spots.

The location of tasks may change over time. Tasks may be dynamically created or removed. We can think of the set of tasks as a classical database with operations for insertions and deletions. The difference to the classical situation here is that the requests for updates do not originate from an external source but from the database itself.

Because of limited memory and computation times the tasks should be distributed evenly across the processors. We will describe a procedure that rebalances a database "on the fly" concurrently with an ongoing application. A kind of orthogonal recursive bisection clustering is used for that purpose. We describe procedures for imbalance detection and for locality preserving rebalancing. The estimated amortized computational time complexity of these procedures is shown to be small for every single dynamic task update.

Section 2 concerns related work to our load balancing approach. In Sect. 3 a computational model is described that is used in our complexity considerations in Sect. 5. A description of the load balancing algorithm and its concurrent execution with a user-defined application is treated in Sect. 4. Section 6 comprises the definition of a concrete simple application and experimental results.

---

* E-mail: robert.garmann@cs.uni-dortmund.de

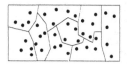

Fig. 1. Every processor gets a different cluster of tasks

2   Related Work

In general parallelizing a scientific application is done by solving the *mapping* problem, i. e. map a guest graph (describing tasks and dependencies) to a host graph (describing the processors and communication channels) thereby minimizing the maximum load, dilation and congestion. Current supercomputers dispose of such powerful routing algorithms that it is justified to ignore the dilation cost. Congestion is hard to estimate since it depends on the specific routing algorithm. Hence mostly it is accounted for only heuristically as below and it remains to solve the load balancing problem.

In applications with local communication a spatial task clustering is used in order to reduce the chance of congestion. Only the marginal tasks of a task-cluster have to communicate with other clusters. Diffusion methods [4, 9] remap tasks while keeping adjacent tasks on neighbouring processors. Neighbouring processors exchange load depending on local load differences. Because of this locality, long distance remote data accesses are difficult, because there is no compact cluster representation that could be replicated on all processors and be used as a directory for remote accesses.

Orthogonal recursive bisection methods [1] have a compact tree representation. Rebalancing can be performed asynchronously and independently in subtrees. In order to minimize communication one would use a cluster shape that minimizes the ratio of surface to volume. Therefore a multidimensional bisection strategy is superior to a onedimensional clustering, which would produce thin long "slices".

Rectangular grid clusterings are often used for matrix-vector-multiplication [5]. Here an exact load balance cannot be achieved by simply moving cutting hyperplanes. Instead the tasks must be permuted randomly in order to achieve balance with high probability [7].

Mapping multidimensional spaces onto a space-filling curve and then partitioning the resulting "pearl necklet" by a onedimensional procedure seems to be a straightforward method to overcome the problem of long, thin clusters. One problem with this are jaggy cluster boundaries, which make statements on the worst case communication complexity of the underlying application difficult.

In this paper we will focus on an orthogonal recursive bisection clustering method. The main reason is that we are concerned with applications consisting of tasks that need to arbitrarily access tasks on distant processors. This is realized most efficiently in an orthogonal recursive bisection clustering, where every processor holds information about the location of other tasks. In the rest of this

```
PQ = part of all initial tasks;
while (not finished)
   while (pending())
      receive(task);
      PQ <- task;
   endwhile
   perform first task of PQ locally;
   if (unbalanced()) then init-rebalance();
endwhile
```

Fig. 2. Scheduling loop

paper we will describe the method and show that this clustering can be adapted dynamically with low amortized overhead for each dynamic task update.

3 Models

We use point-to-point message passing as a programming model. Collective communications such as broadcasts and reductions on m processors can be simulated by many point-to-point communications in $\log(m)$ parallel time in a tree-like manner, if we assume that the communication channels between each pair of processors are independent.

In order to analyze the performance of our algorithm we use a variation of the *LogP* model [2]. This model assumes a set of p processors working asynchronously. Each processor pair is assumed to communicate equally efficient. A processor can communicate with only one other processor at the same time. Communication performance is specified by an upper bound on the *latency L*, a fixed *overhead o* for each transmission and a *gap g* identifying a minimum time interval between consecutive transmissions.

4 Algorithms

We assume that all tasks are equally complex. We gain load balance by clustering the tasks with respect to their associated geometric location. Dynamic updates are monitored by an imbalance detector. If an imbalance occurs, then a simple rebalancing takes place that relocates cluster boundaries thereby preserving locality.

The above technique is similar to *partial rebuilding* [8], which is used in keeping classical dynamic multidimensional databases balanced, and is itself derived from the general concept of dynamizing static data structures [6].

4.1 Embedding of Load Balancing in an Application

On every processor a priority queue PQ stores all tasks that are to be executed on the processor. A scheduling loop that is executed asynchronously on every processor looks for incoming messages and executes the tasks (Fig. 2). Execution of a task may involve dynamical creation or deletion of tasks. This may cause

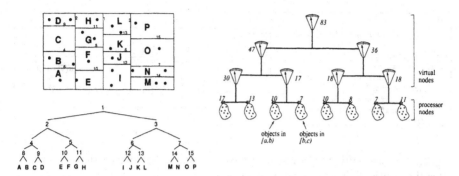

Fig. 3. Left: Tree clustering for $p = 16$ processors. Right: A database as a tree of scales. Each leaf node contains the objects of a processor. Inner nodes measure imbalance between children. The numbers are example values for the number of objects $E(\cdot)$ associated with the nodes.

imbalance of workload and maybe a rebalancing operation is initiated. The call to `init-rebalance` involves sending of messages to other processors and then returns immediately. The messages are processed by the destination processors within their regular scheduling loop. So, rebalancing does not synchronize any processors. It is performed besides the normal computations. This leads to the desirable effect that latency times are overlapped by computations.

4.2 The Clustering Method

The set of points in $\mathbb{R}^k$ representing tasks or objects[1] is stored in a binary search tree of limited depth. We will assume that there exists some integer l such that the number of processors is $p = 2^{kl}$. The real implementation is free of this restriction, but here it simplifies the presentation of the concepts.

We split the objects into 2^l subsets based on the value of the first coordinate (Fig. 3). These subsets are stored in the leaves of a binary tree of height l. At the next l levels we split the subsets with respect to the second coordinate. The last l levels of the tree are formed using the k-th coordinate. We arrive at a binary tree of height $h := kl$.

In theory it is often assumed that the points are *in general position*, because then there exists a perfectly balanced clustering. If this is not true in practice, but the maximum number of points with equal value in some coordinate is much smaller than N, then only small imbalances will occur. Hence, we will neglect the problem and assume that the points are in general position.

4.3 Rebalancing

Let us interpret the database of objects as a tree of scales (Fig. 3), where each inner node measures the imbalance between the two children and each leaf node contains the objects out of an subinterval of $\mathbb{R}^k$.

---

[1] In realistic applications usually tasks are associated with some data stored in an "object". This is why we use the terms task and object synonymically.

```
diff(v, a, b) :=
   ⌊½ ((E(vᵣ) + b) − (E(vₗ) + a))⌋

shift(v, a, b):
   if ( v is leaf )
   then execshift(v, a, b)
   else M :=diff(v, a, b)
        shift(vₗ, a, +M)
        shift(vᵣ, −M, b)
   endif
```

$$\text{diff}(v, a, b) :=$$
$$\text{if } \left(1 - \frac{\beta}{2^{d(v)+1}} \leq \frac{E(v_l) + a}{E(v_r) + b} \leq \frac{1}{1 - \frac{\beta}{2^{d(v)+1}}}\right)$$
$$\text{then return } 0$$
$$\text{else return } \left\lfloor \tfrac{1}{2}\left((E(v_r) + b) - (E(v_l) + a)\right)\right\rfloor$$
$$\text{endif}$$

a)   b)

Fig. 4. a) Strict shifting with respect to external shifts. b) Relaxed shifting. $d(v)$ denotes the length of the path from v to the leaves.

Definition 1. Let $E(v)$ denote the number of objects that are associated with a node v. For inner nodes $E(v)$ is the sum of the $E(\cdot)$ values of the two children v_l and v_r. We call an inner node v perfectly balanced, if $|E(v_l) - E(v_r)| \leq 1$.

The balance of an inner node v is defined as $B(v) := \frac{\min(E(v_l), E(v_r))}{\max(E(v_l), E(v_r))}$. We call an inner node v <u>balanced</u>, if $B(v) \geq 1 - \beta$ for some constant $0 < \beta \leq 1$.

If we could guarantee that at any time every inner node of the tree is balanced, then all processors get roughly the same amount of objects. Assume, that an inner node v of the tree has got out of balance. We can bring v back to perfect balance by shifting $M := \frac{1}{2} |E(v_r) - E(v_l)|$ objects from the overloaded child to the underloaded child. Since inner nodes are virtual nodes not containing real objects, the shift has to take place at the bottom level of the tree. The question, which leaf should deliver the objects to be shifted, is answered by the ordering inherent in our tree clustering, where the objects below v_l are smaller in some fixed coordinate than those below v_r. We treat the one-dimensional case first.

The One-Dimensional Case. Let v denote an unbalanced inner node. E. g. let $\beta = 0.6$, then the node for which $E(v) = 47$ in Fig. 3 is unbalanced. v can be balanced perfectly by shifting $M := \frac{1}{2} |17 - 30| \approx 6$ objects from the left to the right subtree. The source of objects is the rightmost leaf below v_l, the target is the leftmost leaf below v_r. The resulting situation is: $E(v) = 47$, $E(v_l) = 30 - M = 24$, $E(v_r) = 17 + M = 23$, $E(v_{ll}) = 17$, $E(v_{lr}) = 13 - M = 7$, $E(v_{rl}) = 10 + M = 16$, $E(v_{rr}) = 7$. Unfortunately, now the nodes v_l and v_r both are unbalanced. Rebalancing these two nodes can be done in parallel independently of each other in the next step.

We describe a procedure $\text{shift}(v, a, b)$ that is applied recursively to all inner nodes v and is responsible for rebalancing v and all its descendants (Fig. 4a). The arguments a and b represent two shifts that act from the outside to the leftmost or rightmost leaf below v, respectively. The rebalancing of v takes place under the environment conditions that the the leftmost leaf below v has received or will receive a objects from its left neighbour and the rightmost leaf below v gets b objects from its right neighbour. Negative values a, b mean, that the leftmost or rightmost leaves are delivering $|a|$ or $|b|$ objects to their neighbours.

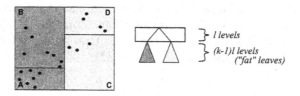

Fig. 5. Higher dimensional rebalancing is reduced to k one-dimensional phases

If v is a leaf, the shifts a and b are executed directly (**execshift**(v, a, b)) by sending up to two messages containing objects to the neighbours. Of course for negative a or b no message is sent, but the leaf will receive objects from its neighbours in the near future. The leaf does not wait for message arrival, but will receive the message in its regular scheduling loop.

If v is an inner node, we calculate the number of objects M that need to be shifted internally between v_l and v_r and then pass M to the children of v in appropriate order. The initial call is **shift**$(r, 0, 0)$, where r denotes the root of the tree.

There are two open questions in how to realize **shift**. The first is, who executes the code of **shift** for the inner nodes v. This can be answered easily by mapping the **shift** calls for an inner node v to the leftmost leaf below v. Then every processor is responsible for not more than h inner nodes. The second question is, what has to be done, if at any leaf w the procedure **execshift**(w, a, b) cannot be performed immediately, because w has to deliver more objects than are available locally, e. g. $a < 0$ and $E(w) < |a|$? Then w has to postpone its **execshift** call until it gets objects from its immediate right neighbour.

The k-Dimensional Case. As in the one-dimensional case we have to move objects between processors if any inner node v has got out of balance. The number of objects to be moved is the same as in the one-dimensional case but not the source of the objects. Consider a tree of height 2 storing points in $\mathbb{R}^2$ as sketched in Fig. 5. Assume that the root r is unbalanced. We can bring r to perfect balance by shifting $M = 3$ objects from $A \cup B$ to $C \cup D$. The objects to be shifted are those with largest value in the first coordinate. The problem with this is that it is not known which of the source processors contain the M largest objects with respect to the first coordinate. In order to determine the M largest objects processors A and B will have to communicate.

The clustering as described above is a tree of height kl with k groups of l levels each dividing the space into slabs with respect to a fixed coordinate. We will isolate the first group of l levels (Fig. 5) and regard the processor sets beneath consisting of subtrees of height $(k - 1)l$ as "fat" leaves. The algorithm of Fig. 4a is applied to this tree of height l. The only difference to the one-dimensional case is the implementation of **execshift**(v, a, b). Here each fat leaf has to determine its M largest or smallest objects in a communication operation before shifting the objects to a neighbouring fat leaf.

After the shifts with respect to the first coordinate have completed, then recursively the subtrees in the fat leaves are balanced independently in the same manner. We need k phases to balance the whole tree. In the k-th phase we will use the ordinary one-dimensional version of **shift**.

4.4 Detecting Imbalance

If once a balanced binary tree is given, how can we detect that any node of the tree gets out of balance? The problem with this is that imbalances result from unpredictable updates that are made independently by the leaves of the tree. The following lemma states a necessary condition for the situation that some inner node gets out of balance.

Lemma 2. *Assume that each node of the binary tree was balanced at time t_0. Let $E^0(v)$ denote the load at an inner node v at time t_0. Let r denote the root of the tree.*

If v is unbalanced at any time t after t_0 then there exists some leaf w below v for which

$$E(w) \notin \left[\frac{E^0(r)}{p} 2 \frac{1-\beta}{2-\beta}, \frac{E^0(r)}{p} 2 \frac{1}{2-\beta} \right] \ .$$

Proof. Assume, that the object number in all leaves below v is covered by the above interval at time t. For $u \in \{v_l, v_r\}$ let $d(u)$ denote the length of the path from u to the leaves. Then $E(u)$ is covered by $2^{d(u)} [\min E(w), \max E(w)]$, where the minimum and maximum covers all leaves w below v. Using Def. 1 we can easily conclude that $B(v) \geq 1 - \beta$, i. e. v is balanced.

The obvious strategy derived from this lemma to maintain balance in the whole tree is the following. Every leaf tracks whether its number $E(\cdot)$ is covered by the above narrowing interval. The tracking does not require any communication. We say a leaf *leaves* the narrowing interval, if its number $E(\cdot)$ is not covered by the interval. If any leaf leaves the interval, we apply the function $\text{shift}(r, 0, 0)$ to the root r in order to rebalance all nodes of the tree. Then we define a new narrowing interval by setting $E^0(r) := E(r)$.

4.5 Improving Average Efficiency

There are update patterns for which the above strategy is best. E. g. consider an update pattern that starts in perfect balance (load is $\frac{E^0(r)}{p}$ everywhere) and proceeds such that all leaves below the left son of the root create objects at the same rate and all leaves below the right son of the root remove objects at this rate. All leaves will leave the narrowing interval at the same time and then the root will be unbalanced. Now, rebalancing the tree perfectly is justified.

But, consider another update pattern where all leaves create objects at the same rate. All leaves will leave the narrowing interval at the same time, but there is no need for any shifts, since all inner nodes are balanced. We will slightly modify the **shift** procedure by replacing the **diff** function as shown in Fig. 4b. Now, a small imbalance at node v is ignored. This prevents slightly unbalanced nodes being rebalanced perfectly which would involve expensive, but useless small messages. Below we will see that using this relaxed version does not worsen the worst case behaviour significantly. But we expect the average behaviour of the relaxed version to perform much better in realistic applications than the strict version. Therefore we used the relaxed version in our experiments in Sect. 6.

5 Complexity Analysis

We will examine the complexity of the above mechanisms for detecting and rebalancing an imbalanced distribution of objects. In sect. 5.1 we calculate the cost of a complete rebalancing operation. Fortunately, we will see that, after rebalancing is complete, several updates are possible without any rebalancing being necessary. We quantify this number separately for the strict case (Sect. 5.3, Lemma 7) and the relaxed case (Sect. 5.4, Lemma 10). We will charge the communication cost of a rebalancing operation to the updates. In Sect. 5.2 we show that every single update has a small *amortized time complexity*. Theorem 8 in Sect. 5.3 and Theorem 11 in Sect. 5.4 specialize the result separately for the strict and the relaxed case.

5.1 Absolute Cost of Rebalancing

Definition 3. The above algorithm uses two types of messages: <u>administrative</u> messages containing small amounts of data and <u>object</u> messages containing objects. For the subsequent estimations we will assume a cost of $C_{\text{adm}} = o + (A-1)g$ for an administrative message where A denotes the maximum memory size of any administrative message. The cost of an object message is assumed to be $C_{\text{obj}}(M) = o + (MS - 1)g$ where S denotes the memory size of an object and M is the number of objects contained in the message.

These formulas assume that the cost of a message consists of a fixed <u>overhead</u> portion and a length-dependent <u>bandwidth</u> portion. We dropped the latency term, since we assume throughout our complexity considerations that latency can be hidden by computation.

Lemma 4. *Let M denote the maximum number of objects contained in any shift between any two nodes during a complete rebalancing operation. Then the total cost of rebalancing the tree perfectly is not larger than*

$$(2k + (k-1)h)C_{\text{obj}}(M) + (k+2)hC_{\text{adm}} .$$

Proof. The first action in order to rebalance the tree perfectly is an administrative message that is cast to all processors, which informs them that the rebalancing starts. This takes hC_{adm} cycles per processor if collective communications on 2^h processors are simulated by h point-to-point communications.

Then a combine operation calculates the array of current numbers $E(\cdot)$ (cost hC_{adm}). After that recursive calls of shift pass the arguments a and b to all nodes of the tree. The cost of all recursive calls is less than hC_{adm}.

The cost $C_{\mathrm{shift}}(d)$ of a single shift between two fat leaves v_{src} and v_{dest} depends on the height d of the involved fat leaves. A shift consists of four phases: extracting objects from the source leaves upwards to v_{src}, moving objects to v_{dest}, inserting below v_{dest} and finally removing below v_{src}. Extraction and insertion are performed independently on different processors and can be implemented as a filtering combine or a cast operation in $dC_{\mathrm{obj}}(M)$ time. Moving takes time for one message $C_{\mathrm{obj}}(M)$. Removing at the leaves below v_{src} involves dC_{adm} for casting a small message containing the actual split coordinate. The total per-processor-cost of the shift is $C_{\mathrm{shift}}(d) = (d+1)C_{\mathrm{obj}}(M) + dC_{\mathrm{adm}}$.

We sum over the heights of all fat leaves involved with shifts. The height of fat leaves varies from $(k-1)l$ in the first phase to 0 in the k-th phase. Every fat leaf is involved in up to two shifts (one to the left, one to the right). The sum of heights is:

$$2 \sum_{0 \leq j \leq k-1} C_{\mathrm{shift}}(jl) = (2k + (k-1)h)C_{\mathrm{obj}}(M) + (k-1)hC_{\mathrm{adm}} .$$

Now, the proof is completed by summing up all above costs:

$$3hC_{\mathrm{adm}} + (2k + (k-1)h)C_{\mathrm{obj}}(M) + (k-1)hC_{\mathrm{adm}}$$
$$= (2k + (k-1)h)C_{\mathrm{obj}}(M) + (k+2)hC_{\mathrm{adm}} .$$

5.2 Amortized Cost of Rebalancing

Lemma 5. *Let t_0 denote a time, when the whole tree was perfectly balanced. Assume for some $t > t_0$ that a rebalancing operation is started and that U updates have happened in the leaves since t_0. Then the maximum number M of objects that have to be shifted between any two nodes during the rebalancing is bounded by $M \leq U$.*

This result is easily seen by considering the border between any two neighbouring leaves. If not more than U updates happened to the left and to the right of the border, then not more than U objects have to move across the border to re-establish perfect balance.

Now we divide the cost of rebalancing (Lemma 4) by the number of updates U between two consecutive rebalancing operations.

Corollary 6. *The amortized time complexity of a single update is not larger than*

$$(2k + (k-1)h)Sg + \frac{((2k+1)(h+1) - 1)C_{\mathrm{adm}}}{U} .$$

Proof. This is easily shown by seeing that $\frac{C_{\mathrm{obj}}(M)}{U} \leq \frac{o+USg}{U} \leq \frac{C_{\mathrm{adm}}}{U} + Sg$.

5.3 The Strict Case

Lemma 7. *Let t_0 denote a time, when the whole tree was perfectly balanced. When a leaf leaves the narrowing interval (Lemma 2) at time $t > t_0$, then at least $U \geq U_{\text{strict}}^{\text{bound}} := \frac{E^0(r)}{p} \frac{\beta}{2-\beta}$ updates have happened in the whole tree since t_0.*

Proof. The minimum number of updates until a leaf may leave the narrowing interval is bounded by the difference of the borders of the narrowing interval (Lemma 2) and the initial number $\frac{E^0(r)}{p}$:

$$U \geq \min\left\{ \frac{E^0(r)}{p} 2 \frac{1}{2-\beta} - \frac{E^0(r)}{p} \quad , \quad \frac{E^0(r)}{p} - \frac{E^0(r)}{p} 2 \frac{1-\beta}{2-\beta} \right\} = \frac{E^0(r)}{p} \frac{\beta}{2-\beta}.$$

Theorem 8. *Consider $p = 2^h$ processors containing a total number of E objects that are rebalanced by the function* shift *(Fig. 4a). Then the amortized update time of a single update is not larger than*

$$(2k + (k-1)h)Sg + O\left(\frac{kph}{\beta E}\right) C_{\text{adm}}.$$

Proof. We assume, that the tree is perfectly balanced at the beginning, i. e. $E(w) = Ep^{-1}$ for all leaves w. Combining Corollary 6 and Lemma 7 we get an amortized update time of

$$(2k + (k-1)h)Sg + ((2k+1)(h+1) - 1)\frac{2-\beta}{\beta}\frac{p}{E}C_{\text{adm}}$$

$$= (2k + (k-1)h)Sg + O\left(\frac{kph}{\beta E}\right) C_{\text{adm}} \ .$$

We interpret the Theorem. If $k = 1$, every update involves sending and receiving the updated object (cost $2Sg$) plus some overhead. If the dimension is increased by 1, additional $2 + h$ send or receive operations are necessary in the worst case, because the object space has no canonical ordering. The overhead of sending and receiving messages is small if $\frac{kph}{\beta E}$ is small. We can reduce the overhead either by increasing β – which clearly increases load imbalances – or by storing many objects on few processors ($p << E$), which opposes to the demand of fast computation. Hence, we have a classical trade-off situation.

5.4 The Relaxed Case

In order to improve the average behaviour of the balancing scheme we will use the relaxed version in Fig. 4b. We will show, that then the number of updates between two successive rebalancing operations is still large.

Lemma 9. *Assume that all inner nodes v of the tree satisfy* diff$(v, 0, 0) = 0$. *Then simulation for $0 \leq \beta \leq 1$ and $p \leq 2^{20}$ shows that the number of objects in every particular leaf is covered by the following interval:*

$$\left[\frac{E(r)}{p} \frac{4 - 3\beta}{2(2 - \beta)}, \frac{E(r)}{p} \frac{4 - \beta}{2(2 - \beta)} \right] \ .$$

Lemma 10. *Let t_0 denote a time where the assumption of Lemma 9 is valid. When a leaf leaves the narrowing interval (Lemma 2) at time $t > t_0$, then at least $U \geq U_{relaxed}^{bound} := \frac{1}{2} \frac{E^0(r)}{p} \frac{\beta}{2-\beta}$ updates happened in the whole tree since t_0.*

Proof. As for Lemma 7.

Theorem 11. *If the relaxed version of* shift *(fig. 4b) is used, then the worst case amortized update time is at most twice as large as the bound, which was stated in Theorem 8.*

Proof. Let C_{strict}^{bound} denote the bound on the cost of the whole rebalancing in the strict case (Lemma 4). Evidently, the relaxed version involves not more communication in the worst case than C_{strict}^{bound}, since the relaxed version simply omits some of the object shifts that are performed in the strict version.

The minimum number of updates in the relaxed case is $U \geq U_{relaxed}^{bound} = \frac{1}{2} U_{strict}^{bound}$ (see Lemma 7 and 10). The amortized update time is the cost C of a whole rebalancing operation divided by the number U of updates between two successive escape events:

$$\frac{C}{U} \leq \frac{C_{strict}^{bound}}{U_{relaxed}^{bound}} = 2 \frac{C_{strict}^{bound}}{U_{strict}^{bound}} \ .$$

6 Experimental Results

We have tested the load balancing algorithm for a simple application that treats a set of points (objects) in $[0, 1]^k$. The application comprises a few loops. Within each loop every object is treated once. The action for each object o consists of creating new objects close to o and/or deleting o. This simple application defines local dependencies between objects, because new objects are created *close* to the original object. Therefore clustering the objects on the processors leads to communication only between neighbouring clusters.

We have examined various load patterns by assigning different *productivity* values to different regions of $[0, 1]^k$. A productivity of 1 means, that after treating an object the expected number of objects is the same as before treating it. A productivity *prod* > 1 means that treating an object involves creating *prod* $- 1$ new objects on the average. If *prod* < 1 this means that the object under treatment is deleted with probability $1 - prod$.

There are many more parameters that affect the behaviour of the application. We experimented with various taskloads, tasknumbers, β-values, and load patterns. The results are lengthy and therefore compiled in a technical report that is available online [3].

Here we present results for two extreme load patterns in two dimensions ($k = 2$). We compared the speedup of the application with and without load balancing. Fig. 6 shows the results obtained on a massively parallel NEC Cenju 3 computer. For the "constant" load pattern – a load pattern where the object density is constant over time everywhere in $[0, 1]^2$ –, the speedup with and without load balancing are approximately equal. The "heavy" pattern – a pattern

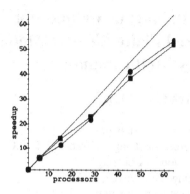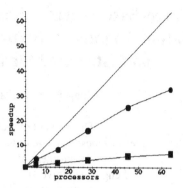

Fig. 6. Speedup for "constant" pattern (left) and "heavy" pattern (right) with (symbol ●) and without (symbol ■) load balancing

with drastically varying object densities in different regions of $[0, 1]^2$ – performs very bad without load balancing (speedup of 6.8 on 64 processors). With load balancing the speedup is improved significantly (33.3 on 64 processors).

References

1. M. J. Berger and S. H. Bokhari. A partitioning strategy for nonuniform problems on multiprocessors. *IEEE Trans. Comp.*, C-36(5), 5 1987.
2. D. Culler, R. Karp, D. Patterson, A. Sahay, K. Schauser, E. Santos, and T. van Eicken. LogP: Towards a realistic model of parallel computation. In *Proc. of the ACM SIGPLAN Symposium on Principles and Practice of Parallel Programming*, pages 1–12, 1993.
3. R. Garmann. Maintaining a dynamic set of geometric objects on parallel processors. Technical Report 646, FB Informatik, Universität Dortmund, D-44221 Dortmund, Germany, ls7-www.cs.uni-dortmund.de, May 1997.
4. A. Heirich and S. Taylor. A parabolic load balancing method. In *Proc. 24th Int. Conf. Par. Proc.*, volume III, pages 192–202, New York, 1995. CRC Press.
5. B. Hendrickson, R. Leland, and S. Plimpton. An efficient parallel algorithm for matrix-vector multiplication. *Intl. J. High Speed Comput.*, 7(1):73–88, 1995.
6. K. Mehlhorn and M. H. Overmars. Optimal dynamization of decomposable searching problems. *Information processing letters*, 12(2):93–98, April 1981.
7. A. T. Ogielski and W. Aiello. Sparse matrix computations on parallel processor arrays. *SIAM J. Sci. Comput.*, 14(3):519–530, May 1993.
8. M. H. Overmars. *The Design of Dynamic Data Structures*. Lecture Notes in Computer Science; 156. Springer-Verlag, Berlin, 1983.
9. C.-Z. Xu and F. C. M. Lau. The generalized dimension exchange method for load balancing in k-ary n-cubes and variants. *J. Par. Distr. Comp.*, 24(1):72–85, January 1995.

Tree-Based Parallel Load-Balancing Methods for Solution-Adaptive Unstructured Finite Element Models on Distributed Memory Multicomputers[*]

Ching-Jung Liao[a] and Yeh-Ching Chung[b]

[a] Department of Accounting and Statistics
The Overseas Chinese College of Commerce, Taichung, Taiwan 407, R.O.C.
[b] Department of Information Engineering
Feng Chia University, Taichung, Taiwan 407, R.O.C
Email: cjliao, ychung@pine.iecs.fcu.edu.tw

Abstract. In this paper, we propose three tree-based parallel load-balancing methods, the MCSTPLB method, the BTPLB method, and the CBTPLB method, to deal with the load unbalancing problems of solution-adaptive finite element application programs. To evaluate the performance of the proposed methods, we have implemented those methods along with three mapping methods, the AE/ORB method, the AE/MC method, and the MLkP method, on an SP2 parallel machine. The experimental results show that (1) if a mapping method is used for the initial partitioning and this mapping method or a load-balancing method is used in each refinement, the execution time of an application program under a load-balancing method is always shorter than that of the mapping method. (2) The execution time of an application program under the CBTPLB method is better than that of the BTPLB method that is better than that of the MCSTPLB method.

1 Introduction

To efficiently execute a finite element application program on a distributed memory multicomputer, we need to map nodes of the corresponding finite element model to processors of a distributed memory multicomputer such that each processor has approximately the same amount of computational load and the communication among processors is minimized. Since this mapping problem is known to be NP-completeness [6], many heuristic methods were proposed to find satisfactory sub-optimal solutions [2-3, 5, 7, 9-11, 16, 18]. If the number of nodes of a finite element graph will not be increased during the execution of a finite element application program, the mapping algorithm only needs to be performed once. For a solution-adaptive finite element application program, the number of nodes will be increased discretely due to the refinement of some finite elements during the execution of a finite element application program. This will result in load unbalancing of processors. A node remapping or a load-balancing algorithm has to be performed many times in

---

[*] The work of this paper was partially supported by NCHC of R.O.C. under contract NCHC-86-08-021.
[b] Correspondence addressee.

order to balance the computational load of processors while keeping the communication cost among processors as low as possible. Since node remapping or load-balancing algorithms were performed at run-time, their execution must be fast and efficient.

In this paper, we propose three tree-based parallel load-balancing methods to efficiently deal with the load unbalancing problems of solution-adaptive finite element application programs on distributed memory multicomputers with fully-connected interconnection networks such as multistage interconnection networks, crossbar networks, etc. They are the *maximum cost spanning tree load-balancing* (MCSTPLB) method, the *binary tree parallel load-balancing* (BTPLB) method, and the *condensed binary tree parallel load-balancing* (CBTPLB) method. To evaluate the performance of the proposed methods, we have implemented those methods along with three mapping methods, the AE/ORB method [3], the AE/MC method [3], and the MLkP method [10], on an SP2 parallel machine. The unstructured finite element graph *truss* is used as test sample. The experimental results show that (1) if a mapping method is used for the initial partitioning and this mapping method or a load-balancing method is used in each refinement, the execution time of an application program under a load-balancing method is always shorter than that of the mapping method. (2) The execution time of an application program under the CBTPLB method is better than that of the BTPLB method that is better than that of the MCSTPLB method.

The paper is organized as follows. The relative work will be given in Section 2. In Section 3, the proposed tree-based parallel load-balancing methods will be described in details. In Section 4, we will present the performance evaluation and simulation results.

2 Related Work

Many methods have been proposed to deal with the load unbalancing problems of solution-adaptive finite element graphs on distributed memory multicomputers in the literature. They can be classified into two categories, the remapping method and the load redistribution method. For the remapping method, many finite element graph mapping methods can be used as remapping methods. In general, they can be divided into five classes, the *orthogonal section* approach [3, 16, 18], the *min-cut* approach [3, 5, 11], the *spectral* approach [2, 9, 16], the *multilevel* approach [2, 10, 17], and others [7, 18]. These methods were implemented in several graph partition libraries, such as Chaco [8], DIME [19], JOSTLE [17], METIS [10], and PARTY [14], etc., to solve graph partition problems. Since our main focus is on the load-balancing methods, we do not describe these mapping methods here.

For the load redistribution method, many load-balancing algorithms can be used as load redistribution methods. The dimension exchange method (DEM) is applied to application programs without geometric structure [4]. Ou and Ranka [13] proposed a linear programming-based method to solve the incremental graph partition problem. Wu [15, 20] propose the tree walking, the cube walking, and the mesh walking run time scheduling algorithms to balance the load of processors on tree-base, cube-base, and mesh-base paradigms, respectively.

3 Tree-Based Parallel Load-Balancing Methods

In the following, the proposed tree-based parallel load-balancing methods, the MCSTPLB method, the BTPLB method, and the CBTPLB method, will be described in details.

3.1 The Maximum Cost Spanning Tree Parallel Load-Balancing (MCSTPLB) Method

The main idea of the MCSTPLB method is to find a maximum cost spanning tree from the processor graph that is obtained from the initial partitioned finite element graph. Based on the maximum cost spanning tree, it tries to balance the load of processors. The MCSTPLB method can be divided into the following four phases.

Phase 1: Obtain a weighted processor graph G from the initial partition.

Phase 2: Use a similar *Kruskal's* [12] algorithm to find a maximum cost spanning tree $T = (V, E)$ from G, where V and E denote the processors and edges of T, respectively. There are many ways to determine the shape of T. In this method, the shape of T is constructed as follows. First, a processor with the largest degree in V is selected as the root of T. For the children of the root, the processor with the largest degree will be the leftmost child of the root. The processor with the second largest degree will be the second leftmost child of the root, and so on. Then, the same construction process is applied to each nonterminal processor until every processor is processed. If the depth of T is greater than $\log P$, where P is the number of processors, we will try to adjust the depth of T. The adjusted method is first to find the longest path (from a terminal processor to another terminal processor) of T. Then, choose the middle processor of the path as the root of the tree and reconstruct the tree according to the above construction process. If the depth of the reconstructed tree is less than that of T, the reconstructed tree is the desired tree. Otherwise, T is the desired tree. The purpose of the adjustment is trying to reduce the load balancing steps among processors.

Phase 3: Calculate the global load balancing information and schedule the load transfer sequence of processors by using the TWA. Assume that there are M processors in a tree and N nodes in a refined finite element graph. We define N/M as the average weight of a processor. In the TWA method, the *quota* and the *load* of each processor in a tree are calculated, where the quota is the sum of the average weights of a processor and its children and the load is the sum of the weights of a processor and its children. The difference of the quota and the load of a processor is the number of nodes that a processor should send to or receive from its parent. If the difference is negative, a processor should send nodes to its parent. Otherwise, a processor should receive nodes from its parent. According to the global load balancing information, a schedule can be determined.

Phase 4: Perform load transfer (send/receive) based on the global load-balancing information, the schedule, and T. The main purposes of load transfer are balancing the computational load of processors and minimizing the communication cost among processors. Assume that processor P_i needs to send m nodes to processor P_j and let N denote the set of nodes in P_i that are adjacent to those of P_j. In order to keep the communication cost as low as possible, in the load transfer, nodes in N are transferred first. If $|N|$ is less than m, then nodes adjacent to those in N are transferred. This process is continued until the number of nodes transferred to P_j is equal to m.

3.2 The Binary Tree Parallel Load-Balancing (BTPLB) Method

The BTPLB method is similar to the MCSTPLB method. The only difference between these two methods is that the MCSTPLB method is based on a maximum cost spanning tree to balance the computational load of processors while the BTPLB method is based on a binary tree. The BTPLB method can be divided into the following four phases.

Phase 1: Obtain a processor graph G from the initial partition.

Phase 2: Use a similar *Kruskal's* algorithm to find a binary tree $T = (V, E)$ from G, where V and E denote the processors and edges of T, respectively. The method to determine the shape of a binary tree is the same as that of the MCSTPLB method.

Phase 3: Calculate the global load balancing information and schedule the load transfer sequence of processors by using the TWA.

Phase 4: Perform load transfer based on the global load-balancing information, the schedule, and T. The load transfer method is the same as that of the MCSTPLB method.

3.3 The Condensed Binary Tree Parallel Load-Balancing (CBTPLB) Method

The main idea of the CBTPLB method is to group processors of the processor graph into meta-processors. Each meta-processor is a hypercube. We call the grouped processor graph as a *condensed processor graph*. From the condensed processor graph, the CBTPLB method finds a binary tree. Based on the binary tree, the global load balancing information is calculated by a similar TWA method. According to the global load balancing information and the current load distribution, a load transfer algorithm is performed to balance the computational load of meta-processors and minimize the communication cost among meta-processors. After the load transfer in performed, a DEM is performed to balance the computational load of processors in a meta-processor. The CBTPLB method can be divided into the following five phases.

Phase 1: Obtain a processor graph G from the initial partition.

Phase 2: Group processors of G into meta-processors to obtain a condensed processor graph G_c incrementally. Each meta-processor of G_c is a hypercube. The meta-processors in G_c are constructed as follows. First, a processor P_i with the smallest degree in G and a processor P_j that is a neighbor processor of P_i and has the smallest degree among those neighbor processors of P_i are grouped into a meta-processor. Then, the same construction is applied to other ungrouped processors until there are no processors can be grouped into a hypercube. Repeat the grouping process to each meta-processor until there are no meta-processors can be grouped into a higher order hypercube.

Phase 3: Find a binary tree $T = (V, E)$ from G_c, where V and E denote the meta-processors and edges of T, respectively. The method of construction a binary tree is the same as that of the BTPLB method.

Phase 4: Based on T, calculate the global load balancing information and schedule the load transfer sequence by using a similar TWA method for meta-processors. Assume that there are M processors in a tree and N nodes in a refined finite element graph. We define N/M as the average weight of a processor. To obtain the global load balancing information, the *quota* and the *load* of each processor in a tree are calculated. The quota is defined as the sum of the average weights of processors in a meta-processor C_i and processors in children processors of C_i. The load is defined as the sum of the weights of processors in a meta-processor C_i and processors in children meta-processors of C_i. The difference of the quota and the load of a meta-processor is the number of nodes that a meta-processor should send to or receive from its parent meta-processor. If the difference is negative, a meta-processor should send nodes to its parent meta-processor. Otherwise, a meta-processor should receive nodes from its parent meta-processor. After calculating the global load balancing information, the schedule is determined as follows. Assume that m is the number of nodes that a meta-processor C_i needs to send to another meta-processor C_j. We have the following two cases.

Case 1: If the weight of C_i is less than m, the schedule of these two meta-processors is postponed until the weight of C_i is greater than or equal to m.

Case 2: If the weight of C_i is greater than or equal to m, a schedule can be made between processors of C_i and C_j. Assume that ADJ denotes the set of processors in C_i that are adjacent to those in C_j. If the sum of the weights of processors in ADJ is less than m, a schedule is made to transfer nodes of processors in C_i to processors in ADJ such that the weights of processors in ADJ is greater than or equal to m. If the sum of the weights of processors in ADJ is greater than or equal to m, a schedule is made to send m nodes from processors in ADJ to those in C_j.

Phase 5: Perform load transfer among meta-processors based on the global load balancing information, the schedule, and T. The load transfer method is similar to that of the BTPLB method. After performing load transfer process among meta-processors, a DEM is performed to balance the computational load of processors in meta-processors. We now give an example to explain the above description.

EXAMPLE 1: An example of the behavior of the CBTPLB method is shown in Figure 1. Part (a) shows an initial partition of a 61-node finite element graph on 7 processors by using the AE/ORB method. In part (a), the number of nodes assigned to processors P_0, P_1, P_2, P_3, P_4, P_5, and P_6 are 8, 9, 9, 8, 9, 9, and 9, respectively. After the refinement, the number of nodes assigned to processors P_0, P_1, P_2, P_3, P_4, P_5, and P_6 are 26, 10, 16, 13, 12, 13, and 10, respectively, as shown in part (b). Part (c) and (d) shows the process of constructing a condensed processor graph from (b). In Part (e), a binary tree can be constructed from the condensed processor graph. The global load balancing information of the condensed binary tree is shown in part (f). From the global load balancing information of part (f), we can decide the following load transfer sequence.

Step 1: $P_2 \rightarrow P_5$;
Step 2: $P_5 \rightarrow P_6$.

Part (g) shows the load transfer of processor in each meta-processor using the DEM. The load transfer sequence is give as follows.

Step 1: $P_0 \rightarrow P_1$, $P_3 \rightarrow P_2$, $P_6 \rightarrow P_4$;
Step 2: $P_0 \rightarrow P_3$, $P_1 \rightarrow P_2$.

Part (h) shows the load balancing result of part (b) after performing the CBTPLB method.

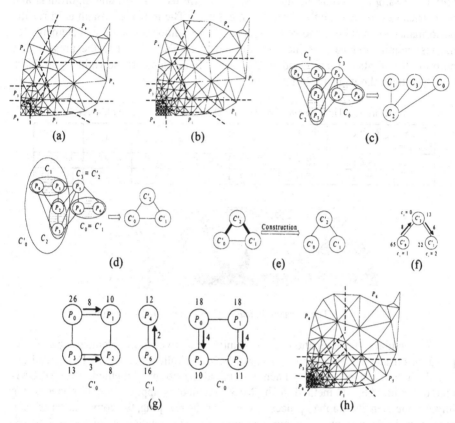

Figure 1: An example of the behavior of the CBTPLB method. (a) The initial partitioned finite element graph. (b) The finite element graph after a refinement. (c) The first grouping process of constructing a condensed processor graph from part (b). (d) The second grouping process of constructing a condensed processor graph from (c). (e) A binary tree constructed from the condensed processor graph. (f) The global load balancing information of the condensed binary tree. (g) The load transfer processors in each meta-processor by using the DEM. (h) The load balancing result of part (b) after performing the CBTPLB method.

4 Performance Evaluation and Experimental Results

To evaluate the performance of the proposed methods, we have implemented the proposed methods along with three mapping methods, the AE/ORB method, the AE/MC method, and the MLkP method, on an SP2 parallel machine. All of the algorithms were written in C with MPI communication primitives. Three criteria, the execution time of mapping/load-balancing methods, the computation time of an application program under different mapping/load-balancing methods, and the speedups achieved by the mapping/load-balancing methods for an application program, are used for the performance evaluation.

In dealing with the unstructured finite element graphs, the distributed irregular mesh environment (DIME) is used. In this paper, we only use DIME to generate the initial test sample. From the initial test graph, we use our refining algorithms and data structures to generate the desired test graphs. The initial test graph used for the performance evaluation is shown in Figure 2. The number of nodes and elements for the test graph after each refinement are shown in Table 1. For the presentation purpose, the number of nodes and the number of finite elements shown in Figure 2 are less than those shown in Table 1.

Table 1. The number of nodes and elements of the test graph *truss*.

Samples	Refinement	Initial (0)	1	2	3	4	5
Truss	node #	18407	23570	29202	36622	46817	57081
	element #	35817	46028	57181	71895	92101	112494

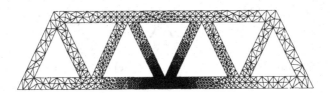

Figure 2. The test sample *truss* (7325 nodes, 14024 elements).

To emulate the execution of a solution-adaptive finite element application program on an SP2 parallel machine, we have the following steps. First, read the initial finite element graph. Then the initial partitioning method, the AE/ORB method or the AE/MC method or the MLkP method, is applied to map nodes of the initial finite element graph to processors. After the mapping, the computation of each processor is carried out. In our example, the computation is to solve Laplace equation (Laplace solver). The algorithm of solving Laplace equation is similar to that of [1]. Since it is difficult to predict the number of iterations for the convergence of a Laplace solver, we assume that the maximum number of iterations executed by our Laplace solver is 1000. When the computation is converged, the first refined finite element graph is read. To balance the computational load of processors, the AE/MC method or the AE/ORB method or the MLkP method or the MCSTPLB method or the BTPLB method or the CBTPLB method is applied. After a mapping/load-balancing method is performed, the computation for each processor is carried out. The procedures of mesh refinement, load balancing, and computation processes are performed in turn until the execution of a solution-adaptive finite element application program is completed. By combining the initial mapping methods and methods for load balancing, there are twelve methods used for the performance evaluation. For examples, the AE/ORB method uses AE/ORB to perform the initial mapping and AE/ORB to balance the computational load of processors in each refinement. The AE/ORB/BTPLB method use AE/ORB to perform the initial mapping and BTPLB to balance the computational load of processors in each refinement.

4.1 Comparisons of the Execution Time of Mapping/Load-Balancing Methods

The execution time of different mapping/load-balancing methods for the test unstructured finite element graph *truss* on SP2 with 10, 30, 50, and 70 processors are shown in Table 2. From Table 2, we can observe that the execution time of the AE/MC method ranges from tens of seconds to thousands of seconds. The execution time of the AE/ORB method and the ML*k*P method are about a few seconds. The execution time of the MCSTPLB method, the BTPLB method, and the CBTPLB method are less than two seconds. Obviously, the execution time of the proposed methods is much less than that of the mapping methods. The main reason is that the load-balancing methods use the current load distribution of processors to do the local load transfer task while the mapping methods need to repartition the finite element graph and redistribute nodes to processors. The overheads of the load-balancing methods are much less than those of the mapping methods. For the proposed methods, the execution time of the CBTPLB method is less than the execution time of the MCSTPLB method and the BTPLB method. This is because that the CBTPLB method can reduce the size of a tree with a large ratio so that the overheads to do the load transfer among meta-processors are less than those of the MCSTPLB method and the BTPLB method. Therefore, it can shorten the load transfer time efficiently. The disadvantage of the MCSTPLB method is when all of the processors except the root want to send nodes to their parents, the bottleneck will be occurred in the root. The BTPLB method can avoid this situation since the degree of the root in a binary tree is two.

4.2 Comparisons of the Execution Time of the Test Sample under Different Mapping/Load-Balancing Methods

The time of a Laplace solver to execute one iteration (computation + communication) for the test sample under different mapping/load-balancing methods on a SP2 parallel machine with 10, 30, 50, and 70 processors are shown in Figure 3. Since we assume a synchronous mode of communication in our model, the total time for a Laplace solver to complete its job is the sum of the computation time and the communication time. We can see that the execution time of a Laplace solver under the proposed load-balancing methods (for example AE/ORB/BTPLB) is less than that of their counterparts (AE/ORB). For the proposed methods, the execution time of a Laplace solver under the CBTPLB method is always less than that of the MCSTPLB and the BTPLB methods for all cases (Assume that the same initial mapping method is used).

4.3 Comparisons of the Speedups under the Mapping/Load-Balancing Methods for the Test Sample

The speedups and the maximum speedups under the mapping/load-balancing methods on an SP2 parallel machine with 10, 30, 50, and 70 processors for the test sample are shown in Table 3 and Table 4, respectively. The maximum speedup is defined as the ratio of the execution time of a sequential Laplace solver to the execution time of a parallel Laplace solver. In Table 3, the MLkP/CBTPLB method has the best performance among the mapping/load-balancing methods for all cases. The speedups produced by the AE/MC method and its counterparts (AE/MC/MCSTPLB, AE/MC/BTPLB, and AE/MC/CBTPLB) are always much smaller than other mapping/load-balancing methods. The main reason is that the execution of the AE/MC method is time consuming. However, if the number of iterations executed by a Laplace solver is set to ∞, the AE/MC method (its counterparts), in general, produces better speedups than those of the AE/ORB (its counterparts) and the MLkP (its counterparts) methods for all cases. We can see this situation from Table 4. Therefore, a fast and efficient mapping/load-balancing method is of great important to deal with the load unbalancing problems of solution-adaptive finite element application programs on distributed memory multicomputers.

5 Conclusions

In this paper, we have proposed three tree-based parallel load-balancing methods to deal with the load unbalancing problems of solution-adaptive finite element application programs. To evaluate the performance of the proposed methods, we have implemented those methods along with three mapping methods, the AE/ORB method, the AE/MC method, and the MLkP method, on an SP2 parallel machine. The unstructured finite element graph *truss* is used as test sample. Three criteria, the execution time of mapping/load-balancing methods, the execution time of a solution-adaptive finite element application program under different mapping/load-balancing methods, and the speedups under mapping/load-balancing methods for a solution-adaptive finite element application program, are used for the performance evaluation. The experimental results show that (1) if a mapping method is used for the initial partitioning and this mapping method or a load-balancing method is used in each refinement, the execution time of an application program under a load-balancing method is always shorter than that of the mapping method. (2) The execution time of an application program under the CBTPLB method is better than that of the BTPLB method that is better than that of the MCSTPLB method.

References

1. I.G. Angus, G.C. Fox, J.S. Kim, and D.W. Walker, *Solving Problems on Concurrent Processors*, Vol. 2, N. J.: Prentice-Hall, Englewood Cliffs, 1990.
2. S.T. Barnard and H.D. Simon, "Fast Multilevel Implementation of Recursive Spectral Bisection for Partitioning Unstructured Problems," *Concurrency: Practice and Experience*, Vol. 6, No. 2, pp. 101-117, Apr. 1994.
3. Y.C. Chung and C.J. Liao, "A Processor Oriented Partitioning Method for Mapping Unstructured Finite Element Graphs on SP2 Parallel Machines," Technical Report, Institute of Information Engineering, Feng Chia University, Taichung, Taiwan, Sep. 1996.

4. G. Cybenko, "Dynamic Load Balancing for Distributed Memory Multiprocessors," *Journal of Parallel and Distributed Computing*, Vol. 7, No. 2, pp. 279-301, Oct. 1989.

5. F. Ercal, J. Ramanujam, and P. Sadayappan, "Task Allocation onto a Hypercube by Recursive Mincut Bipartitioning," *Journal of Parallel and Distributed Computing*, Vol. 10, pp. 35-44, 1990.

6. M.R. Garey and D.S. Johnson, Computers and Intractability, A Guide to Theory of NP-Completeness. San Francisco, CA: Freeman, 1979.

7. J.R. Gilbert, G.L. Miller, and S.H. Teng, "Geometric Mesh Partitioning: Implementation and Experiments," *Proceedings of 9th International Parallel Processing Symposium*, Santa Barbara, California, pp. 418-427, Apr. 1995.

8. B. Hendrickson and R. Leland, "The Chaco User's Guide: Version 2.0," Technical Report SAND94-2692, Sandia National Laboratories, Albuquerque, NM, Oct. 1994.

9. B. Hendrickson and R. Leland, "An Improved Spectral Graph Partitioning Algorithm for Mapping Parallel Computations," *SIAM Journal on Scientific Computing*, Vol. 16, No.2, pp. 452-469, 1995.

10. G. Karypis and V. Kumar, "Multilevel *k*-way Partitioning Scheme for Irregular Graphs," Technical Report 95-064, Department of Computer Science, University of Minnesota, Minneapolis, 1995.

11. B.W. Kernigham and S. Lin, "An Efficient Heuristic Procedure for Partitioning Graphs," *Bell Syst. Tech. J.*, Vol. 49, No. 2, pp. 292-370, Feb. 1970.

12. J.B. Kruskal, "On the Shortest Spanning Subtree of a Graph and the Traveling Salesman Problem," In *Proceeding of the AMS*, Vol. 7, pp. 48-50, 1956.

13. C.W. Ou and S. Ranka, "Parallel Incremental Graph Partitioning," *IEEE Trans. Parallel and Distributed Systems*, Vol. 8, No. 8, pp. 884-896, Aug. 1997.

14. R. Preis and R. Diekmann, "The PARTY Partitioning – Library User Guide – Version 1.1," HENIZ NIXDORF INSTITUTE Universität Paderborn, Germany, Sep. 1996.

15. W. Shu and M.Y. Wu, "Runtime Incremental Parallel Scheduling (RIPS) on Distributed Memory Computers," *IEEE Trans. Parallel and Distributed Systems*, Vol. 7, No. 6, pp. 637-649, June 1996.

16. H.D. Simon, "Partitioning of Unstructured Problems for Parallel Processing," *Computing Systems in Engineering*, Vol. 2, No. 2/3, pp. 135-148, 1991.

17. C.H. Walshaw and M. Berzins, "Dynamic Load-Balancing for PDE Solvers on Adaptive Unstructured Meshes," *Concurrency: Practice and Experience*, Vol. 7, No. 1, pp. 17-28, Feb. 1995.

18. R.D. Williams, "Performance of Dynamic Load Balancing Algorithms for Unstructured Mesh Calculations," *Concurrency: Practice and Experience*, Vol. 3, No. 5, pp. 457-481, Oct. 1991.

19. R.D. Williams, *DIME: Distributed Irregular Mesh Environment*, California Institute of Technology, 1990.

20. M.Y. Wu, "On Runtime Parallel Scheduling," *IEEE Trans. Parallel and Distributed Systems*, Vol. 8, No. 2, pp. 173-186, Feb. 1997.

Table 2: The execution time of different mapping/load-balancing methods for test samples on different numbers of processors.

# of Processors	Refine # Method	Initial (0)	1	2	3	4	5	Total
10	AE/ORB	0.633	0.78	1.261	1.645	1.713	2.094	8.126
	AE/MC	46.89	347.45	591.91	567.32	349.47	1263.41	3166.39
	MLkP	0.573	0.673	0.72	0.92	1.21	1.431	5.527
	AE/ORB/MCSTPLB	0.614	0.689	0.683	0.635	1.135	1.076	4.832
	AE/MC/MCSTPLB	46.83	0.676	0.667	0.621	1.14	0.726	50.659
	MLkP/MCSTPLB	0.548	0.436	0.372	0.776	1.004	0.889	4.025
	AE/ORB/BTPLB	0.614	0.161	0.215	0.167	0.243	0.183	1.583
	AE/MC/BTPLB	46.83	0.163	0.206	0.157	0.207	0.178	47.74
	MLkP/BTPLB	0.548	0.157	0.159	0.153	0.216	0.163	1.396
	AE/ORB/CBTPLB	0.614	0.081	0.11	0.084	0.17	0.126	1.185
	AE/MC/CBTPLB	46.83	0.082	0.103	0.079	0.147	0.119	47.359
	MLkP/CBTPLB	0.548	0.079	0.087	0.078	0.124	0.092	1.008
30	AE/ORB	0.637	0.776	1.005	1.269	1.642	2.021	7.35
	AE/MC	23.33	174.64	296.49	284.62	186.25	587.36	1552.69
	MLkP	0.62	0.701	0.79	0.973	1.28	1.504	5.868
	AE/ORB/MCSTPLB	0.614	0.618	0.601	0.836	1.212	1.251	5.132
	AE/MC/MCSTPLB	23.27	0.621	0.619	0.887	1.225	1.439	28.061
	MLkP/MCSTPLB	0.566	0.647	0.688	0.799	1.217	1.388	5.305
	AE/ORB/BTPLB	0.614	0.243	0.186	0.131	0.197	0.313	1.684
	AE/MC/BTPLB	23.27	0.236	0.186	0.141	0.199	0.297	24.329
	MLkP/BTPLB	0.566	0.193	0.127	0.177	0.198	0.261	1.522
	AE/ORB/CBTPLB	0.614	0.118	0.068	0.098	0.212	0.199	1.309
	AE/MC/CBTPLB	23.27	0.106	0.093	0.073	0.109	0.169	23.82
	MLkP/CBTPLB	0.566	0.093	0.078	0.089	0.099	0.185	1.11
50	AE/ORB	0.742	0.805	1.071	1.292	1.75	2.02	7.68
	AE/MC	57.37	26.75	64.53	163.83	182.87	202.05	697.4
	MLkP	0.72	0.795	0.933	1.04	1.515	1.676	6.679
	AE/ORB/MCSTPLB	0.586	0.281	0.382	0.538	0.627	1.059	3.473
	AE/MC/MCSTPLB	57.28	0.331	0.448	0.659	0.682	0.812	60.212
	MLkP/MCSTPLB	0.621	0.266	0.379	0.624	0.893	1.293	4.076
	AE/ORB/BTPLB	0.586	0.234	0.283	0.34	0.503	0.608	2.554
	AE/MC/BTPLB	57.28	0.188	0.239	0.252	0.353	0.509	58.821
	MLkP/BTPLB	0.621	0.193	0.208	0.378	0.542	0.634	2.576
	AE/ORB/CBTPLB	0.586	0.178	0.224	0.248	0.32	0.501	2.057
	AE/MC/CBTPLB	57.28	0.161	0.201	0.232	0.274	0.474	58.622
	MLkP/CBTPLB	0.621	0.072	0.192	0.291	0.3	0.434	1.91
70	AE/ORB	0.855	0.873	1.105	1.399	1.572	1.989	7.793
	AE/MC	56.96	78.38	91.19	105.56	116.12	134.69	582.9
	MLkP	0.73	0.819	1.026	1.146	1.505	1.743	6.969
	AE/ORB/MCSTPLB	0.714	0.256	0.346	0.46	0.673	0.916	3.365
	AE/MC/MCSTPLB	56.87	0.201	0.249	0.408	0.518	0.819	59.065
	MLkP/MCSTPLB	0.674	0.249	0.284	0.515	0.797	1.284	3.803
	AE/ORB/BTPLB	0.714	0.253	0.245	0.567	0.669	0.889	3.337
	AE/MC/BTPLB	56.87	0.196	0.247	0.364	0.501	0.752	58.93
	MLkP/BTPLB	0.674	0.193	0.263	0.499	0.462	0.864	2.955
	AE/ORB/CBTPLB	0.714	0.169	0.187	0.32	0.435	0.671	2.496
	AE/MC/CBTPLB	56.87	0.121	0.157	0.229	0.408	0.609	58.394
	MLkP/CBTPLB	0.674	0.146	0.167	0.21	0.335	0.485	2.017

Time unit: seconds

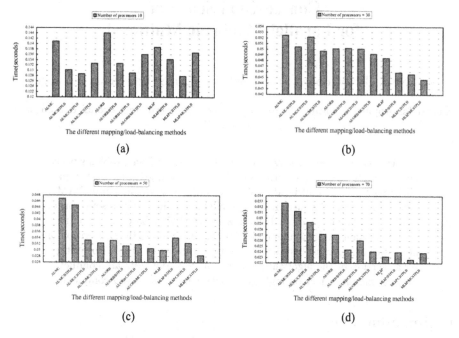

Figure 3: The time for Laplace solver to execute one iteration (computation + communication) for the test sample under different mapping/load-balancing methods on 10, 30, 50, and 70 processors (Time unit: 1×10^{-3} seconds).

Table 3: The speedups under the mapping/load-balancing methods for the test sample on an SP2 parallel machine.

# of processors Method	10	30	50	70
AE/ORB	6.55	16.65	18.24	24.60
AE/MC	0.30	0.62	1.35	1.63
MLkP	6.91	17.69	19.40	26.08
AE/ORB/MCSTPLB	7.07	17.39	27.12	30.54
AE/MC/MCSTPLB	5.44	12.81	10.77	11.55
MLkP/MCSTPLB	7.07	17.96	26.78	32.29
AE/ORB/BTPLB	7.42	19.17	28.62	34.07
AE/MC/BTPLB	5.60	13.36	11.05	11.95
MLkP/BTPLB	7.35	19.29	28.92	34.35
AE/ORB/CBTPLB	7.65	19.73	30.55	37.50
AE/MC/CBTPLB	5.66	13.79	11.25	12.22
MLkP/CBTPLB	7.73	20.15	30.56	38.69

Table 4: The maximum speedups under the mapping/load-balancing methods for the test sample on an SP2 parallel machine.

# of processors Method	10	30	50	70
AE/ORB	6.92	18.99	21.22	30.46
AE/MC	7.07	19.79	23.38	35.54
MLkP	7.18	19.75	22.3	31.9
AE/ORB/MCSTPLB	7.32	19.09	29.95	34.05
AE/MC/MCSTPLB	7.51	20.04	30.85	36.66
MLkP/MCSTPLB	7.28	19.86	30.07	36.83
AE/ORB/BTPLB	7.51	19.82	30.88	38.46
AE/MC/BTPLB	7.65	19.83	31.74	40.78
MLkP/BTPLB	7.43	19.88	31.26	38.24
AE/ORB/CBTPLB	7.71	20.26	32.61	41.38
AE/MC/CBTPLB	7.74	20.57	33.23	43.01
MLkP/CBTPLB	7.79	20.61	32.4	41.97

Coarse-Grid Selection
for Parallel Algebraic Multigrid

Andrew J. Cleary, Robert D. Falgout, Van Emden Henson, Jim E. Jones

Center for Applied Scientific Computing, Lawrence Livermore National Laboratory, Livermore, CA

Abstract. The need to solve linear systems arising from problems posed on extremely large, unstructured grids has sparked great interest in parallelizing algebraic multigrid (AMG). To date, however, no parallel AMG algorithms exist. We introduce a parallel algorithm for the selection of coarse-grid points, a crucial component of AMG, based on modifications of certain parallel independent set algorithms and the application of heuristics designed to insure the quality of the coarse grids. A prototype serial version of the algorithm is implemented, and tests are conducted to determine its effect on multigrid convergence, and AMG complexity.

1   Introduction

Since the introduction of algebraic multigrid (AMG) in the 1980's [4, 2, 3, 5, 19, 16, 18, 17] the method has attracted the attention of scientists needing to solve large problems posed on unstructured grids. Recently, there has been a major surge of interest in the field, due in large part to the need to solve increasingly larger systems, with hundreds of millions or billions of unknowns. Most of the current research, however, focuses either on improving the standard AMG algorithm [9, 7], or on dramatic new algebraic approaches [20, 6]. Little research has been done on parallelizing AMG. The sizes of the modern problems, however, dictate that large-scale parallel processing be employed.

Methods for parallelizing geometric multigrid methods have been known for some time [10], and most of the AMG algorithm can be parallelized using existing technology. Indeed, much of the parallelization can be accomplished using tools readily available in packages such as PETSc or ISIS++. But, the heart of the AMG setup phase includes the coarse-grid selection process, which is inherently sequential in nature.

In this note we introduce a parallel algorithm for selecting the coarse-grid points. The algorithm is based on modifications of parallel *independent set* algorithms. Also, we employ heuristics designed to insure the quality of the coarse grids. A prototype serial code is implemented, and we examine the effect the algorithm has on the multigrid convergence properties.

In Section 2 we outline the basic principles of AMG. Section 3 describes our parallelization model and the underlying philosophy, while the details of the parallel algorithm are given in Section 4. Results of numerical experiments with the serial prototype are presented and analyzed in Section 5. In Section 6 we make concluding remarks and indicate directions for future research.

2 Algebraic Multigrid

We begin by outlining the basic principles of AMG. Detailed explanations may be found in [17]. Consider a problem of the form $A\mathbf{u} = \mathbf{f}$, where A is an $n \times n$ matrix with entries a_{ij}. For AMG, a "grid" is simply a set of indices of the variables, so the original grid is denoted by $\Omega = \{1, 2, \ldots, n\}$. In any multigrid method, the central idea is that error $\mathbf{e}$ not eliminated by relaxation is eliminated by solving the residual equation $A\mathbf{e} = \mathbf{r}$ on a coarser grid, then interpolating $\mathbf{e}$ and using it to correct the fine-grid approximation. The coarse-grid problem itself is solved by a recursive application of this method. Proceeding through all levels, this is known as a multigrid cycle. One purpose of AMG is to free the solver from dependence on geometry (which may not be easily accessible, if it is known at all). Hence, AMG fixes a relaxation method, and its main task is to determine a coarsening process that approximates error the relaxation cannot reduce.

Using superscripts to indicate level number, where 1 denotes the finest level so that $A^1 = A$ and $\Omega^1 = \Omega$, the components that AMG needs are: "grids" $\Omega^1 \supset \Omega^2 \supset \ldots \supset \Omega^M$; grid operators $A^1, A^2, \ldots, A^M$; interpolation operators $I_{k+1}^k, k = 1, 2, \ldots M - 1$; restriction operators $I_k^{k+1}, k = 1, 2, \ldots M - 1$; and a relaxation scheme for each level. Once these components are defined, the recursively defined multigrid cycle is as follows:

> **Algorithm:** $MV^k(\mathbf{u}^k, \mathbf{f}^k)$. The (μ_1, μ_2) V-cycle.
> If $k = M$, set $\mathbf{u}^M = (A^M)^{-1}\mathbf{f}^M$.
> Otherwise:
> 　　Relax μ_1 times on $A^k\mathbf{u}^k = \mathbf{f}^k$.
> 　　Perform coarse grid correction:
> 　　　　Set $\mathbf{u}^{k+1} = 0, \mathbf{f}^{k+1} = I_k^{k+1}(\mathbf{f}^k - A^k\mathbf{u}^k)$.
> 　　　　"Solve" on level $k + 1$ with $MV^{k+1}(\mathbf{u}^{k+1}, \mathbf{f}^{k+1})$.
> 　　　　Correct the solution by $\mathbf{u}^k \leftarrow \mathbf{u}^k + I_{k+1}^k\mathbf{u}^{k+1}$.
> 　　Relax ν_2 times on $A^k\mathbf{u}^k = \mathbf{f}^k$.

For this to work efficiently, two principles must be followed:

P1: *Errors not efficiently reduced by relaxation must be well-approximated by the range of interpolation.*

P2: *The coarse-grid problem must provide a good approximation to fine-grid error in the range of interpolation.*

AMG satisfies **P1** by automatically selecting the coarse grid and defining interpolation, based solely on the algebraic equations of the system. **P2** is satisfied by defining restriction and the coarse-grid operator by the *Galerkin* formulation [14]:

$$I_k^{k+1} = \left(I_{k+1}^k\right)^T \quad \text{and} \quad A^{k+1} = I_k^{k+1} A^k I_{k+1}^k. \tag{1}$$

Selecting the AMG components is done in a separate preprocessing step:

AMG Setup Phase:

1. Set $k = 1$.
2. Partition Ω^k into disjoint sets C^k and F^k.
   (a) Set $\Omega^{k+1} = C^k$.
   (b) Define interpolation I^k_{k+1}.
3. Set $I^{k+1}_k = \left(I^k_{k+1} \right)^T$ and $A^{k+1} = I^{k+1}_k A^k I^k_{k+1}$.
4. If Ω^{k+1} is small enough, set $M = k + 1$ and stop. Otherwise, set $k = k + 1$ and go to step 2.

2.1 Selecting Coarse Grids and Defining Interpolation

Step 2 is the core of the AMG setup process. The goal of the setup phase is to choose the set C of coarse-grid points and, for each fine-grid point $i \in F \equiv \Omega - C$, a small set $C_i \subset C$ of interpolating points. Interpolation is then of the form:

$$\left(I^k_{k+1} \mathbf{u}^{k+1} \right)_i = \begin{cases} \mathbf{u}^{k+1}_i & \text{if } i \in C, \\ \displaystyle\sum_{j \in C_i} \omega_{ij} \mathbf{u}^{k+1}_j & \text{if } i \in F. \end{cases} \tag{2}$$

We do not detail the construction of the interpolation weights ω_{ij}, instead referring the reader to [17] for details.

An underlying assumption in AMG is that smooth error is characterized by small residuals, that is, $Ae \approx 0$, which is the basis for choosing coarse grids and defining interpolation weights. For simplicity of discussion here, assume that A is a symmetric positive-definite M-matrix, with $a_{ii} > 0, a_{ij} \leq 0$ for $j \neq i$, and $\sum a_{ij} \geq 0$.

We say that point i *depends* on point j if a_{ij} is "large" in some sense, and hence, to satisfy the ith equation, the value of u_i is affected more by the value of u_j than by other variables. Specifically, the *set of dependencies* of i is defined by

$$S_i \equiv \left\{ j \neq i : -a_{ij} \geq \alpha \max_{k \neq i}(-a_{ik}) \right\}, \tag{3}$$

with α typically set to be 0.25. We also define the set $S^T_i \equiv \{j : i \in S_j\}$, that is, the set of points j that depend on point i, and we say that S^T_i is the set of *influences* of point i.

A basic premise of AMG is that relaxation smoothes the error in the direction of influence. Hence, we may select $C_i = S_i \cap C$ as the set of interpolation points for i, and adhere to the following criterion while choosing C and F:

P3: *For each $i \in F$, each $j \in S_i$ is either in C or $S_j \cap C_i \neq \emptyset$.*

That is, if i is a fine point, then the points influencing i must either be coarse points or must depend on the coarse points used to interpolate u_i.

The coarse grid is chosen to satisfy two criteria. We enforce **P3** in order to insure good interpolation. However, we wish to keep the size of the coarse-grid as small as possible, so we desire that

P4: *C is a maximal set with the property that no C-point influences another C-point.*

It is not always possible to enforce both criteria. Hence, we enforce **P3** while using **P4** as a guide in coarse-point selection.

AMG employs a two-pass process, in which the grid is first "colored", providing a tentative C/F choice. Essentially, a point with the largest number of influences ("influence count") is colored as a C point. The points depending on this C point are colored as F points. Other points influencing these F points are more likely to be useful as C points, so their influence count is increased. The process is repeated until all points are either C or F points. Next, a second pass is made, in which some F points may be recolored as C points to ensure that **P3** is satisfied. Details of the coarse-grid selection algorithm may be found in [17], while a recent study of the efficiency and robustness of the algorithm is detailed in [7].

Like many linear solvers, AMG is divided into two main phases, the *setup* phase and the *solve* phase. Within each of these phases are certain tasks that must be parallelized to create a parallel AMG algorithm. They are:

- **Setup phase:**
  - Selecting the coarse grid points, Ω^{k+1}.
  - Construction of interpolation and restriction operators, I_{k+1}^k, I_k^{k+1}.
  - Constructing the coarse-grid operator $A^{k+1} = I_k^{k+1} A^k I_k^{k+1}$.
- **Solve phase:**
  - Relaxation on $A^k \mathbf{u}^k = \mathbf{f}^k$.
  - Calculating the residual $\mathbf{r}^k \leftarrow \mathbf{f}^k - A^k \mathbf{u}^k$.
  - Computing the restriction $\mathbf{f}^{k+1} = I_k^{k+1} \mathbf{r}^k$.
  - Interpolating and correcting $u^k \leftarrow \mathbf{u}^k + I_{k+1}^k \mathbf{u}^{k+1}$.

3 Parallelization Model

In this work we target massively parallel distributed memory architectures, though it is expected that the method will prove useful in other settings, as well. Currently, most of the target platforms support shared memory within clusters of processors (typically of size 4 or 8), although for portability we do not utilize this feature. We assume explicit message passing is used among the processors, and implement this with MPI [15]. The equations and data are distributed to the processors using a *domain-partitioning* model. This is natural for many problems of physics and engineering, where the physical domain is partitioned by subdomains. The actual assignment to the processors may be done by the application code calling the solver, by the gridding program, or by a subsequent call to a graph partitioning package such as Metis [12]. The domain-partitioning strategy should not be confused with *domain decomposition*, which refers to a family of solution methods.

We use object-oriented software design for parallel AMG. One benefit of this design is that we can effectively employ kernels from other packages, such as

PETSc [1] in several places throughout our code. Internally, we focus on a *matrix object* that generalizes the features of "matrices" in widely-used packages. We can write AMG-specific routines once, for a variety of matrix data structures, while avoiding the necessity of reinventing widely available routines, such as matrix-vector multiplication.

Most of the required operations in the solve phase of AMG are standard, as are several of the core operations in the setup phase. We list below the standard operations needed by AMG:

- *Matrix-vector multiplication*: used for residual calculation, for interpolation, and restriction (both use rectangular matrices; restriction multiplies by the transpose). Some packages provide all of the above, while others may have to be augmented, although the coding is straightforward in these cases.
- *Basic iterative methods*: used for the smoothing step. Jacobi or scaled Jacobi are most common for parallel applications, but any iterative method provided in the parallel package could be applied.
- *Gathering/scattering processor boundary equations*: used in the construction of the interpolation operators and in the construction of coarse-grid operators via the Galerkin method. Each processor must access "processor-boundary equations" stored on neighboring processors. Because similar functionality is required to implement additive Schwarz methods, parallel packages implementing such methods already provide tools that can be modified to fulfill this requirement.

4 Parallel Selection of Coarse Grids

Designing a parallel algorithm for the selection of the coarse-grid points is the most difficult task in parallelizing AMG. Classical AMG uses a two-pass algorithm to implement the heuristics, **P3** and **P4**, that assure grid quality and control grid size. In both passes, the algorithm is inherently sequential. The first pass can be described as:

1) Find a point j with maximal measure $w(j)$. Select j as a C point.
2) Designate neighbors of j as F points, and update the measures of other nearby points, using heuristics to insure grid quality.

Repeat steps 1) and 2) until all points are either C or F points.

This algorithm is clearly unsuitable for parallelization, as updating of measures occurs after each C point is selected. The second pass of the classical AMG algorithm is designed to enforce *P3*, although we omit the details and refer the reader to [17]. We can satisfy **P3** and eliminate the second pass through a simple modification of step 2).

Further, we may allow for parallelism by applying the following one-pass algorithm. Begin by performing step 1) globally, selecting a *set* of C points, D, and then perform step 2) locally, with each processor working on some portion of the set D. With different criteria for selecting the set D, and armed with various heuristics for updating the neighbors in **2)**, a family of algorithms may be developed. The overall framework is:

> Input: the $n \times n$ matrix A^k (level k).
> Initialize:
> $F = \emptyset$, $C = \emptyset$.
> $\forall i \in \{1...n\}$,
> $w(i) \leftarrow$initial value.
> Loop until $|C| + |F| = n$:
>   Select an independent set of points D.
> $\forall j \in D$:
> $C = C \cup j$.
> $\forall k$ in set local to j, update $w(k)$.
>   if $w(k) = 0$, $F = F \cup k$.
> End loop

4.1   Selection of the set D

For the measure $w(i)$, we use $|S_i^T| + \sigma(i)$, the number of points influenced by the point i plus a random number in $(0,1)$. The random number is used as a mechanism for breaking ties between points with the same number of influences. The set D is then selected using a modification of a parallel maximal independent set algorithm developed in [13, 11, 8].

A point j will be placed in the set D if $w(j) > w(k)$ for all k that either influence or depend on j. By construction, this set will be independent. While our implementation selects a maximal set of points possessing the requisite property, this is not necessary, and may not be optimal. An important observation is that this step can be done entirely in parallel, provided each processor has access to the w values for points with influences that cross its processor boundary.

4.2   Updating $w(k)$ of neighbors

Describing the heuristics for updating $w(k)$ is best done in terms of graph theory. We begin by defining S, the auxiliary *influence matrix*:

$$S_{ij} = \begin{cases} 1 & \text{if } j \in S_i, \\ 0 & \text{otherwise.} \end{cases} \qquad (4)$$

That is, $S_{ij} = 1$ only if i depends on j. The ith row of S gives the dependencies of i while the ith column of S gives the influences of i. We can then form the directed graph of S, and observe that a directed edge from vertex i to vertex j exists only if $S_{ij} \neq 0$. Notice that the directed edges point in the direction of dependence. To update the $w(k)$ of neighbors, we apply the following pair of heuristics.

P5: Values at C points are not interpolated; hence, neighbors that influence a C point are less valuable as potential C points themselves.

P6: If k and j both depend on c, a given C point, and j influences k, then j is less valuable as a potential C point, since k can be interpolated from c.

The details of how these heuristics are implemented are:

$$\forall\, c \in D,$$

	P5:
$\forall\, j \mid S_{cj} \neq 0,$	(each j that influences c)
$\quad w(j) \leftarrow w(j) - 1$	(decrement the measure)
$\quad S_{cj} \leftarrow 0$	(remove edge cj from the graph)
	P6:
$\forall\, j \mid S_{jc} \neq 0$	(each j that depends on c),
$\quad S_{jc} \leftarrow 0$	(remove edge jc from the graph)
$\quad \forall\, k \mid S_{kj} \neq 0,$	(each k that j influences),
$\qquad$ if $S_{kc} \neq 0$	(if k depends on c),
$\qquad\quad w(j) \leftarrow w(j) - 1$	(decrement the measure)
$\qquad\quad S_{kj} \leftarrow 0$	(remove edge kj from the graph)

The heuristics have the effect of lowering the measure $w(k)$ for a set of neighbors of each point in D. As these measures are lowered, edges of the graph of S are removed to indicate that certain influences have already been taken into account. Frequently the step $w(j) \to w(j) - 1$ causes $\lfloor w(j) \rfloor = 0$. When this occurs j is flagged as an F point.

Once the heuristics have been applied for all the points in D, a global communication step is required, so that each processor has updated w values for all neighbors of all their points. The entire process is then repeated. C points are added by selecting a new set, D, from the vertices that still have edges attached in the modified graph of S. This process continues until all n points have either been selected as C points or F points.

5 Numerical Experiments

To test its effect on convergence and algorithmic scalability, we include a serial implementation of the parallel coarsening algorithm in a standard sequential AMG solver. Obviously, this does not test parallel efficiency, which must wait for a full parallel implementation of the entire AMG algorithm.

Figure 1 shows the coarse grid selected by the parallel algorithm on a standard test problem, the 9-point Laplacian operator on a regular grid. This test is important because the grid selected by the standard sequential AMG algorithm

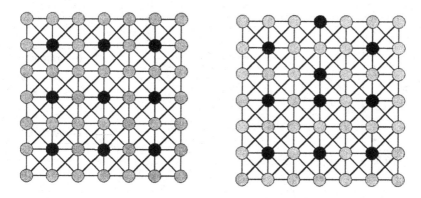

Fig. 1. *Coarse grids for the structured-grid 9-point Laplacian operator. The dark circles are the C points. Left: Grid selected by the standard algorithm. Right: Grid selected by the parallel algorithm.*

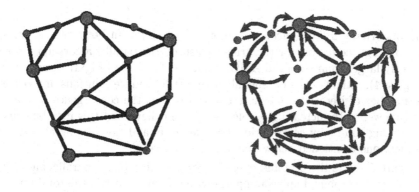

Fig. 2. *Coarse grids for an unstructured grid. The large circles are the C points. Left: Grid selected by the standard algorithm. Right: Grid selected by the parallel algorithm. Graph connectivity is shown on the left, while the full digraph is shown on the right.*

is also the optimal grid used in geometric multigrid for this problem. Examining many such test problems on regular grids, we find that the parallel coarsening algorithm typically produces coarse grids with 10–20% more C points than the sequential algorithm. On unstructured grids or complicated domains, this increase tends to be 40–50%, as may be seen in the simple example displayed in Figure 2.

The impact of the parallel coarsening algorithm on convergence and scalability is shown in two figures. Figure 3 shows the convergence factor for the 9-point Laplacian operator on regular grids ranging in size from a few hundred to nearly a half million points. Several different choices for the smoother and the parameter α are shown. In Figure 4 the same tests are applied to the 9-point Laplacian operator for anisotropic grids, where the aspect ratios of the under-

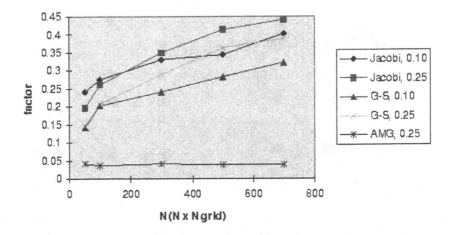

Fig. 3. *Convergence factors for parallel AMG for the 9-point Laplacian*

lying quadrilateral finite elements are extremely high. In both figures, we see that convergence factors for the grids chosen by the parallel algorithm are significantly larger than standard AMG (shown as "AMG" in Figure 3; not shown in Figure 4), although the parallel algorithm still produce solutions in a reasonable number of iterations. Of more concern is that the convergence factors do not scale well with increasing problem size. We believe that this may be caused by choosing too many coarse grid points at once, and that simple algorithmic modifications mentioned below may improve our results.

Figure 5 shows the grid and operator complexities for the parallel algorithm applied to the 9-point Laplacian operator. Grid complexity is the total number of grid points, on all grids, divided by the number of points on the original grid. Operator complexity is the total number of non-zeros in all operators $A^1, A^2, \ldots$ divided by the number of non-zeros in the original matrix. Both the grid and operator complexities generated using by the parallel algorithm are essentially constant with increasing problem size. While slightly larger than the complexities of the sequential grids, they nevertheless appear to be scalable.

The framework described in Section 4 permits easy modification of the algorithm. For example, one may alter the choice of the set D of C points. We believe that the convergence factor degradation shown in our results may be due to selecting too many coarse grid points. One possibility is to choose the minimal number of points in D, that is, one point per processor. This amounts to running the sequential algorithm on each processor, and there a number of different ways to handle the interprocessor boundaries. One possibility is to coarsen the processor boundary equations first, using a parallel MIS algorithm, and then treat each domain independently. Another option is to run the sequential algorithm on each processor ignoring the nodes on the boundary, and then patch up the grids on the processor boundaries.

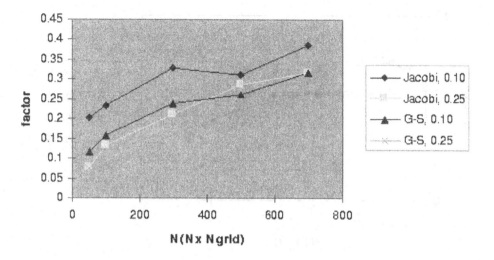

Fig. 4. *Convergence rates for parallel AMG for the anisotropic grid problem.*

6 Conclusions

Modern massively parallel computing requires the use of scalable linear solvers such as multigrid. For unstructured-grid problems, however, scalable solvers have not been developed. Parallel AMG, when developed, promises to be such a solver. AMG is divided into two main phases, the setup phase and the solve phase. The solve phase can be parallelized using standard techniques common to most parallel multigrid codes. However, the setup phase coarsening algorithm is inherently sequential in nature.

We develop a family of algorithms for selecting coarse grids, and prototype one member of that family using a sequential code. Tests with the prototype indicate that the quality of the selected coarse grids are sufficient to maintain constant complexity and to provide convergence even for difficult anisotropic problems. However, convergence rates are higher than for standard AMG, and do not scale well with problem size. We believe that this degradation may be caused by choosing too many coarse grid points at once, and that simple algorithmic modifications may improve our results. Exploration of these algorithm variants is the subject of our current research.

References

1. S. BALAY, W. GROPP, L. C. McINNES, AND B. SMITH, *Petsc 2.0 user's manual*, Tech. Rep. ANL-95/11, Argonne National Laboratory, Nov. 1995.

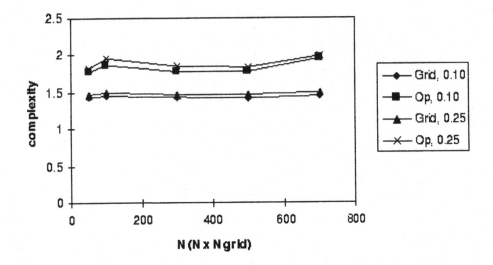

Fig. 5. *Operator complexity for parallel AMG on example problem*

2. A. BRANDT, *Algebraic multigrid theory: The symmetric case*, in Preliminary Proceedings for the International Multigrid Conference, Copper Mountain, Colorado, April 1983.

3. A. BRANDT, *Algebraic multigrid theory: The symmetric case*, Appl. Math. Comput., 19 (1986), pp. 23–56.

4. A. BRANDT, S. F. MCCORMICK, AND J. W. RUGE, *Algebraic multigrid (AMG) for automatic multigrid solutions with application to geodetic computations*. Report, Inst. for Computational Studies, Fort Collins, Colo., October 1982.

5. ———, *Algebraic multigrid (AMG) for sparse matrix equations*, in Sparsity and Its Applications, D. J. Evans, ed., Cambridge University Press, Cambridge, 1984.

6. M. BREZINA, A. J. CLEARY, R. D. FALGOUT, V. E. HENSON, J. E. JONES, T. A. MANTEUFFEL, S. F. MCCORMICK, AND J. W. RUGE, *Algebraic multigrid based on element interpolation (AMGe)*. Submitted to the SIAM Journal on Scientific Computing special issue on the Fifth Copper Mountain Conference on Iterative Methods, 1998.

7. A. J. CLEARY, R. D. FALGOUT, V. E. HENSON, J. E. JONES, T. A. MANTEUFFEL, S. F. MCCORMICK, G. N. MIRANDA, AND J. W. RUGE, *Robustness and scalability of algebraic multigrid*. Submitted to the SIAM Journal on Scientific Computing special issue on the Fifth Copper Mountain Conference on Iterative Methods, 1998.

8. R. K. GJERTSEN, JR., M. T. JONES, AND P. E. PLASSMAN, *Parallel heuristics for improved, balanced graph colorings*, Journal of Parallel and Distributed Computing, 37 (1996), pp. 171–186.

9. G. GOLUBOVICI AND C. POPA, *Interpolation and related coarsening techniques for the algebraic multigrid method*, in Multigrid Methods IV, Proceedings of the Fourth European Multigrid Conference, Amsterdam, July 6-9, 1993, vol. 116 of ISNM, Basel, 1994, Birkhäuser, pp. 201–213.

10. J. E. JONES AND S. F. MCCORMICK, *Parallel multigrid methods*, in Parallel Numerical Algorithms, D. E. Keys, A. H. Sameh, and V. Venkatakrishnan, eds., Dordrecht, Netherlands, 1997, Kluwer Academic Publications.

11. M. T. JONES AND P. E. PLASSMAN, *A parallel graph coloring heuristic*, SIAM Journal on Scientific Computing, 14 (1993), pp. 654–669.

12. G. KARYPIS AND V. KUMAR, *A coarse-grain parallel multilevel k-way partitioning algorithm*, in Proceedings of the 8th SIAM Conference on Parallel Processing for Scientific Computing, 1997.

13. M. LUBY, *A simple parallel algorithm for the maximal independent set problem*, SIAM Journal on Computing, 15 (1986), pp. 1036–1053.

14. S. F. MCCORMICK, *Multigrid methods for variational problems: general theory for the V-cycle*, SIAM J. Numer. Anal., 22 (1985), pp. 634–643.

15. MPI FORUM, *MPI: A message-passing interface standard*, International J. Supercomputing Applications, 8(3/4) (1994), pp. 654–669.

16. J. W. RUGE AND K. STÜBEN, *Efficient solution of finite difference and finite element equations by algebraic multigrid (AMG)*, in Multigrid Methods for Integral and Differential Equations, D. J. Paddon and H. Holstein, eds., The Institute of Mathematics and its Applications Conference Series, Clarendon Press, Oxford, 1985, pp. 169–212.

17. ———, *Algebraic multigrid (AMG)*, in Multigrid Methods, S. F. McCormick, ed., vol. 3 of Frontiers in Applied Mathematics, SIAM, Philadelphia, PA, 1987, pp. 73–130.

18. K. STÜBEN, *Algebraic multigrid (AMG): experiences and comparisons*, Appl. Math. Comput., 13 (1983), pp. 419–452.

19. K. STÜBEN, U. TROTTENBERG, AND K. WITSCH, *Software development based on multigrid techniques*, in Proc. IFIP-Conference on PDE Software, Modules, Interfaces and Systems, B. Enquist and T. Smedsaas, eds., Sweden, 1983, Söderköping.

20. P. VANĚK, J. MANDEL, AND M. BREZINA, *Algebraic multigrid based on smoothed aggregation for second and fourth order problems*, Computing, 56 (1996), pp. 179–196.

Overlapping and Short-Cutting Techniques in Loosely Synchronous Irregular Problems

Ernesto Gomez[1] and L. Ridgway Scott[2]

[1] Department of Computer Science and Texas Center for Advanced Molecular Computation, University of Houston, Texas, USA

[2] Department of Mathematics and Texas Center for Advanced Molecular Computation, University of Houston, Texas, USA

Abstract. We introduce short-cutting and overlapping techniques that, separately and in combination show promise of speedup for parallel processing of problems with irregular or asymmetric computation. Methodology is developed and demonstrated on an example problem. Experiments on an IBM SP-2 and a workstation cluster are presented.

1 Introduction

We are interested in parallel programs for irregular and asymmetric problems. We have the same program on all processors, but data dependent logic may evaluate differently at different nodes, causing different code to execute, or resulting in different amounts of work by the same code. Such problems present challenges for parallel processing because synchronization delays can be of the order of the computation time.

There are different kinds of irregularity in parallel codes. Data structure irregularity is one common case that has been addressed previously [1]. In other applications, data among parallel tasks can be regular, but access unpredictable [2] or asymmetric [3]. Another large class of problems are very regular in sequential execution, but when executed in parallel, exhibit irregularities in time. We propose a technique we call short-cutting to reduce the execution time of such problems. These codes can have parallel efficiency exceeding one in some cases.

Short-cutting gains are typically of the same order as the synchronization delays in the problems to which they may be applied. We propose a protocol to tolerate or hide synchronization delays by overlapping intervals between variable definition and use at each parallel process. Both techniques should be employed together, to prevent losses from synchronization offsetting gains from short-cutting.

2 Irregular Problems and Synchronization

For any non-trivial computation carried out in parallel there is some communication cost added to the actual computational cost. This is at least the cost of moving data between processes. In asymmetric or irregular computation, however, we have an added cost due to time or control asymmetry. Note that this time is in addition to the communications required to synchronize processes. In the following we will refer to "synchronization time" as this waiting time, and include any message passing required to synchronize in the communications time. Suppose two processes, p1 and p2 are initially synchronized and must communicate after performing some computation. Suppose that p1 executes n_1 instructions and p2 executes n_2 instructions, and that each instruction takes a time t to execute. Then there will be a synchronization cost $T_s = t(n_2 - n_1)$ added to the actual message times. One measure of the irregularity or asymmetry of the problem is be given by:

$$\iota = \frac{|n_2 - n_1|}{max(n_i)} \tag{1}$$

Let T_c be the time it takes for signals to travel between processes (including any signals required for process coordination), and T_n be the computation time equal to $t*max(n_i)$. Normally we would want to compute in parallel in situations where $T_n >> T_c$. In highly skewed, irregular problems, the difference $|n_2 - n_1|$ can be of the same order as $max(n_i)$; the synchronization time T_s is therefore much greater than T_c and the irregularity T_s/T_n can be a large fraction of unity.

We want to emphasize that we are not talking about badly load-balanced codes. We have in mind codes that are well balanced at a coarser scale but irregular at a fine scale.

3   Short-Cutting

In some problems, a parallel execution may give dramatic improvement over its serial counterpart. For example, suppose the computation involves testing various cases and terminates as soon as any one case passes a test. In a parallel execution, as soon as one process finds a "passing" case, it can "short-cut" the other processes by telling them to quit. The parallel algorithm may do less work if it finds the answer on one node before finishing the computation on other nodes. We have so far applied this technique to a minimization problem, but we believe its applicability can be extended to more general cases.

Consider a backtracking algorithm. Conceptually we are performing different computations on the same data, by following first one possible computational branch, then another if the first fails, and so on. Suppose we parallelize this algorithm by going down multiple branches simultaneously. (This is a parallel emulation of a non-deterministic choice algorithm, except that, lacking unlimited parallelism we may either have to sequence multiple branches on each node or limit the approach to problems with a fixed branching factor). Suppose further that we have some criterion that tells us when a solution has been reached. Since each branch follows a different computational path, the amount of work to be done along each branch is different.

Let us consider a simple case in which we have P processors and P branches must be explored. Let W_i be the work required on branch i. If all branches are explored one after another, the total amount of work done serially is:

$$W_s = \sum_i W_i + O_s \qquad (2)$$

where O_s is the serial overhead.

Let $W_{avg} = Avg(W_i)$. Then we may replace this by:

$$W_s = P * W_{avg} + O_s \qquad (3)$$

Assume that not all branches lead to solutions, but that out of P branches there are $s < P$ solutions, and that we have some criterion that tells us when a solution has been reached along a particular branch without comparing to other branches (for example, this could be the case for a search with a fixed goal state where all we care about is the goal, not the path). Then we can halt as soon as a solution has been reached. When this will happen depends on the order in which we take the branches. Assume for the sake of simplicity that there is a single solution: s=1. Also assume that the units of work and time are the same, so that one unit of work executes in one unit of time. On the average we would expect to find a solution after trying about half the branches, and we should have an average case sequential time:

$$T_{avg.case} = .5(P * W_{avg} + O_s). \qquad (4)$$

Here we also assume that if we can stop after taking half the branches, we only incur half the overhead. We are also assuming that the algorithm halts on all branches; we can always force this by some artificial criterion like a counter).

Now consider the parallel case. In this situation, we can halt as soon as any one of our processes has found a solution. We would then have that parallel work done on P branches is:

$$W_p = P * (min(W_i) + O_p). \qquad (5)$$

Take the case of small overhead. Then we have:

$$T_{avg.case} = .5P * W_{avg} \qquad (6)$$

and the parallel execution time T_p is

$$T_p = W_p/P = min(W_i) = W_{min}. \qquad (7)$$

The average case parallel efficiency would be:

$$E_p = \frac{T_{avg.case}}{PT_p} = \frac{W_{avg}}{2W_{min}} \qquad (8)$$

where we know that $W_{min} \leq W_{avg}$. If it should happen that $W_{min} < .5W_{avg}$, then we would have a parallel efficiency greater than 1 in the average case.

In the worst case for parallel efficiency, we would find the solution on the first sequential trial. Then the total sequential work would be W_{min}, much less than the parallel work of $P * W_{min}$. But the parallel task would still complete at the same clock-time. In the best case, the sequential algorithm finds the solution on the last branch; this would give us a best case parallel efficiency greater than 2.

4 The Overlap Protocol

The possible gains from short-cutting are of the order of the computation time. As we have seen, this is also the case for synchronization costs in highly irregular problems. It is therefore possible that synchronization delays will cancel out all or most of the potential gains from short-cutting. Synchronization cost is due to the properties of the code, or to random factors in the runtime environment, and therefore may not be eliminated.

A standard way to hide communications cost is to overlap computations with communications. However, this will not be beneficial to processes that finish their execution in less time than others, and in any case the benefits of such overlap are only possible if special hardware allows communications to be executed in parallel.

We propose therefore to separate sends and receives and overlap the periods between generation and use of data in different processes. If this period at a producer process overlaps the period between references to data at a consumer process, and the data can be sent during this overlap, then neither process will have to wait for the other. Depending on which process is faster we generally only need to have either sender or receiver wait, but rarely both at once. It is difficult for the programmer to implement such overlap, so we propose here a protocol for overlapping communications that may be implemented automatically at the level of a compiler or preprocessor, or with a library with suitable run-time environment.

Consider a simple case of point to point communication between sender P1 and receiver P2 for data item X. We have the following possibilities: X.1, X.2, ...X.n are the first, second and nth values in memory location X. In general we have statements such as:

S1: X = (expression)

...

S2: Z = (expression using X)

...

(S3: X at P2 = X at P1 ; original communication statement)

...

S4: Y = (expression using X)

...

S5: X = (expression) ; maybe next iteration of a loop

...

S6: Y = (expression using X)

Table 1. Sender(P1)-receiver(P2) timing

time	P1	P2
case 1: sender is slightly faster		
1	write X.1	—
2	—	read X.1
3	write X.2	—
case 2: sender is much faster		
1	write X.1	—
2	wait X.2	read X.1
3	write X.2	—
case 3: receiver is faster		
1	—	wait X.1
2	write X.1	wait X.1
3	write X.2	read X.1

All references to X are in S1 ... S6. Referring to Table 1; S1 is a write at P1 and S4 is a read at P2. Since the processes are not synchronized, later statements at one process can occur before earlier statements at another; e.g. case 3, t=1 is the situation where P2 reaches S4 before P1 reaches S1; this forces P2 to wait before it can read. Recall that both P1 and P2 are executing the above sequence of statements, but in the given case we need the value of Y only in P2, and we are only using the value of X produced at P1.

P1 has data which it may send at S1, it must send it before S4 when it needs to update the local value of X. P2 may read X any time after S2; it must read it at S4. Similarly, the value of X that P2 uses in statement S6 must be read after S5 (where X is once more updated - P2 executes this statement even though this value of X is not used). X must be read at S6 at the latest.

P1 is in a state MAY-SEND (with respect to a given data item and communication statement) as soon as it calculates the value of X at S1; it is in a state MUST-SEND at S5, because here it cannot proceed to recalculate X until it sends the old value. P2 is in a state MAY-READ after S2 and before S4. It can't read X before S1, because here it executes code to recalculate X, even if we don't want this value. Once P2 reaches S4, it is in a state MUST-READ, and here it waits until P1 is ready to send.

In case 3, if we start at t=2 we have sender writes X.1 when receiver wants it, that is, synchronized. If sender is much faster than receiver, it starts waiting for X.1 earlier, the end result is same as case 3.

Processes in MAY states (MAY-SEND, MAY-READ) can continue processing, until they reach a statement that puts them in a MUST state (MUST-SEND, MUST-READ). In a MUST state we either need to use a value from another process to calculate something, or we need to replace a value that some other process needs. In either case, we have to wait for the data transfer to take place. We consider a process to be in a QUIET state if it is neither ready to send nor to receive.

Table 2. States of a sender-receiver pair

	P1 state	P2 state	results
00	QUIET	QUIET	both processes continue
01	QUIET	MAY-READ	both processes continue
02	QUIET	MUST-READ	P2 waits, P2 continues
10	MAY-SEND	QUIET	both processes continue
11	MAY-SEND	MAY-READ	P1 sends to P2, both continue
12	MAY-SEND	MUST-READ	P1 sends to P2, P2 waits for data
20	MUST-SEND	QUIET	P1 waits, P1 continues
21	MUST-SEND	MAY-READ	P1 sends to P2, P1 waits for data
22	MUST-SEND	MUST-READ	P1 sends to P2, both wait

Data is transferred in states 11, 12, 21 and 22. In cases 12 and 21, one process must wait for one data copy, but the other (the one in the MAY state) can continue. In case 11, neither process has to wait, computation can completely overlap communications. In case 22, both processes wait, but they wait at the same time so total wait is only for one copy.

All states are with respect to some specific data item. For example, a process may have calculated X and so be in MAY-SEND with respect to it, and need a value Z from someplace else in the current statement being executed, which puts it in MUST-READ with respect to Z.

5 The Overlap FSM

Table 2 defines a Finite State Machine (FSM) that implements a protocol that is sufficient to handle simple point to point communications. However, we also wish to handle more complex operations, such as reductions and limited broadcasts, including reductions among a sub-group of processors. We refer to this general class of operations as *group communications*. Figure 1 defines a Finite State Machine (FSM) that implements our protocol, with added states to permit short-cutting and group communications. A complete description of this FSM is beyond the scope of this paper and will be presented elsewhere.

We assign a separate FSM to each variable being communicated at each node. Group communications, to be efficient, require cooperative action, in which some nodes will pass information on, both sending and receiving. We use a *message count* to determine completion, implemented by adding a counter to our FSM, which emits a "zero" message when there are no pending operations. Messages define the transitions in the FSM: each message corresponds to a particular event, and is a letter in the language accepted.

States in Fig. 1 are divided in 3 general paths corresponding to sends, receives and group communications. Any particular communication is defined as a set of point to point sends between nodes, such that the receiver sends a CS message when ready to receive, and the sender then sends the data message. One or more sends from a node may start with Send, allow multiple RH and CS (each CS

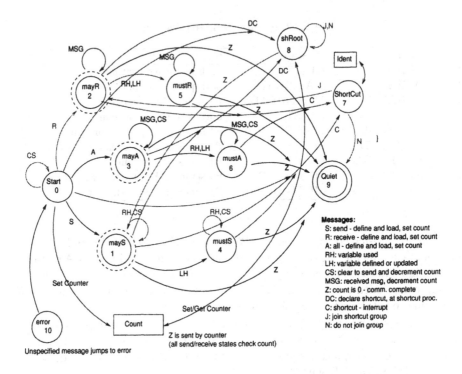

Fig. 1. Overlapping and Short-cutting FSM

corresponds to a message send and a decrement of the count, each RH is a use of the variable without redefinition), terminating in Z or blocking at a redefinition (LH), followed by at least one CS (send and decrement), terminating in Z. (That is, $S(RH|CS)*Z$ or $S(RH|CS)*LHCSCS*Z$ is a word accepted by the FSM.)

One or more receives at a node allow multiple message (MSG) arrival (and decrement count), terminating in Z or blocking on first use or definition (RH or LH), followed by at least one MSG (and decrement), terminating in Z. (That is, $RMSG*Z$ or $RMSG*(LH|RH)MSGMSG*Z$)

A group communication allows multiple message (MSG, decrement) arrival, multiple CS (message sends, decrement), terminating in Z or blocking on first use or definition (RH or LH), followed by at least one MSG (and decrement) or CS (send and decrement), terminating in Z. (That is, $A(MSG|CS)*Z$ or $A(MSG|CS)*(LH|RH)(MSG|CS)(MSG|CS)*Z$)

We have defined protocols for overlap between definition and next definition at the sender, and use and next use at the receiver; processes being in a may send/receive state during the overlap period, and entering a must send/receive state just before the next use/definition. Note, however, even if we are in an overlap section, communication can not occur until the senders and receivers are identified. Therefore we consider the 'may communicate' interval to begin when senders and receivers are defined, and to end (at sender) before the next definition and (at receiver) before the next use.

Suppose we have a group of processes $\{Pi \mid i \in G\}$ indexed by a group G with $|G|$ elements. For example, if $G = \{1..n\}$ then $|G| = n$. We are presently able to handle the following cases (others are treated as compositions of these):

1. Point to point: send x at j, receive y at i; a single send matched by a single receive. (In this case, we define the count to be 1.)

2. Send to a group G: (count of $n = |G|$ at sender, count of 1 at each receiver). It is possible to define it as a collaborative broadcast, using e.g. a spanning tree rooted in j. Doing this costs implicit synchronization and ordering

3. Gather: Y at i is the result of applying some operator to pairs of values x at all nodes in G. (count of $n = |G|$ at receiver, count of 1 at each sender). Order is arbitrary, therefore correctness requires operations to be commutative and associative. This can also be defined as a collaborative operation involving sets of sends and receives at particular nodes.

All communications are resolved into sets of point to point sends and receives (note that this does not preclude collaborative communications, since a particular node may receive a message from another and forward it to yet another node, optionally adding information of its own).

We assume a correct program in which communication statements express a transfer of information and include senders and receivers; this is the style of MPI [4] group communications and of Pfortran [5], to name two examples. Processes each see the communications in which they participate in the same order. As a result, we know the communication pattern before communicating (that is, we know the group of processes involved, and what the initial and final situation should be). We can therefore resolve any communication into a pattern of point to point sends ahead of time. Therefore for every send of a variable at a given process, there is a matching receive (of possibly a different variable) at the destination process, and vice-versa. Assume the message includes an identifier that indicates what variable is being sent (this is necessary because, if several messages are received by a given node, it must know which buffers to use for the received data).

Note that communications proceeds by alternating CS (receiver→sender) and MSG (sender→receiver). A sender state may not accept without receiving a correct count of CS messages; a receiver state may not accept without receiving a correct count of MSG, and an all-gather state may not accept without receiving a correct count of both CS and MSG.

Therefore communications are ordered with respect to each variable, each communication completes before the next communication of the given variable is required. It can be shown that the correct value is transmitted. For collaborative communication, we must also prove the correctness of the group algorithm. This issues will be addressed in subsequent publications.

6   Implementation

The test program was written in Fortran77 for serial processing. It was parallelized using Pfortran and MPI, and a one node version without communications was used for speed benchmarks running on a single node.

We chose to implement a real algorithm rather than a synthetic benchmark; although a test on one algorithm is clearly not an exhaustive proof, it should be more indicative of real world performance than an artificial benchmark program, which we could write to respond arbitrarily well to our overlap synchronization strategy. We picked a multidimensional function minimization technique called the Downhill Simplex Method in Multi-dimensions from [6] for use in our test.

We parallelized by projecting through all faces of the simplex at the same time rather than just through a single face: speedup was obtained through convergence in fewer steps. The parallelism and branching factor in this case is limited to the number of dimensions in the problem.

At each stage we do the following:

- Input: simplex
- P: parallel tasks on each vertex:
  - P1: project through opposite base
  - P2: if medium good project farther
  - P3: if nothing worked contract simplex
- compare all nodes and pick best simplex
- C: if simplex has converged stop; else go back to Input.

The problem is irregular because the evaluation at each node is different.

We then modified the algorithm by short-cutting as follows: If at any stage P1, P2 or P3 we have a better simplex than the original (possibly by some adjustable criterion of enough better) then go directly to C, the convergence test, interrupting all the other parallel tasks. We then have:

- Input: simplex
- P: parallel tasks on each vertex:
  - P1: project through opposite base; if good goto C
  - P2: if medium good project farther; if good goto C
  - P3: if nothing worked contract simplex
- compare all nodes and pick best simplex
- C: halt parallel tasks
- if simplex has converged stop; else go back to Input.

Potentially at each stage some process could short-cut the computation after doing only 1/3 the work, at P1. Therefore it is possible that the criterion $T_{min} < T_{avg}$ is met, and there is a potential in the modified algorithm for improved parallel efficiency.

Although in the serial version it is sufficient to take any good enough new value as the base for the next iteration, in the parallel version we must consider the possibility that more than one node has a good enough value at the same time (traces while running the program showed this circumstance in fact occurred). If two nodes think they have a good value, and both try to broadcast in the same place, the program will deadlock.

The protocol definition allows for definition of separate streams of execution corresponding to each of multiple short-cutting nodes, but this is not fully implemented. The present implementation simply forces a jump to the final comparison section when either a node has a good enough value or it has received a message that tells it some other node has one. The evaluation then proceeds as in the standard algorithm.

Note that the short-cutting version of the program is nondeterministic but not random; the short-cutting process is the one that first (by clock time) finds a good enough solution, and this may vary from one run to the next.

It may be argued that the parallel short-cut works better in this case than the serial short-cut because it evaluates the best partial solution, whereas the serial algorithm merely takes the first one it finds that is good enough. But in fact, the first good enough solution found will be the best value the serial algorithm has; in order to find a better value it must continue trials which would cause it to evaluate more functions.

7 Tests

For testing we used a simple seven dimensional paraboloid requiring fifteen floating point operations to compute, embedded within a loop that repeated it a number of times to simulate a more complicated function. We took an initial simplex with a corner at (-209,-209,-209,-209,-209,-209,-209) and then changed the starting point along a diagonal in increments of +80 to 451; the minimum was at (3,0,0,0,0,0,0).

We ran six variations of the program:

A0 - the original serial algorithm. From a starting point of -209 this required 11,513 function evaluations.

A1 - the parallelized algorithm, run serially. From the same starting point, 9,899 function calls.

A2 - the short-cutting algorithm, run serially. 9543 function calls.

A3 - the parallel algorithm. 9899 function calls.

A4 - the parallel short-cutting algorithm. 8000-11000 function calls.

A5 - the parallel overlapping algorithm. 9899 function calls.

A6 - the parallel overlapping and short-cutting algorithm. 8000-11000 function calls.

Figure 2 shows work required for convergence of the deterministic algorithm (det - corresponds to A1 serially or A5 in parallel); for the serial short-cutting algorithm A2 (short); and for several runs of the non-deterministic short-cutting algorithm in parallel, A4 (10 runs, n0 .. n9, equivalent to A6). Although the short-cutting algorithm executes differently each time, it converged correctly to the same value. These tests were run on a cluster of four 200 MHz PC's running linux and communicating via MPICH over 10 Mbit ethernet.

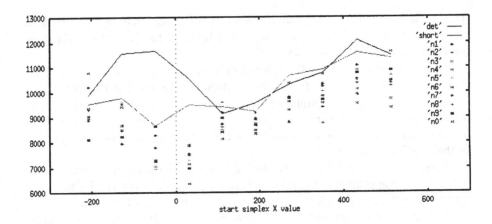

Fig. 2. Function Evaluations Required for Convergence

The level of irregularity in the program may be seen in the variation in number of function evaluations when different paths are taken to a solution. There is variation of up to about one tenth of the function evaluations, indicating an irregularity of the order of .1.

Timings were done on the SP2 for A3, the standard parallel algorithm, A4 the short-cutting algorithm, and and A5, the short-cutting algorithm with overlap, generating a starting simplex from (-209,-209,-209,-209,-209,-209,-209). We found that the SP2, using IBM's "load leveler," exhibited large variations in time depending on what other jobs were running on it (differences by factors of as much as 2 or 3). The following results were taken when the machine was lightly loaded.

A2 deterministic: Average = 48.6 sec, range between 47.9-49.8 sec.

A3 short-cutting: Average = 47.8 sec, range between 46.2-48.9 sec.

A4 short-cutting with overlap: Average = 46.9 sec, range between 46.3-48.3 sec.

In the tests, each function evaluation was set to 15 million floating point operations, by setting the loop index to 1000000. Reducing the loop index to 1, program run times were about 10 seconds; we took this to correspond to the base communication time, so the actual computation time is about 40 seconds. The time irregularity should be about 4 seconds, and the combination of short-cutting and overlapping appears to be capturing almost half of the possible savings.

Some of the time variation is probably due to differences in the SP2 environment. The two non-deterministic short-cutting programs exhibit more variation in time, as expected. We see that the short-cutting program has a slight advantage over the standard program, and the overlapping program performs better than short-cutting alone.

8 Conclusions

We have shown two techniques that, separately and in combination show promise of speedup for parallel processing of irregular programs.

The overlapping of regions between data definition and use in different processes lightens the synchronization requirements by spreading the work of synchronization over an interval of time. This is particularly important for irregular problems in which we want to not only hide the communications cost, but also minimize the time processes must spend waiting for others to reach synchronization points.

Where applicable, the short-cutting method allows the implementation of parallel algorithms that are inherently better than their serial counterparts in that they do less work to reach the same solution. Note that this means increasing advantage with increasing problem size, since there will be a greater difference between the work done in parallel and that done serially. In addition, short-cutting dynamically load balances parallel execution of irregular problems since all processes will take the same clock time when short-cut.

Problems to which short-cutting is applicable are extreme cases of irregular problems; this in our view justifies the development and application of overlapping technology together with short-cutting, for maximum benefits.

References

1. R. Ponnusamy, J. Saltz, and A. Choudhary; Runtime-Compilation Techniques for Data Partitioning and Communication Schedule Reuse. Proceedings Supercomputing '93, pages 361-370, Portland, Oregon, November 15-19, 1993
2. L. R. Scott and Xie Dexuan; Parallel linear stationary iterative methods. Research Report UH/MD 239, Dept. Math., Univ. Houston, 1997.
3. G. A. Geist and C. H. Romine; LU factorization algorithms on distributed-memory multiprocessor architectures. *SIAM J. Sci. Stat. Comput.*, 9:639–649, 1988.
4. Message Passing Interface Forum; MPI: A message Passing Interface Standard. June 12, 1995, Version 1.1, http://www.mcs.anl.gov/mpi/
5. B. Bagheri, T. W. Clark and L. R. Scott; Pfortran: A Parallel Dialect of Fortran. Fortran Forum, ACM Press, 11, pp. 3-20, 1992.
6. W. H. Press, B. P.Flannery, A. A. Teulkosky, W. T. Vetterling; Numerical Methods: The Art of Scientific Computing. Cambridge 1986

Control Volume Meshes Using Sphere Packing

Gary L. Miller

School of Computer Science, Carnegie Mellon University, Pittsburgh, Pennsylvania 15213.

Abstract. We present a sphere-packing technique for Delaunay-based mesh generation, refinement and coarsening. We have previously established that a bounded radius of ratio of circumscribed sphere to smallest tetrahedra edge is sufficient to get optimal rates of convergence for approximate solutions of Poisson's equation constructed using control volume (CVM) techniques. This translates to Delaunay meshes whose dual, the Voronoi cells diagram, is well-shaped. These meshes are easier to generate in 3D than finite element meshes, as they allow for an element called a *sliver*.

We first support our previous results by providing experimental evidence of the robustness of the CVM over a mesh with slivers. We then outline a simple and efficient sphere packing technique to generate a 3D boundary conforming Delaunay-based mesh. We also apply our sphere-packing technique to the problem of automatic mesh coarsening. As an added benefit, we obtain a simple 2D mesh coarsening algorithm that is optimal for finite element meshes as well.

This is a joint work with Dafna Talmor (CMU), Shang-Hua Teng (UIUC), Noel Walkington (CMU), and Han Wang (CMU).

The control volume method (CVM) is a popular method for solving partial differential equations, especially when the underlying physical problem has some conservation properties, such as for heat or flow problems. The domain is partitioned into small "control" volumes, and the approximation is derived using the conservation properties over each control volume. The Voronoi diagram, and its dual, the Delaunay triangulation, are particularly fitted for the control volume method, with the Voronoi cells acting as the control volumes, and the Delaunay triangulation as the neighborhood structure of interacting Voronoi cells.

The quality of a triangular or simplicial mesh is measured in terms of both the shape of its elements and their number. Unstructured Delaunay triangulation meshes are often used in conjunction with the finite element method (FEM). In the context of the finite element meshes, the element shape quality is measured by its aspect ratio. Several definitions for the element aspect ratio exist, and they are all equivalent up to a constant factor: the ratio of the element height to the length of the base, the ratio of the largest inscribed sphere to the smallest circumscribed sphere, or the size of the smallest element angle.

The problem of generating good aspect ratio meshes has been discussed extensively in the literature (see the survey by Bern and Eppstein [2]). CVM mesh generation was assumed to pose the same requirements as FEM mesh generation, and did not merit its own discussion. However, we have shown recently [10] that the error estimates of the CVM depend on a shape criteria weaker than the aspect-ratio. In particular, CVM meshes require elements with a bounded ratio of radius of circumscribed sphere to smallest element edge, among other requirements mentioned below. In 2D, this *radius-edge ratio* is equivalent to the standard aspect ratio, but in 3D it allows a very flat element with arbitrarily small dihedral angles, and edges proportional to the circumscribed sphere radius. This type of element is referred to as a *sliver*.

Generating a mesh of good radius-edge ratio appropriate for the CVM is the motivation behind this paper. In 3D, it is an easier task than generating a mesh with a good aspect ratio. Our approach is to use an approximate sphere-packing technique. In this paper we analyze the conditions under which a sphere-packing yields a bounded radius-edge ratio mesh that conforms to a 3D domain description.

These are what we consider to be the main contributions of this paper:

- Comparing and contrasting the FEM and the CVM over a mesh with slivers. We provide experimental evidence that the CVM is insensitive to slivers.
- Characterizing the requirements of CVM mesh generation.
- Proposing a new sphere-packing approach: we analyze the conditions under which a sphere-packing yields a good radius-edge mesh that conforms to a 3D domain description. This also provides some insight into the workings of other related mesh generation methods, such as advancing front methods and Ruppert's method, and also into the development of efficient 3D mesh generation.
- Applying the new sphere-packing technique to 2D provably optimal automatic mesh coarsening. In 2D, this automatic mesh coarsening applies to good aspect-ratio (FEM) meshes as well.

Previous Work The study of mesh generation abounds in the literature. Bern, Eppstein and Gilbert [4] had the first provably optimal 2D good aspect-ratio mesh generation algorithm. The mesh they generated is based on a quad-tree. Ruppert [15] presented a simpler algorithm that produces smaller meshes in practice. Ruppert's algorithm is based on a Delaunay triangulation, and extends the approach Chew [6] suggested for quasi-uniform meshes to general unstructured meshes.

3D mesh generation is still a mostly uncharted research area for good aspect-ratio meshes. Mitchell and Vavasis [11] extended to 3D the quad-tree approach of Bern, Eppstein and Gilbert [4]. The simpler Delaunay triangulation approach of Ruppert and Chew has not been successfully generalized. The work of Dey, Bajaj and Sugihara [7] is a generalization of Chew's algorithm into 3D. It is targeted at quasi-uniform meshes only, and addresses only simple convex boundaries. Even for that simpler case their 3D generalization of Chew's approach fails to generate good aspect ratio Delaunay triangulation. It does, however, generate good bounded radius-edge ratio meshes.

The failure of the Dey, Bajaj and Sugihara algorithm to produce good aspect-ratio meshes preempted any discussion on how to generalize Ruppert's related algorithm to 3D. Now that the weaker condition of radius-edge ratio is proven useful, the question of generalization of Ruppert's algorithm regains its relevance. Even if this generalization were straight-forward, it would merit a lengthy discussion so that the issues of conforming to 3D boundaries are thoroughly investigated.

However, the obvious generalization of Ruppert's algorithm to 3D suffers from some drawbacks. Recall that the basis of Ruppert's scheme is to iteratively produce the target point set by maintaining a Delaunay triangulation of the point set, picking any bad aspect-ratio triangle, adding its circumcenter to the point set or splitting a segment if appropriate, retriangulating the point set and repeating until all triangles are well-shaped. This scheme is simple and efficient in 2D in practice, though is not theoretically guaranteed to be efficient. In 3D, two facts hinder us: (1) The intermediate 3D triangulation maintained is potentially of size $O(|P|^2)$, P being the intermediate point set, whereas in 2D the triangulation is always linear. (2) In 2D the triangulation is repaired by an edge flipping algorithm; the complexity of a repair can be $O(n)$ but in practice is smaller. In 3D each edge flip is more complex, and furthermore the total number of flips necessary to update the Delaunay diagram can be of a worst case $O(n^2)$. Its complexity in practice remains to be seen. (Note however that in this case, as the starting point of the incremental construction is a DT, the edge flips algorithm is guaranteed to converge to a Delaunay diagram. This has been shown separately by Joe [8], and Rajan [14]).

A different technique is therefore necessary to produce an efficient generalization of Ruppert's algorithm to 3D. In this paper we concentrate on one such technique: sphere-packing. It has been observed before [16] that a pattern of tightly packed spheres led in nature to well shaped Delaunay triangulations and Voronoi diagrams.

We are aware of at least two instances where sphere packing was used as a basis for mesh generation, but with a different goal and underlying technique: (1) Shimada's bubble packing: spheres are packed in the domain in some initial configuration, and then a physically based iterative method is used to smooth the mesh and obtain a better shaped mesh. Our method is not an iterative method, and its emphasis is on laying down the spheres such that the corresponding Delaunay triangulation carries radius-edge ratio guarantees. As such, it is orthogonal to Shimada's method. (2) Bern, Mitchell and Ruppert [3] used sphere packing to generate an $O(n)$ non-obtuse triangulation of a polygonal domain in 2D. The elements produced in that instance could be of bad aspect ratio and radius-edge ratio.

References

1. Chanderjit L. Bajaj and Tamal K. Dey. Convex decomposition of polyhedra and robustness. *SIAM J. Comput.*, 21(2):339–364, April 1992.
2. M. Bern and D. Eppstein. Mesh generation and optimal triangulation. Tech. Report CSL-92-1, Xerox PARC, Palo Alto, CA, 1992.
3. M. Bern, S. Mitchell, and J. Ruppert. Linear-size nonobtuse triangulation of polygons. In *Proc. 10th Annu. ACM Sympos. Comput. Geom.*, pages 221–230, 1994.
4. Marshall Bern, David Eppstein, and John Gilbert. Provably good mesh generation. In *FOCS90*, pages 231–241, St. Louis, October 1990. IEEE.
5. A. Brandt. Multi-level adaptive solutions to boundary value problems. *Mathematics of Computation*, 31:333–390, 1977.
6. L.P. Chew. Guaranteed-quality triangular meshes. CS 89-983, Cornell, 1989.
7. Tamal K. Dey, Chanderjit L. Bajaj, and Kokichi Sugihara. On good triangulations in three dimensions. In *Proceedings of the ACM symposium on solid modeling and CAD/CAM Applications 1991*, pages 431–441. ACM, 1991.
8. B. Joe. Construction of three-dimensional Delaunay triangulations using local transformations. *Comput. Aided Geom. Design*, 8(2):123–142, May 1991.
9. R. H. MacNeal. An asymmetrical finite difference network. *Quarterly of Applied Math*, 11:295–310, 1953.

10. G. L. Miller, D. Talmor, S.-H. Teng, and N. Walkington. A Delaunay based numerical method for three dimensions: generation, formulation, and partition. In *Proc. 27th Annu. ACM Sympos. Theory Comput.*, pages 683–692, 1995.

11. S. A. Mitchell and S. A. Vavasis. Quality mesh generation in three dimensions. In *Proceedings of the ACM Computational Geometry Conference*, pages 212–221, 1992. Also appeared as Cornell C.S. TR 92-1267.

12. R. A. Nicolaides. Direct discretization of planar div–curl problems. *SIAM J. Numer. Anal.*, 29(1):32–56, 1992.

13. Carl F. Ollivier-Gooch. Upwind acceleration of an upwind euler solver on unstructured meshes. *AIAA Journal*, 33(10):1822–1827, 1995.

14. V. T. Rajan. Optimality of the Delaunay triangulation in R^d. In *Proc. 7th Annu. ACM Sympos. Comput. Geom.*, pages 357–363, 1991.

15. Jim Ruppert. A new and simple algorithm for quality 2-dimensional mesh generation. In *Third Annual ACM-SIAM Symposium on Discrete Algorithms*, pages 83–92, 1992.

16. Kenji Shimada and David C. Gossard. Bubble mesh: Automated triangular meshing of non-manifold geometry by sphere packing. In *Proc. 3rd Sympos. on Solid Modeling and Applications.*, pages 409–419, 1995.

Using Multithreading for the Automatic Load Balancing of Adaptive Finite Element Meshes

Gerd Heber[1], Rupak Biswas[2], Parimala Thulasiraman[1], and Guang R. Gao[1]

[1] CAPSL, University of Delaware, 140 Evans Hall, Newark, DE 19716, USA
[2] MRJ, NASA Ames Research Center, MS T27A-1, Moffett Field, CA 94035, USA

Abstract. In this paper, we present a multithreaded approach for the automatic load balancing of adaptive finite element (FE) meshes. The platform of our choice is the EARTH multithreaded system which offers sufficient capabilities to tackle this problem. We implement the adaption phase of FE applications on triangular meshes, and exploit the EARTH token mechanism to automatically balance the resulting irregular and highly nonuniform workload. We discuss the results of our experiments on EARTH-SP2, an implementation of EARTH on the IBM SP2, with different load balancing strategies that are built into the runtime system.

1 Introduction

In this paper, we present and examine a multithreaded approach for the automatic load balancing of adaptive finite element (FE) meshes. During a FE adaption phase, the unstructured mesh is refined/coarsened (according to some application-specific criteria), and the workload may become (seriously) unbalanced on a multiprocessor system. This significantly affects the overall efficiency of parallel adaptive FE calculations. Some of the difficulties encountered when using the traditional approach to resolve the load imbalance problem can be summarized as follows:

- It is necessary to assemble global information to get an accurate estimate of the load distribution – a step that often creates serious bottlenecks in practice.
- Significant effort is usually required to preserve data locality while making the load balancing decisions – an optimization which is computationally difficult.
- To make matters worse, the evolution of computational load and data locality requirements are dynamic, irregular, and unpredictable at compile time.

Multithreaded architectures, such as the EARTH (*Efficient Architecture for Running THreads*) system [8], offer new capabilities and opportunities to tackle this problem. They strive to hide long latency operations by overlapping computation and communication with the help of *threads*. A thread is a (small) sequence of instructions. EARTH provides mechanisms to enable automatic load balancing for applications that do not allow a good static (compile time) task distribution. The programmer can simply encapsulate a threaded function invocation as a *token*. The advantage of this token mechanism is that they can flexibly migrate over processors, by means of a load balancer.

Given our experience with EARTH, we decided to pursue a novel approach to implement a two-dimensional version of the dynamic mesh adaption procedure, called 3D_TAG [1]. Our decisions were based on the following observations:

- EARTH provides a runtime load balancing mechanism at a very fine-grain level, i.e., the token can be an ultra light-weight function invocation. Therefore, as long as the computation generates a large number of such tokens, it is likely that the EARTH runtime system can automatically distribute them to the processors to keep them usefully busy. It is our hypothesis that we can thus eliminate an explicit repartitioning/remapping phase (in the first order) and the gathering of global mesh information.

- Locality is always an important issue for efficiency. However, since EARTH has the ability to tolerate latency using multithreading, its performance is less sensitive to the amount of non-local accesses by each processor. In other words, the optimality of locality or data distribution is less critical. Instead of optimally trying to repartition/remap the data to minimize communication, the user should take advantage of the EARTH token mechanism and structure the algorithm/code such that the input arguments to a token can be migrated easily with the token itself. The token should thus have good *mobility*.

- The difficulty of handling irregular and unpredictable load evolution in traditional systems is partly due to the limitations of the data parallel programming model and the SPMD execution paradigm where the computation in each processor should be loosely synchronous at a coarse-grain level. The EARTH execution model is fundamentally asynchronous and does not rely on the SPMD model. Function-level parallelism can be naturally expressed at a fine grain in the EARTH programming paradigm. The possible association of a parallel function invocation with a token at such a fine grain enables the runtime load balancing mechanism to work smoothly with the evolving computational load.

We implement the adaption phase of a simulated FE calculation and exploit the EARTH token mechanism to automatically balance the highly nonuniform workload. To achieve good token mobility, we apply a new indexing methodology for FE meshes that does not assume a particular architecture or programming environment. During mesh adaption, tokens that migrate to another processor to be executed leave the processed data in that processor instead of transferring the data back to the processor that generated the token. This is a novel approach in solving this problem: we avoid doing work twice and do a load-driven remapping without explicit user intervention. A major part of this paper is devoted to a discussion of the *quality* of such an approach which can be achieved using different load balancing strategies (built into the EARTH runtime system).

2 The EARTH System

EARTH (*Efficient Architecture for Running THreads*) [8] supports a multithreaded program execution model in which user code is divided into threads that are scheduled atomically using dataflow-like synchronization operations. These "EARTH operations" comprise a rich set of primitives, including remote loads and stores, synchronization operations, block data transfers, remote function calls, and dynamic load balancing. EARTH operations are initiated by the threads themselves. Once a thread is started, it runs to completion, and instructions within it are executed sequentially.[1] Therefore, a conventional processor can execute a thread efficiently even when the thread is purely sequential. It is thus possible to obtain single-node performance close to that of a purely sequential implementation, as was shown in our earlier work [10].

Conceptually, each EARTH node consists of an *Execution Unit* (EU), which executes the threads and a *Synchronization Unit* (SU), which performs the EARTH operations requested by the threads (cf. Fig. 1). The EARTH runtime system is presently available on the MANNA architecture [7], the IBM SP2 multiprocessor [3], and the Beowulf workstation clusters.

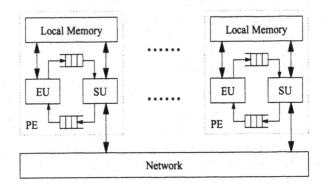

Fig. 1. The EARTH architecture

Currently, programs running on EARTH are written in Threaded-C [13], a C extension containing multithreading instructions. It is clean and powerful enough to be used as a user-level, explicitly parallel programming language. The keyword THREADED specifies that the function's frame should be allocated from the heap and that the function may contain multiple threads.[2] Each thread, except for the first, begins with a label THREAD_n; the first thread begins at the start of the function, and is automatically executed when the function is invoked using an INVOKE or TOKEN instruction. Each thread ends with either END_THREAD(), which simply executes a **fetch_next** instruction, or END_FUNCTION(), which executes an **end_function** instruction to deallocate the frame before executing a **fetch_next**.

In addition, for those applications for which a good task distribution cannot be determined statically by the programmer and communicated to the compiler, EARTH provides an automatic load balancing mechanism. The programmer can simply encapsulate a function invocation as a token. The token is sent to the SU, which puts it on top of the local token queue. When there are no more threads in the ready queue, the SU removes a token from the top of the token queue and

---

[1] Instructions may be executed out of order, as on a superscalar machine, as long as the semantics of the sequential ordering are obeyed.

[2] Non-threaded functions are specified as regular C functions.

invokes the function as specified locally. This load balancing technique is derived from the token management strategy used in the ADAM architecture [9]. Note that putting locally-generated tokens on top of the queue and then removing tokens from the top leads to a *depth-first* traversal of the call-graph. This generally leads to better control of functional parallelism, i.e., it diminishes the likelihood of parallelism explosion that can exhaust the memory resources of a node.

When both the ready queue and the token queue are empty, the SU sends a message to a neighboring processor requesting a token, in effect performing *work stealing*. The neighboring processor, if it has tokens in its queue, extracts one from the bottom of its queue and sends it back to the requester. In this manner, a *breadth-first* traversal of the call-graph is implemented across processors, hopefully resulting in a better distribution of tasks. If the neighboring processor does not have any tokens to satisfy a request, the neighbor's neighbor is queried, and so on. Either a token will be found, or the request cannot be fulfilled. Idle processors periodically query their neighbors for work.

3 An Indexing Technique for FE Meshes

Dynamic mesh adaption is a common and powerful technique for efficiently solving FE problems, and is achieved by *coarsening* and/or *refining* some elements of the computational mesh. In 2D FE meshes, triangular elements are the most popular. Once a triangle is targeted for refinement, it is generally subdivided by bisecting its three edges to form four congruent triangles as shown in Fig. 2(a). This type of subdivision is called *isotropic* (or 1:4) and the resulting triangles are referred to as being *red*. A problem is that the refined mesh will be *nonconforming* unless all the triangles are isotropically subdivided. To get a consistent triangulation without global refinement, a second type of subdivision is allowed. For example, a triangle may be subdivided into two smaller triangles by bisecting only one edge as shown in Fig. 2(b). This type of subdivision is called *anisotropic* (or 1:2 in this case) and the resulting triangles are referred to as being *green*. The process of creating a consistent triangulation is defined as a *closure* operation. Note that several iterations may be necessary to achieve closure [1].

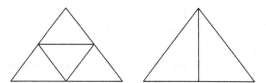

Fig. 2. (a) Isotropic and (b) anisotropic subdivision of a triangle

A couple of additional rules are applied, primarily to assure that the quality of the adapted mesh does not deteriorate drastically with repeated refinement:
1. All triangles with exactly two bisected edges have their third edge also bisected. Thus, such triangles are isotropically refined.

2. A green triangle cannot be further subdivided. Instead, the previous subdivision is discarded and isotropic subdivision is first applied to the (red) ancestor triangle [2].

It is the task of an *index scheme* to properly name or label the various objects (vertices, edges, triangles) of a mesh. We prefer the term index scheme instead of *numbering* to stress that the use of natural numbers as indices is not sufficient to meet the naming requirements of the FE objects on parallel architectures.

We give a brief description of our indexing scheme for the sake of completeness; refer to [6] for a detailed discussion. Our technique is a combination of *coarse* and *local* schemes, and is intended for hierarchical meshes. The coarse scheme labels the initial mesh objects in such a way that the incidence relations can be easily derived from the labels. For the example in Fig. 3, the set of vertices for the coarse triangulation consists of the following numbers:

$$\text{vertices} = \{1, 2, 3, 4, 5, 6, 7, 8\}.$$

The edges of the coarse triangulation are indexed by ordered pairs of integers corresponding to the endpoints of the edges. The ordering is chosen so that the first index is less than the second:

$$\text{edges} = \{(1,2),\ (1,3),\ (1,5),\ (2,5),\ (3,4),\ (3,5),\ (3,6),$$
$$(4,6),\ (5,6),\ (5,7),\ (5,8),\ (6,7),\ (7,8)\}.$$

The same principles are applied to index the coarse triangles. They are denoted by the triple consisting of their vertex numbers in ascending order:

$$\text{triangles} = \{(1,2,5),\ (1,3,5),\ (3,4,6),\ (3,5,6),\ (5,6,7),\ (5,7,8)\}.$$

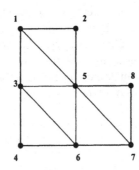

Fig. 3. An L-shaped domain and its coarse triangulation

The local scheme exploits the regularity (and the finiteness) of the refinement rules to produce names for the objects at subsequent refinement levels [6]. We use (scaled) natural coordinates as indices in the local scheme. Again, this is done in a way such that the incidence relations are *encoded* in the indices of the objects. For example, the set of vertices at level k in the local model is given by:

$$V_k = \{(a,b,c) \in \mathbb{N}^3 \mid a + b + c = 2^k\}.$$

We do not know which of these will actually be present; however, we already have names for them. Figure 4 shows the local indices for the vertices and the triangles on the first two refinement levels (for isotropic subdivision). The coarse and local schemes are combined by taking the union of the Cartesian products of the coarse mesh objects with their corresponding local schemes. Ambiguities are resolved by using a *normal* form of the index.

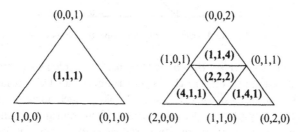

Fig. 4. Examples of the local index scheme for triangular elements (the vertices and triangles are denoted by integer triples, triangles in bold)

The key features of such a scheme are:
- Each object is assigned a *global name* that is independent of any architectural considerations or implementation choices.
- Combinatorial information is translated into simple arithmetic.
- It is *well-behaved* under (adaptive) refinement. No artificial synchronization/serialization is introduced.
- It can be extended (with appropriate modifications) to three dimensions [6].

4 Description of the Test Problems

4.1 Mesh, Partitioning, and Mapping

As the initial mesh for our experiments, we choose a rectangular mesh similar to, but much larger than, the one shown in Fig. 5. Mesh generation simplicity was an overriding concern. However, we never, neither in the algorithms nor in the implementation, exploit this simple connectivity. Recall that geometry information is irrelevant for our index scheme. Mesh adaption on even a simple case like this can fully exercise several aspects of the EARTH multithreaded system and its dynamic load balancing capabilities.

Fine-grain multithreading would allow us to have one thread per element. However, to control granularity and to prevent *fragmentation*[3] of the triangulation, we choose partitions as our smallest migratable unit. A good overview of common partitioning techniques and mapping strategies is given in [4]. Based on observations made there, we decided to use a $P \times Q$ block partitioning.

Initially, the entire mesh resides on node 0. A marking procedure (an *error estimator/indicator* in real applications) marks edges/elements for refinement or

---

[3] The EARTH runtime system does not support a restricted migration of tokens.

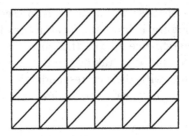

Fig. 5. A small sample of our test mesh

coarsening (on higher levels). Based on this marking, tokens (one for each partition) are generated that perform the actual refinement. This situation immediately serves as a *stress* test for the EARTH token mechanism to automatically balance the highly nonuniform workload. If a token migrates to and is executed on another processor, it transfers the partition and refinement information to the executing node. The actual refinement/coarsening is done there. It is thus absolutely crucial to *compress* the *metacontext* as much as possible. The property of our index scheme to provide global indices becomes extremely useful here: it is not necessary to transform names if processor/memory boundaries are crossed. Furthermore, only the indices of the elements have to be transferred since the indices of all other objects (edges, vertices) can be easily recomputed [6]. (This can be further optimized since a parent element can, in conjunction with the refinement information, rapidly calculate the indices of its children.)

For mesh refinement, one can distinguish between two strategies. In the *first* case, parent elements are deleted from the data structure when they are refined. In the *second* case, the refinement history is maintained. The advantage of the second strategy is that parent objects do not have to be reconstructed from scratch during mesh coarsening. We have implemented this second strategy [1].

When all elements in a FE mesh are subdivided isotropically, it is referred to as *uniform* refinement and automatically leads to a conforming triangulation. In the case of *nonuniform* refinement, a (iterative) consistency procedure is required since adaptive coarsening/refinement generally leads to inconsistent triangulations. Moreover, to guarantee mesh quality using our adaption algorithm, the parents of green triangles are reinstated before further refinement [2].

In real applications, mesh adaption is driven by an error indicator that is obtained either geometrically or from the numerical solution. The computational mesh is refined in regions where the indicator is high, and coarsened in regions where it is low. In our experiments, elements are randomly selected for adaption. The behavior of an error indicator can generally be mimicked by suitably adjusting the distribution pattern of the random numbers.

4.2 Load Balancing Strategies

For a detailed description of the load balancing strategies that have been implemented in the EARTH-SP[4] runtime system, we refer the reader to [3]. This is

---

[4] This refers to the version of Threaded-C available for the IBM SP2 and SP3.

a comprehensive study that provides an excellent overview of the implemented strategies and their qualitative behavior under several stress tests.

Some questions we would like to answer for our particular simulation are:

- What portion of the total number of elements are finally located on nodes other than node 0? How do the numbers of nodes and tokens generated per node affect this behavior? Is the load acceptably balanced outside of node 0?
- How does the token generation process affect the runtime? Is there a measurable system overhead?
- Does an increased number of tokens decrease the *variance* of the load balancer (as one would expect)?
- How does the system behave if there are less tokens than processors?

5 Results

The data structure used to keep the indices of our FE mesh objects is a red-black tree [5], as it is used in the implementation of set container in C++ STL. The code for the uniform refinement case (including that for the red-black tree) consists of about 2000 lines of Threaded-C. In the nonuniform case, closure of the triangulation process is added, and the code is about 2700 lines.

The program contains certain optimizations for the case when a token is executed on the node it originates. Executing the program on one node implies that all tokens are both generated and executed on the same node. In that case, no remote operations or tree compression/expansion are necessary.

The initial mesh is the same for all the experiments and consists of 2048 vertices, 5953 edges, and 3906 triangular elements. At the beginning of each experiment, the entire mesh resides on node 0. It is the responsibility of the EARTH runtime system to scatter partitions across the available nodes. Given the space limitations, we analyze only the uniform refinement case in this paper. A total of four mesh refinement steps were performed. The final mesh contained 999,936 triangles since each triangle was refined isotropically. All measurements were done upon completion of these four steps.

Currently, there are eight different load balancers available with EARTH-SP, and we tested all of them. As a representative for the discussions in this section, we chose the DUAL load balancer. It is one of the simplest load balancers (a *virtual ring* [3]) and does not provide the best quantitative results; however, its general behavior is comparable with the other load balancers in EARTH-SP.

Discussion of Table 1: Runtime versus the number of tokens generated per node.

- Even on one node, the program runs faster as the number of tokens increases! Recall that the number of tokens corresponds to the number of partitions. Hence, for a fixed mesh size, more partitions imply less elements per partition. Since the elements in a partition are organized in a tree, it follows that the trees will be shallower. This, in turn, accelerates the tree operations. The entries for one node in the table confirm that the underlying data structure must be of logarithmic complexity.

Table 1. Execution time (in seconds) versus the number of tokens generated per node

Nodes	Tokens									
	4	8	16	32	64	128	256	512	1024	2048
1	20.78	19.77	18.73	17.74	16.74	15.68	14.91	14.39	13.71	12.96
2	18.13	14.85	13.79	12.37	11.53	11.92	10.91	10.89	10.38	9.94
3	11.50	11.97	9.72	9.44	9.14	7.90	8.54	8.03	7.98	8.06
4	19.48	10.25	7.36	7.99	7.08	6.82	7.25	6.06	6.35	6.87
5	11.30	10.12	8.06	6.69	5.81	5.47	6.24	5.40	5.92	5.35
6	11.96	10.56	10.13	6.94	5.90	5.40	5.84	5.78	5.36	4.89
7	26.83	9.97	5.25	6.89	5.20	5.26	5.49	4.94	4.69	5.07
8	22.62	10.76	7.91	5.91	5.55	5.44	5.05	5.15	5.00	5.04

- Reasonable speedup is obtained for up to five nodes. However, we must consider the additional quality information (discussed below) for our evaluation, since quality has rather strong implications for other parts (solver/preconditioner) of a complete FE application. It might therefore be useful to invest some more resources that do not necessarily increase the speedup but improve quality.
- A certain instability is observed in the region where the number of tokens is not much greater than the number of processors.

Table 2. Migration (as a percentage of the total number of triangles) from node 0 versus the number of tokens generated per node

Nodes	Tokens									
	4	8	16	32	64	128	256	512	1024	2048
2	24	37	37	31	32	25	27	25	25	24
3	50	49	61	50	53	51	44	45	45	41
4	52	75	61	58	60	58	52	58	58	49
5	100	88	68	75	70	66	59	63	58	60
6	75	100	81	75	71	68	62	61	66	65
7	75	88	75	80	75	68	68	68	68	73
8	100	100	75	84	68	67	71	66	66	69

Discussion of Table 2: Migration from node 0 versus the number of tokens generated per node.
- Migration is measured as the percentage of all triangles that are, at the end, owned by nodes other than node 0. This measure is somewhat imprecise since a partition may migrate back and forth.[5]
- Migration increases with the number of nodes, but might decrease with the number of tokens if not enough processors are available. This is probably because the DUAL load balancer is unable to handle the load within these resource limitations.

---

[5] Such fluctuations occur especially if there are only a few processors and/or tokens.

- It appears that there is a region of stabilization at about 70 percent of migration. Asymptotically, the migration percentage from node 0 when using P nodes is $100 \times (P-1)/P$. Migration can be increased toward this asymptotic by using more processors or more tokens. Generating more tokens implies downsizing partitions, which would consist of only one triangle in the extreme case. However, we would then inevitably be faced with a fragmentation problem (and a total loss of locality).
- Once again, there is a region of instability when the number of tokens is not much greater than the number of processors.

Table 3. Total variance (as a percentage of the total number of triangles) versus the number of tokens generated per node

Nodes	Tokens									
	4	8	16	32	64	128	256	512	1024	2048
2	26	13	13	19	18	25	23	25	25	26
3	12	24	7	14	16	12	15	15	16	19
4	17	9	10	11	9	10	14	10	10	15
5	10	6	7	4	5	7	10	9	11	10
6	12	9	5	5	7	7	10	10	8	9
7	18	5	7	4	4	8	8	7	8	5
8	13	9	7	4	8	8	7	9	9	7

Table 4. Unbiased variance (as a percentage of the total number of triangles) outside of node 0 versus the number of tokens generated per node. Note that results are not shown for two nodes since node 0 is excluded when computing this variance

Nodes	Tokens									
	4	8	16	32	64	128	256	512	1024	2048
3	1	24	7	8	7	7	5	6	6	5
4	12	10	8	5	2	3	3	4	4	3
5	1	5	5	4	3	2	1	3	3	3
6	12	6	5	4	3	3	3	3	3	4
7	19	5	6	4	3	2	3	2	3	3
8	12	8	6	4	2	3	4	4	4	3

Discussion of Tables 3 and 4: Total variance and unbiased variance outside of node 0 versus the number of tokens generated per node.

- Total variance is defined as the expected deviation from the mean value of the number of triangles per node measured as a percentage of the total number of triangles. Since this measure is biased towards node 0, the variance outside of node 0 is computed to evaluate the uniformity of the migration (cf. Table 4).
- The variances oscillate rapidly if not enough tokens are available. If not enough

processors are available, increasing the number of tokens tends to overload the load balancer.

- A large number of tokens seems to guarantee the stability of, and possibly a slight improvement in, the variances.
- Perhaps the most important result is the quality of load balancing. Experimental values of the unbiased variance presented in Table 4 show that the load is extremely well balanced (between 3 and 6 percent). This is generally acceptable for actual adaptive FE calculations.

5.1 Summary of Results

We summarize by addressing each of the three observations made in Sec. 1:

- The experimental results confirm our hypothesis outlined in the introduction: although we did not have an explicit repartitioning/remapping phase, the processors appear to have a good and balanced utilization as long as there are enough tokens being processed in the system (cf. Tables 2 and 3).
- The experiments also show that the natural indexing method used to enhance token mobility does appear to work; i.e., the remote communication due to token migration does not seem to have a major impact on the overall performance (see the region of good speedup and balanced workload implied in Table 1). Of course, the FE solution phase is usually more sensitive to data locality which is not included in the scope of this paper.
- Finally, the experimental results have shown that the load balancing works smoothly as long as there are enough computational resources available, even when the workload evolution is irregular. This can be observed in Table 4, where we see a stable and small variance in the expected workload (measured by the number of triangles).

6 Conclusions

The ability to dynamically adapt a finite element mesh is a powerful tool for solving computational problems with evolving physical features; however, an efficient parallel implementation is rather difficult. To address this problem, we examined a multithreaded approach for performing the load balancing automatically.

Preliminary results from experiments with 2D triangular meshes using the EARTH multithreaded system are extremely promising. The EARTH token mechanism was used to balance the highly irregular and nonuniform workload. To achieve good token mobility, a new indexing methodology was used. During mesh adaption, tokens that migrate to another processor to be executed leave the processed data in the destination processor. This is a novel approach that allows simultaneous load balancing and load-driven data remapping. Results showed that the load is extremely well balanced (3–6% of ideal) when a sufficient number of tokens is generated (at least 512 tokens per node). Token generation (up to 2048 tokens per node) does not significantly increase system overhead or degrade performance, independent of the number of nodes. Having a large number of tokens decreases the total variance of the load balancer and stabilizes the uniformity of the migration. Finally, an underloaded system (one that has few tokens per node) causes instability and leads to unpredictable behavior by the load balancer.

These conclusions do not seem very surprising, but qualitatively confirm what was expected. Given the fact that very minimal programming effort (on the user level) was necessary to obtain these results and that this is one of the first multithreaded approaches to tackle unstructured mesh adaption, our findings and observations become extremely valuable. More extensive experiments will be done in the future, and the results compared critically with an explicit message-passing implementation [12] and a global load balancer [11]. One must remember however that dynamic mesh adaption comprises a small though significant part of a complete application. Further investigations are needed to determine whether the functionality of such an approach is viable for real applications.

References

1. Biswas, R., Strawn, R.: A new procedure for dynamic adaption of three-dimensional unstructured grids. Appl. Numer. Math. **13** (1994) 437–452
2. Biswas, R., Strawn, R.: Mesh quality control for multiply-refined tetrahedral grids. Appl. Numer. Math. **20** (1996) 337–348
3. Cai, H., Maquelin, O., Gao, G.: Design and evaluation of dynamic load balancing schemes under a multithreaded execution model. CAPSL Technical Memo TM-05 (1997)
4. Chrisochoides, N., Houstis, E., Rice, J.: Mapping algorithms and software environment for data parallel PDE iterative solvers. J. Par. Dist. Comput. **21** (1994) 75–95
5. Cormen, C., Leiserson, C., Rivest, R.: Introduction to Algorithms. MIT Press (1990)
6. Gerlach, J., Heber, G.: Fundamentals of natural indexing for simplex finite elements in two and three dimensions. RWCP Technical Report 97-008 (1997)
7. Hum, H., Maquelin, O., Theobald, K., Tian, X., Gao, G., Hendren, L.: A study of the EARTH-MANNA multithreaded system. Intl. J. of Par. Prog. **24** (1996) 319–347
8. Hum, H., Theobald, K., Gao, G.: Building multithreaded architectures with off-the-shelf microprocessors. Intl. Par. Proc. Symp. (1994) 288–297
9. Maquelin, O.: Load balancing and resource management in the ADAM machine. Advanced Topics in Dataflow Computing and Multithreading, IEEE Press (1995) 307–323
10. Maquelin, O., Hum, H., Gao, G.: Costs and benefits of multithreading with off-the-shelf RISC processors. Euro-Par'95 Parallel Processing, Springer-Verlag LNCS **966** (1995) 117–128
11. Oliker, L., Biswas, R.: PLUM: Parallel load balancing for adaptive unstructured meshes. NAS Technical Report NAS-97-020 (1997)
12. Oliker, L., Biswas, R., Strawn, R.: Parallel implementation of an adaptive scheme for 3D unstructured grids on the SP2. Parallel Algorithms for Irregularly Structured Problems, Springer-Verlag LNCS **1117** (1996) 35–47
13. Theobald, K., Amaral, J., Heber, G., Maquelin, O., Tang, X., Gao, G.: Overview of the Threaded-C language. CAPSL Technical Memo TM-19 (1998)

Dynamic Load Balancing for Parallel Adaptive Mesh Refinement*

Xiang-Yang Li[1] and Shang-Hua Teng[1]

Department of Computer Science and Center for Simulation of Advanced Rockets,
University of Illinois at Urbana-Champaign, Urbana, IL 61801.

Abstract. Adaptive mesh refinement is a key problem in large-scale numerical calculations. The need of adaptive mesh refinement could introduce load imbalance among processors, where the load measures the amount of work required by refinement itself as well as by numerical calculations thereafter. We present a dynamic load balancing algorithm to ensure that the work load are balanced while the communication overhead is minimized. The main ingredient of our method is a technique for the estimation of the size and the element distribution of the refined mesh before we actually generate the refined mesh. Base on this estimation, we can reduce the dynamic load balancing problem to a collection of static partitioning problems, one for each processor. In parallel each processor could then locally apply a static partitioning algorithm to generate the basic units of submeshes for load rebalancing. We then model the communication cost of moving submeshes by a condensed and much smaller subdomain graph, and apply a static partitioning algorithm to generate the final partition.

1 Introduction

Many problems in computational science and engineering are based on unstructured meshes in two or three dimensions. An essential step in numerical simulation is to find a proper discretization of a continuous domain. This is the problem of *mesh generation* [1, 9], which is a key component in computer simulation of physical and engineering problems. The six basic steps usually used to conduct a numerical simulation include, 1: **Mathematical modeling**, 2: **Geometric modeling**, 3: **Mesh generation**, 4: **Numerical approximation**, 5: **Numerical solution**, 6: **Adaptive refinement**.

In this paper, we consider issues and algorithms for adaptive mesh refinement, (cf, Step 6 in the paradigm above). The general scenario is the following. We partition a well shaped mesh M_0 and its numerical system and map the submeshes and their fraction of the numerical system onto a parallel machine. By solving the numerical system in parallel, we obtain an initial numerical solution S_0. An error-estimation of S_0 generates a refinement spacing-function h_1 over the domain. Therefore, we need properly refine M_0 according to h_1 to generate a well shaped mesh M_1. The need of refinement introduces load imbalance among processors in the parallel machine.

* Supported in part by the Academic Strategic Alliances Program (ASCI) of U.S. Department of Energy, and by an NSF CAREER award (CCR-9502540) and an Alfred P. Sloan Research Fellowship.

The work-load of a processor is determined by refining its submesh and solving its fraction of the numerical system over the refined mesh M_1. In this paper, we present a dynamic load balancing algorithm to ensure that the computation at each stage of the refinement is balanced and optimized. Our algorithm estimates the size and distribution of M_1 before it is actually generated. Based on this estimation, we can compute a quality partition of M_1 before we generate it. The partition of M_1 can be projected back to M_0, which divides the submesh on each processor into one or more subsubmeshes. Our algorithm first moves these subsubmeshes to proper processors before performing the refinement. This is more efficient than moving M_1 because M_0 is usually smaller than M_1. In partitioning M_1, we take into account of the communication cost of moving these subsubmeshes as well as the communication cost in solving the numerical system over M_1.

This paper is organized as the following. Section 2 introduces an abstract problem to model parallel adaptive mesh refinement. Section 3 presents an algorithm to estimate the size and distribution of the refined mesh before its generation. It also presents a technique to reduce dynamic load balancing for mesh refinement to a collection of static partitioning problems. Section 4 extends ours algorithm from the abstract problem to unstructured meshes. Section 5 concludes the paper with a discussion of some future research directions.

2 Dynamic Balanced Quadtrees

In this section, we present an abstract problem to model the process of parallel adaptive mesh refinement. This abstract problem is general enough; it uses balanced quadtrees and octrees to represent well-shaped meshes; it allows quadtrees and octrees to grow dynamically and adaptively to approximate the process of adaptive refinement of unstructured meshes. This model is also simple enough geometrically to provide a good framework for the design of mesh refinement algorithms.

2.1 Balanced Space Decomposition

The basic data structure for quad-/oct-tree based adaptive mesh refinement is a *box*. A box is a d-dimensional cube embedded in an axis-parallel manner in $\mathbb{R}^d$ [9]. Initially, there is a large d-dimensional box, we call it the *top-box*, which contains the interior of the domain, and a neighborhood around the domain. The box may be split, meaning that it is replaced by 2^d equal-sized boxes. These smaller boxes are called the *children boxes* of the original box. A sequence of splitting starting at the *top-box* generates a 2^d-tree, i.e., a *quadtree* in two dimensions, and an *octree* in three dimensions. *Leaf-boxes* of a 2^d-tree are those that have no child. Other boxes are *internal-boxes*, i.e., have children. The size of 2^d-tree T, denoted by $size(T)$, is the number of the leaf-boxes of T. The *depth* of a box b in T, denoted by $d(b)$, is the number of splittings needed to generate b from the top box. The depth of the top box, hence is 0. Two leaf-boxes of a 2^d-tree are *neighbors* if they have a $(d-1)$ dimensional intersection. A 2^d-tree is balanced iff for any two neighbor leaf-boxes b_1, b_2, $|d(b_1) - d(b_2)| \leq 1$. For convenience, leaf-boxes b and b_1 of a quadtree are called *edge neighbors* if they intersect on an edge; and they are called *corner neighbors* if they share only a vertex.

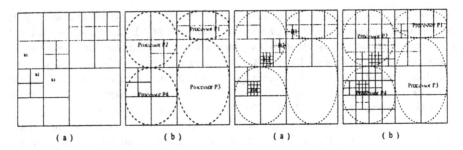

Fig. 1. A balanced quadtree T in two dimensions, a 4-way partition, refinement and balancing. Assume the splitting depth of b_1, b_2, b_3 and b_4 are 1, 1, 2 and 2 respectively.

2.2 Modeling Adaptive Refinement with Dynamic Quadtree

A balanced 2^d-tree can be viewed as a well-shaped mesh in $\mathbb{R}^d$. In fact, most quad-/oct-tree based mesh generation algorithms first construct a balanced 2^d-tree over an input domain, and then apply a local warping procedure to build the final triangular mesh [1, 9]. When the accuracy requirement of a problem is changed during the numerical simulation, we need refine the mesh accordingly. In particular, an error estimation of the computation from the previous stage generates a new spacing-function over the input domain. The new spacing-function defines the expected size of mesh elements in a particular region for the formulation in the next stage. In the context of a 2^d-tree, the refinement requires that some leaf-boxes be split into a collection of boxes of a certain size while globally maintains that the resulting 2^d-tree is still balanced. We model the refinement of a 2^d-tree as following:

Definition 1 Adaptive Refinement of 2^d-trees. A balanced 2^d-tree T and a list of non-negative integers δ, one for each leaf-box, i.e., associated with each leaf-box b is an integer $\delta(b)$, generate a new balanced tree T^* as the following.

1. Construct T' by dividing each leaf-box b of T into $2^{d\delta(b)}$ equal sized boxes.
2. Construct T^* by balancing T'.

The refinement most likely introduces load imbalance among processors, which reflects in the work for both refinement and computations thereafter. Therefore, as an integral part of parallel adaptive computation, we need to dynamically repartition the domain for both refinement and computations of the next stage. To balance the work for refinement, we need to partition a 2^d-tree before we actually refine it. In the next section, we present an efficient method to estimate the size and element distribution of a refined 2^d-tree without actually generating it.

3 Reduce Dynamic Load Balancing to Static Partitioning

The original 2^d-tree T is distributed across a parallel machine based on a partition of T. Assume we have k processors, and we have divided T into k-subdomains $S_1 \ldots, S_k$, and have mapped S_i onto processor i. A good partition is in general *balanced*, i.e., the sizes of $S_1, \ldots, S_k$ are approximately the same size. In addition, the number of *boundary boxes*, the set of leaf-boxes that have neighbors located at different processors should be small.

A simple minded way to refine a 2^d-tree (or a mesh in general) for a new spacing-function is to have each processor refine its own subdomain to collectively construct T^*. Note that the construction of T^* from T' needs communication among processors. The original k-way partition of T naturally defines a k-way partition $(S_1', \ldots, S_k')$ for T' and a k-way partition $(S_1^* \ldots, S_k^*)$ for T^*. However, these partitions may not longer be balanced. In addition, the computation of the next stage will no longer be balanced either. Note also that the set of boundary boxes will change during the construction of T' and T^*. The number of the boundary boxes may not be as small as it should be. In this approach, to balance the computation for the next stage, we could repartition T^* and distribute it according to the new partition. One shortcoming of this approach is that T^* could potentially be larger than T, and hence the overhead for redistributing T^* could be more expensive.

We would like to have a mechanism to simultaneously balance the work for refinement and for the computation of the next stage. To do so, we need to properly partition T^* before we actually generate it. Furthermore, we need a dynamic load balancing scheme that is simple enough for efficient parallel implementation. In this section, we present an algorithm that effectively reduce the dynamic load balancing problem to a collection of static partitioning problems. We first give a high-level description of our approach. Details will be given in subsequent subsections.

Algorithm Dynamic Repartitioning and Refinement
Input (1) a balanced 2^d-tree T that is mapped onto k processors according to a k-way partition $S_1, \ldots, S_k$, and (2) a list of non-negative integers δ, one for each leaf-box.

1. In parallel, processor i estimates the size and the element distribution of its subdomains S_i' and S_i^* without constructing them.

2. Collectively, all processors estimate the size of T^*. Assume this estimation is N. Let $W = \alpha(N/k)$ for a predefined positive constant $\alpha < 1$.

3. In parallel, if the estimated size of S_i^* is more than W, then processor i applies the geometric partitioning algorithm of Miller-Teng-Thurston-Vavisis [6, 3] to implicitly partition S_i^* into a collection of subsubdomains $S_{i,1}^*, \ldots, S_{i,L_i}^*$. We can naturally project this partition back to S_i to generate subsubdomains $S_{i,1}, \ldots, S_{i,L_i}$.

4. We remap these subsubdomains to generate a partition of T so that the projected work for the refinement and computations thereafter at each processor is balanced. We would also like to minimize the overhead in moving these subsubdomains. We introduce a notion of subdomain graph over these subsubdomains to do dynamic balancing.

5. We construct a k-way partition of the subdomain graph using a standard static graph partitioning algorithms such as provided in Chaco and Metis [4, 5]. The partition of the subdomain graph defines a new distribution for T^* before its refinement.

6. Move each subsubdomain to its processor given by the partition of the subdomain graph.

7. In parallel, each processor refines and balances its subdomain.

8. Solve the resulting numerical system for the next stage.

3.1 Subdomain Estimation

We now estimate the size of the quadtree T^* after refining and balancing quadtree T. Our technique can be directly extended to general 2^d-trees. For each leaf-box b of a balanced 2^d-tree, the *effect region* of b, denoted by $region(b)$, is defined to be the set of all boxes that share at least one point with b. We mainly concern about the two dimensions case during the later discussion.

Lemma 2. *For any leaf-box b of a quadtree T, the size of $region(b)$ satisfies $|region(b)| \leq 12$. The region of b can be computed in a constant time.*

The proof is omitted here.

The *pyramid* of a leaf-box b of a quadtree T, denoted by $pyramid(b)$, is defined as the following. (1):if a leaf-box $b_1 \notin region(b)$ shares a $(d-1)$ dimensional face with a box $b_2 \in region(b)$, $\delta(b_1)=0$, and $d(b_2) = d(b_1) + 1$, then $b_1 \in pyramid(b)$. (2):if a leaf-box $b_3 \notin region(b)$ shares a $(d-1)$ dimensional face with a leaf-box b_1 in $pyramid(b)$, $\delta(b_3) = 0$, and $d(b_1) = d(b_3) + 1$, then $b_3 \in pyramid(b)$.

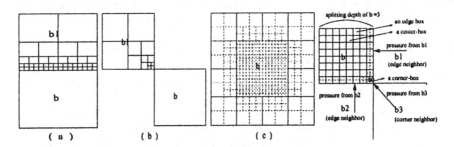

Fig. 2. The two templates for splitting boxes in $region(b)$. Fig (a): the refinement of b causes the edge neighbor b_1 to be split, Fig (b): it causes corner neighbor b_1 to be split, Fig (c): how the refinement of b influence the splitting of leaf-boxes in $region(b)$.

Lemma 3. *The set of boxes that we need to split in the construction of T^* due to the refinement of a leaf-box b is contained in $region(b) \cup pyramid(b)$.*

If b_1 is a leaf-box contained in $pyramid(b)$, then we need to split b_1 at most once. If b_1 is a leaf-box in $region(b)$, then b_1 can only be split *geometrically away* from the face shared between b_1 and b (see Figure 2 (a)) or geometrically from a corner shared by b_1 and b (see Figure 2 (b)). The depth of splitting is a function of $\delta(b), \delta(b_1), d(b)$ and $d(b_1)$. As shown in Figure2(c), after the refinement, the impact of b geometrically weakens away from b in $region(b)$.

The *unit boxes* of leaf-box b are the smaller box uniformly split in b by depth $\delta(b)$, i.e., the side length of this kind boxes is $2^{-\delta(b)}$ of that of b. There are three kinds of unit boxes: *corner-boxes*, which locates at the four corners of b; *edge-boxes*, which intersect b with only one edge; and *center-boxes*, which are all other boxes . Figure 2 shows an example of these unit boxes. Only the edge boxes and corner boxes need to be split again for balancing.

Let $\theta(b, b_1) = \max([\delta(b)-\delta(b_1)], 0)$. Let $\beta(b, b_1) = d(b)-d(b_1)$. Let $pressure(b, b_1)$ be the needed splitting depth of an unit box of b_1, due to the imbalance caused by the refinement of leaf-box b. Let $\gamma(b, b_1) = 1$ or 2 if b and b_1 are edge neighbors or corner neighbors. Then we have $pressure(b, b_1) = \theta(b, b_1) - \beta(b, b_1) - \gamma(b, b_1)$.

We now consider how the refinement of one leaf-box may influence the splitting of its neighboring leaf-boxes. Suppose b and b_1 are two edge neighbors in T. We have three cases based on the value of $d(b) - d(b_1)$: $1, 0, -1$. Assume the refinement of b_1 causes an imbalance between b and b_1. Note that there are $2^{\delta(b)} - 2$ edge-boxes need be split along one edge of b. Let a_k be the number of smaller boxes introduced in splitting an edge-box of b into depth k using the template shown at Figure 2(a). Then a_k can be computed as following.

Lemma 4. *The number of smaller boxes introduced in splitting an edge-box of b into depth k is $a_k = 3 * 2^k - 3$.*

Proof: We have $a_0 = 0$. Splitting each small box will introduce four smaller boxes. Then we have $a_k = a_{k-1} + 3 * 2^{k-1}$, which implies that $a_k = 3 * 2^k - 3$. □

We now consider the splitting of a corner-box of b to eliminate the imbalance caused by the refinement of boxes in $region(b)$. Let b_3 be a corner neighbor of b. Let b_0 be the corner-box of b which shares a vertex v with b_3, as shown in Figure 2. Let s_k be the number of boxes introduced in b_0 if we split it by depth k according to the pressure of b_3. Then we have the following lemma to compute s_k.

Lemma 5. *The number of smaller boxes introduced in splitting b_0 by depth k from pressure of b_3 is $s_k = 3 * k$.*

Proof: Clearly $s_0 = 0$. And $s_k = s_{k-1} + 3$, which implies that $s_k = 3 * k$. □

We now consider the case that two edge neighbors of b cause the corner-box to be split. W.l.o.g., let b_1 and b_2 be the two neighbors of b. Boxes b, b_1 and b_2 intersect at the vertex v of b. Let b_0 be the corner-box which also contains v. Let $k_1 = pressure(b_1, b)$. Let $k_2 = pressure(b_2, b)$. Let $c(k_1, k_2)$ be the number of smaller boxes split in b_0 by pressure k_1 and pressure k_2. We do not consider the corner-pressure from b_3 in $c(k_1, k_2)$ now. Then we have the following lemma to compute $c(k_1, k_2)$.

Lemma 6. *The number of smaller boxes split in b_0 by pressure k_1 and pressure k_2, is $c(k_1, k_2) = 3 * (2^{k_1} + 2^{k_2}) - 3k_2 - 6$. where we assume $k_1 \geq k_2$.*

Proof: Clearly $c(k_1, k_2) = c(k_2, k_1)$. If $k_1 = k_2$, then we have $c(0, 0) = 0, c(k, k) = c(k - 1, k - 1) + 3 * (2^k - 1)$. Hence, we have $c(k, k) = 3 * 2^{k+1} - 3k - 6$.

If $k_1 \neq k_2$, then w.l.o.g, we assume $k_1 > k_2$. The splitting of the corner unit box b_0 can be viewed as two steps: (1) split it by depth k_2 in both directions of b_1 and b_2, (2) split the much smaller boundary boxes generated in (1) into depth $k_1 - k_2$ along the common edge of b and b_1. Note that there are 2^{k_2} much smaller boundary boxes needed to be split in step (2). The number of total smaller boxes split is $c(k_1, k_2) = c(k_2, k_2) + 2^{k_2} * a_{k_1-k_2}$. Generally, we have $c(k_1, k_2) = 3 * (2^{k_1} + 2^{k_2}) - 3k_2 - 6$. $\square$

We now take into account the corner pressure from corner neighbor b_3 and the edge pressures from two edge neighbors b_1 and b_2 together, as shown in Figure 2. Let k_1 and k_2 be the edge pressure from two edge neighbors b_1 and b_2 respectively. And let k_3 be the corner pressure from a corner neighbor b_3 of b. And let v be the common vertex of b, b_1, b_2 and b_3. Let $g(k_1, k_2, k_3)$ be the number of much smaller boxes introduced in corner-box b_0 of b due to the refinement of b_1, b_2 and b_3. Then from the above lemmas, we have the following lemma.

Lemma 7. *The number of smaller boxes introduced in b_0 due to the corner pressure from corner neighbor b_3 and the edge pressures from two edge neighbors b_1 and b_2 is $g(k_1, k_2, k_3) = c(k_1, k_2) + s_{k_3-k_1}$ if $k_3 \geq k_1$, otherwise $g(k_1, k_2, k_3) = c(k_1, k_2)$.*

The proof is omitted here. Let $F(T, \delta) = \cup_{b, \delta(b)>0} pyramid(b) - \cup_{b, \delta(b)>0} region(b)$. Let $f(T, \delta) = |F(T, \delta)|$. In other words, $f(T, \delta)$ counts the leaf-boxes that have to be split only due to the imbalance introduced by the refinement of some boxes which are not their neighbors. Then by computing $pyramid(b_1)$ for $\delta(b_1) > 0$, we can compute $f(T, \delta)$ in worst case $size(T)$ time. The idea is as following. For every box b of T, if there is a box $b_1 \in region(b)$ such that $delta(b_1) > 0$, then $b \notin F(T, \delta)$. Otherwise, if there is a box $b_2 \in region(b)$ such that b_2 belongs to pyramid of some box b_3, and $d(b) = d(b_2) - \gamma(b, b_2)$, then $b \in F(T, \delta)$.

For a leaf-box b, let $V(b) = \{v_1, v_2, v_3, v_4\}$ be the vertex set of b. Let $Intr(v_i)$ be the number of smaller boxes introduced in corner-box b_0 of b, which shares v_i with b. Let $E(b) = \{e_1, e_2, e_3, e_4\}$ be the edge set of b. Let $Intr(e_i)$ be the number of smaller boxes introduced in edge-boxes of b which intersect b on e_i. $Intr(v_i)$ and $Intr(e_i)$ can be computed at constant time according to analysis before. Then we have the following theorem to compute $size(T^*)$.

Theorem 8. *Suppose T is a balanced quadtree and δ is a list of non-negative integers for refining its leaf-boxes. Then after balancing and refining T,*

$$size(T^*) = \sum_{b \in T} \left(\sum_{v_i \in V(b)} Intr(v_i) + \sum_{e_i \in E(b)} Intr(e_i) + 2^{2\delta(b)} \right) + 4 * f(T, \delta).$$

Proof: The elements of T^* has four resources. The first contribution is from the refinement of each box b (there are $2^{2\delta(b)}$ small boxes constructed). The second is from the re-refinement of edge neighbors in $region(b)$. There are $Intr(e_i)$ introduced from pressure an edge neighbor sharing e_i with b. The third is from the refinement of the corner neighbors in $region(b)$. There are $Intr(v_i)$ introduced from pressure of a corner neighbor sharing v_i with b. And there are no overlaps when we do $\sum_b Intr(e_i) + Intr(v_i)$. The last is from the re-refinement of boxes in $F(T, \delta)$. There are $4 * f(T, \delta)$ small boxes introduced in these boxes, because each box in $F(T, \delta)$ is split to 4 smaller boxes. Note that if $b_1 \in region(b)$ then $b \in region(b_1)$. $\square$

3.2 Sampling Boxes from T^*

According to the size computation of mesh T^*, we can approximately sample a random leaf-box of T^*. For a leaf-box b of mesh T, let k_e, k_s, k_w and k_n be the pressure of b from four edge-neighbor leaf-boxes respectively. Let k_{se}, k_{sw}, k_{nw} and k_{ne} be the pressure of b from four corner-neighbor leaf-boxes respectively. Then according to the lemmas of size estimation, we can compute the number of introduced splitting boxes in b due to the refinement of the neighbor leaf-boxes. Let c_1, c_2, c_3 and c_4 be the number of splitting boxes introduced in four corner-boxes of b respectively. Let c_5, c_6, c_7 and c_8 be the number of splitting boxes introduced in edge-boxes of b along four edges respectively. And let c_9 be the number of center-boxes splitting in b. The set of small boxes, which c_i is counted from, is called *block i*. Let (x, y) be the geometric center point of b. Let h be the side length of b. Let $h_0 = h/2^{\delta(b)}$, the side length of the split boxes in b according to the refinement of depth $\delta(b)$.

For sampling a leaf-box in T^*, we uniformly generate a random positive integer r, which is not larger than $\sum_{i=1}^{9} c_i$. W.l.o.g. we assume that $\sum_{j=1}^{i-1} c_j < r \leq \sum_{j=1}^{i} c_j$. The value r specifies the block i at which the random small box will be located. First consider the case when the object block i is center block, i.e., $r > \sum_{i=1}^{8} c_i$. Let $t = r - \sum_{i=1}^{8} c_i$. Let $e = 2^{\delta(b)} - 2$, the number of small split boxes per row at the center block c_9. Let m,n be the integer such that $m \leq e - 1$, $n \leq e - 1$ and $t = m * e + n$. In other words, the object small box will locate the mth row and nth column at the center block of b. The coordinates of left-up corner point of center block of b is $(x - h/2 + h_0, y + h/2 - h_0)$. Then the center point of the object small box is $(x - h/2 + (n + 3/2)h_0, y + h/2 - (m + 3/2)h_0)$. The side length of the sampled small box is h_0.

Now consider the case when object block is an edge-block, i.e., $\sum_{i=1}^{4} c_i < r \leq \sum_{i=1}^{8} c_i$. Similarly let $t = r - \sum_{i=1}^{j-1} c_i$, where j and t satisfy that $0 \leq t < c_j$ and $5 \leq j \leq 8$. Let k_j be the corresponding pressure to b from neighbor leaf-box. For t, we similarly compute m and n as following. If $(\sum_{i=1}^{k_j} 2^i) * e \leq t < c_j$ then let $n = t - \sum_{i=1}^{k_j} 2^i$. The center point of the object small box is $(x - h/2 + h_0 + (n + 1/2) * h_0/(2^{k_j}), y + h/2 - (1/2) * h_0/(2^{k_j}))$. And the side length of the small box is $h_0/(2^{k_j})$. If $t < e * (\sum_{i=1}^{k_j} 2^i)$, then assume $t = \sum_{i=0}^{m}(e * 2^i) + n$, where $n < e * 2^{m+1}$. The center point of the object small box is $(x - h/2 + h_0 + (n + 1/2) * h_0 * 2^{-m}, y + h/2 - 3 * h_0/2^{m+1})$. The side length of the small boxes is $h_0/2^m$. Note that, we only consider the case that the block is the north block. For other 3 case blocks, we can use similar way to sample the small boxes.

For the cases that the object block is a corner-block, we have similar sampling methods. Detail of the sampling is omitted here.

3.3 Subdomain Partitioning

We first review the basic concepts of graph partitioning. Suppose we have a weighted graph $G = (V, E, w)$, where V is the set of vertices and E is the set of edges, and w assigns a positive weight to each vertex and each edge. A *k-way* partition of G is a division of its vertices into k subsets $V_1, \ldots, V_k$. The set of edges whose endpoints are in two different subsets are call the *edge-separator* of the partition. The goal of graph partitioning is to find a k-way partition such that (1) V_i has approximately equal total weight, and (2) the separator is small. There are several available software for graph partition [4, 5]. However, most of these algorithms require the full combinatorial description of an input graph.

To partition each subdomain according to its size and distribution in T^*, we do not have its final combinatorial structure available before the refinement is actually performed. What do we have is a geometric approximation of its size and element distribution. Fortunately, the geometric information is sufficient for us to use the geometric partitioning algorithm of Miller-Teng-Thurston-Vavisis [6, 7].

Recall that the original k-way partition of T defines a k-way partition $(S'_1, \ldots, S'_k)$ for T' and a k-way partition $(S^*_1, \ldots, S^*_k)$ for T^*. However, these partitions may not longer be balanced. What we are going to do is to use the estimation of the element distribution of T^*, to implicitly divide each subdomain from $(S^*_1, \ldots, S^*_k)$ into subsubdomains of approximately equal size. The subsubdomain decomposition is described explicitly using T and its initial partition $S_1, \ldots, S_k$. The subsubdomains will be the units for the final partition.

In particularly, we use the size estimation algorithm presented in the previous section to estimate the size and element distribution of each leaf-box in T. This estimation allows us to sample a random leaf-box of T^* in each leaf-box of T. By doing so, we can obtain a sample of random leaf-boxes of S^*_i. We then apply the geometric mesh partitioning algorithm to this sample to obtain a proper multiway partition of S^*_i. This multiway partition is described as a partition tree of separating spheres and hence we can use this set of separating spheres to build a multiway partition $(S_{i,1}, \ldots, S_{i,L_i})$ of S_i. Details of the geometric mesh partitioning algorithm uses samples can be found in [6].

After the size estimation of each subdomain, we use the sampling technique to uniformly and randomly select leaf-boxes in T^*. We can use the sphere based technique to partition the sampled leaf-boxes. The partition of the sampling leaf-boxes in T^* implied a partition of subdomain in T.

3.4 Subdomain Redistribution

After we have divided each subdomain of T into a collection of subsubdomains, we need to redistribute them to proper processors to balance the load and minimize the communication requirement. We introduce a *subdomain graph SG* to model the redistribution of these subsubdomains. This graph is a weighted graph and its node set contains two parts. The first part has one node for each subsubdomain that we have generated. These nodes will be referred as *subdomain nodes*. The weight of each subdomain node is equal to the estimated size of the subsubdomain in T^*. The second part has one node for each processor. We will call these nodes *processor nodes*. We will discuss the weight of processors nodes later.

Two subdomain nodes are connected in SG if they are directly connected by boundary boxes. The weight of the edge between them is equal to the the number of shared boundary leaf-boxes times a scaler which is determined by the communication cost in solving the numerical system in the parallel computer.

Each processor node is connected in SG with all subdomain nodes of its subsubdomains. The weight on the edge between a processor node and a subdomain node is the cost of moving the subsubdomain to any other processor.

We now come back to the issue of the weight of a processor node. Let W be the total weight of all subsubdomain nodes in SG. Let $w = (1/2 + \epsilon) * W/k$ for a predefined positive constant ϵ. The constant ϵ is also a function of the constant α used in the repartition method of Section 3. For example $\epsilon = \alpha/2$. The choice of the weight of processor nodes is to ensure that in the subsequent partition of SG, no two processor nodes will be assigned to the same partition. That is why we choose the weight larger than $0.5W/k$. However, if the weight is too large, then it might disturb the balance of the final partition of some static partitioning algorithm. An example of a subdomain partition and its subdomain graph is given in Figure 3.

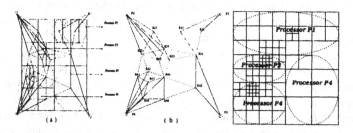

Fig. 3. constructing subdomain graph and redistribution of subdomain.

In Figure 3, we use the following notation. Node p_i denotes the processor i. Node S_{ij} denotes the jth subsubdomain of subdomain associated with processor i, generated by the subdomain partition algorithm.

Note that the subdomain graph is very small. It usually has $\Theta(k)$ nodes. So the cost for partitioning subdomain graph will be small as well.

We can use any static graph partitioning algorithm such as those provided in Chaco and Metis on SG to divide its nodes into k subsets of roughly equal total weights. It follows from the weight that we assigned to processor nodes, in the k-way partition, each subset contains exactly one processor node. Hence this partition generates a redistribution map of subsubdomains among processors in the parallel computer: a subsubdomain will be moved to the processor whose processor node is in the same subset in the k-way partition. Therefore the weight on the edge between a processor node and a subdomain node faithfully includes the communication cost in the partition.

Figure 3 gives an example of a quadtree T and its redistribution over the four processors. After the redistribution, each processor then refines and balances its new subdomain and solves its fraction of the numerical system for the next stage.

4 Remeshing Unstructured Meshes

Our parallel adaptive 2^d-tree refinement algorithm can be extended to general unstructured meshes. In this section, we outline our approach. It follows from a series of work by Bern-Eppstein-Gilbert [1], Mitchell-Vavasis [9], and Miller-Talmor-Teng-Walkington [8] that given a well-shaped mesh M in $\mathbb{R}^d$, there is a balanced 2^d-tree

T_M that approximates M. In particularly, T_M has the property that there are three positive constants $c > 1$, $\beta_1 < 1$ and $\beta_2 > 1$ such that (1) For each element e in M, the number of leaf-boxes of T_M that intersect e is at most c. (2) For each leaf-box b in T_M, the number of elements of M that intersect b is at most c. (3) In addition, if a leaf-box b of T_M intersects e, then $\beta_1 area(e) \leq area(b) \leq \beta_2 area(e)$.

Given a well-shaped mesh M, we can construct T_M in time linear in the size of M. Moreover, such computation can be optimally speeded-up if we have a multiple number of processors. We can then use the following strategy to design a parallel adaptive refinement algorithm for unstructured meshes.

Algorithm Parallel Refinement

Input (1) a well-shaped mesh M that is mapped onto k processors according to a k-way partition $M_1, \ldots, M_k$, and (2) a spacing-function f defining the new spacing at each vertices of M.

In parallel, we generate T_M. We project the k-way partition $M_1, \ldots, M_k$ to T_M to obtain a k-way partition of T_M. For each vertex $v \in M$, we compute the ratio r_m that is equal to the ratio of the current spacing at v to $f(v)$. For each box $b \in T_M$, let $\delta(b)$ be the logarithm of the average ratio of all vertices of M that lies inside b. We then apply our 2^d-tree load balancing algorithm to compute a k-way partition of T and project it back to M to obtain a new k-way partition of M. This k-way partition will be balanced for M^*, the refined mesh for M. Then permute M according this new partition and each processor applies a sequential mesh refinement algorithm to their own submesh and collaboratively refine the boundary elements among the submeshes.

In the full version of this paper, we will review some of sequential mesh refinement algorithms. We can support any sequential mesh refinement algorithm.

5 Final Remarks

In this paper, we present a dynamic load balancing algorithm for parallel adaptive mesh refinement. The main objective of this research is to develop effective algorithms that are simple enough for implementation. We focus on reducing dynamic load balancing to static partitioning in a black-box fashion and on reducing parallel mesh refinement to a collection of traditional sequential mesh refinements. We show how the estimation of the size and element distribution of a refined mesh can be used for this objective. There are several directions that we can extend and improve the method presented in this paper.

In our abstract model for adaptive mesh refinement, we assume that each leaf-box will be uniformly split in T'. In practice, we may need to split each leaf-box according to a given pattern.

The scheme developed in this paper for unstructured mesh refinement first builds a balanced 2^d-tree to approximate the unstructured mesh. This could be cumbersome. It is desirable to have a more direct method to estimate the size and element distribution of unstructured meshes other than 2^d-tree.

In our current model for adaptive refinement, we assume that the mesh will be made finer at every region. For certain applications, some regions will be "de-refined", i.e., will be coarsened. We need to extend our adaptive refinement scheme to handle mixed adaptive refinement and coarsening.

We are in the process of implementing ideas and methods developed in this paper. Experimental results will be presented in subsequent writings. The full version of the paper is available at request.

References

1. M. Bern, D. Eppstein and J. R. Gilbert. Provably good mesh generation. In *31st Annual Symposium on Foundations of Computer Science, IEEE*, 231–241, 1990.
2. L. Paul Chew, Nikos Chrisochoides, Florian Sukup. Parallel constrained delaunay meshing. *Trends in Unstructured Mesh Generation* edited by S.A.Canann and S.Saigal, pp89-96, 1997.
3. J. R. Gilbert, G. L. Miller, and S.-H. Teng. Geometric mesh partitioning: Implementation and experiments. *SIAM J. Scientific Computing*, to appear, 1998.
4. B. Hendrickson and R. Leland. The Chaco user's guide, Version 1.0. Technical Report SAND93-2339, Sandia National Laboratories, Albuquerque, NM, 1993.
5. G. Karypis and V. Kumar. A fast and high quality multilevel scheme for partitioning irregular graphs. *SIAM J. Scientific Computing* to appear, 1997.
6. G. L. Miller, S.-H. Teng, W. Thurston, and S. A. Vavasis. Automatic mesh partitioning. In A. George, J. Gilbert, and J. Liu, editors, *Sparse Matrix Computations: Graph Theory Issues and Algorithms*, IMA Volumes in Mathematics and its Applications. Springer-Verlag, pp57-84, 1993.
7. G. L. Miller, S.-H. Teng, W. Thurston, and S. A. Vavasis. Geometric separators for finite element meshes. *SIAM J. Scientific Computing*, to appear, 1998.
8. G. L. Miller, D. Talmor, S.-H. Teng, and N. Walkington. A Delaunay based numerical method for three dimensions: generation, formulation, and partition. In *Proc. 27th Annu. ACM Sympos. Theory Comput.*, pages 683–692, 1995.
9. S. A. Mitchell and S. A. Vavasis. Quality mesh generation in three dimensions. Proc. ACM Symposium on Computational Geometry, pp 212-221, 1992.
10. T.Okusanya, J.Peraire. 3D parallel unstructured mesh generation. *Trends in Unstructured Mesh Generation* edited by S.A.Canann and S.Saigal, pp109-116, 1997.
11. M. L.Staten and S. A. Canann. Post refinement element shape improvement for quadrilateral meshes. *Trends in Unstructured Mesh Generation* edited by S.A.Canann and S.Saigal, pp9-16, 1997.
12. G. Strang and G. J. Fix. *An Analysis of the Finite Element Method*, Prentice-Hall, 1973.
13. S.-H. Teng. A geometric approach to parallel hierarchical and adaptive computing on unstructured meshes. In *Fifth SIAM Conference on Applied Linear Algebra*, pages 51-57, June 1994.

A Robust and Scalable Library for Parallel Adaptive Mesh Refinement on Unstructured Meshes

John Z. Lou, Charles D. Norton, and Tom Cwik

Jet Propulsion Laboratory

California Institute of Technology, Pasadena, CA 91009-8099

Abstract

The design and implementation of a software library for parallel adaptive mesh refinement in unstructured computations on multiprocessor systems are described. This software tool can be used in parallel finite element or parallel finite volume applications on triangular and tetrahedral meshes. It contains a suite of well-designed and efficiently implemented modules that perform operations in a typical P-AMR process. This includes mesh quality control during successive parallel adaptive mesh refinement, typically guided by a local-error estimator, and parallel load-balancing. Our P-AMR tool is implemented in Fortran 90 with a Message-Passing Interface (MPI) library, supporting code efficiency, modularity and portability. The AMR schemes, Fortran 90 data structures, and our parallel implementation strategies are discussed in the paper. Test results of our software, as applied to a few selected engineering finite element applications, will be demonstrated. Performance results of our code on Cray T3E, HP/Convex Exemplar systems, and on a PC cluster (a Beowulf-class system) will also be reported.

Keywords: parallel adaptive mesh refinement, unstructured mesh, Fortran 90

1. Introduction

Adaptive mesh refinement (AMR) represents a class of numerical techniques that has demonstrated great effectiveness for a variety of computational applications including computational physics, structural mechanics, electromagnetics, and semiconductor device modeling. However, the development of an efficient and robust adaptive mesh refinement component for an application, particularly on unstructured meshes on multiprocessor systems, involves a level of complexity that is beyond the technical domain of most computational scientists and engineers. The motivation for our work is to provide an efficient and robust parallel AMR library that can be easily integrated into unstructured parallel applications.

Research on parallel AMR for unstructured meshes has been previously reported [2,10]. Most efforts are C++-based, and many realize that mesh quality control issues during successive adaptive refinements is an active research topic. The features of our work include the design of a complete Fortran 90 data structure for parallel AMR on unstructured triangular and tetrahedral meshes, the implementation of a robust scheme for mesh quality control during a parallel AMR process, and a "plug-in" component for stiffness matrix construction in finite element applications. We selected Fortran 90 for our implementation mainly because it provides adequate facilities for parallel unstructured AMR development while simplifying interface concerns with scientific application codes, many of which were developed in Fortran 77.

Mesh quality control is an important issue that should be addressed in any AMR algorithm since it has a direct impact on the accuracy and efficiency of numerical schemes. When adaptive refinements on a mesh are guided by a local-error estimator [1] and AMR is performed repeatedly, a straightforward AMR scheme would usually result in serious degradations of mesh quality, generating elements with poor aspect-ratios in the refined mesh. Several strategies have been proposed to improve mesh quality after an AMR step. Among them are global mesh-smoothing schemes after each adaptive refinement [7], selection of element refinement patterns based on the element shape, and other approaches (e.g. [4]). A mesh-smoothing scheme, in our view, should be both efficient and preserving local mesh characteristics. We will discuss a robust approach to mesh quality control during successive mesh refinements and our efficient implementation of the scheme. Results of applying this scheme to a few test problems will be presented.

Our parallel adaptive refinement scheme consists of two steps: a logical step in which the information needed to refine each element in the coarse mesh is constructed and stored, and a physical refinement step in which the coarse mesh is actually refined to produce a new mesh. The decomposition of an adaptive refinement process into a logical refinement phase and a physical refinement phase offers several advantages in a parallel AMR implementation. It makes the AMR code highly modular, and makes the actual refinement local to each element. It also makes it possible to perform parallel load-balancing by migrating only the coarse mesh instead of the refined mesh, thus with a much reduced communication cost. Such a refinement strategy also confines the interprocessor communication required by the parallel adaptive refinement scheme to the logical refinement phase. The code for the logical phase is small compared to the actual refinement phase, with the latter basically an operation local to each processor.

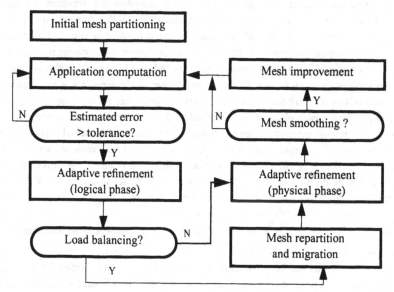

Figure 1. Parallel AMR process for unstructured meshes

2. The Parallel AMR Algorithm

Our parallel AMR framework comprises the following components: parallel adaptive refinement schemes for triangular and tetrahedral meshes, a parallel mesh partitioner for partitioning meshes with weighted elements, and a parallel mesh migration scheme that redistributes a partitioned mesh to their assigned processors. In a typical parallel AMR application scenario, the input mesh is randomly distributed among the processors after being read in from a disk file. This mesh is then partitioned with the mesh partitioner and redistributed among the processors by the mesh migration module according to the new partitioning, so the partitions of the input mesh will generally be connected and load-balanced. After an application computation step and a local error estimate step, a logical parallel AMR scheme is applied to the distributed (input) mesh using a local-error estimate in each mesh element. Since the outcome from the logical AMR step fully defines the refined mesh, thus also indicating the distribution of computational load among all processors for the refined mesh, a load balancing decision can be made, and load-balancing can be performed if needed. The physical AMR is performed after the load-balancing step, where a new mesh is created by consistently refining a subset of elements in the coarse mesh. As the final (and optional) step in the parallel AMR loop, the refined mesh may be checked for element quality, and some global mesh smoothing scheme [7] can be applied to the refined mesh. The control diagram of our parallel AMR algorithm is shown in Figure 1.

2.1 Logical and Physical Refinements

When a mesh is adaptively refined on a subset of elements, refinement also needs to be performed on "transitional elements" to produce a consistently refined mesh. On a triangular mesh, for example, an adaptive mesh can be constructed with refinement patterns shown in Figure 2. The logical refinement step uses an iterative procedure that traverses through elements of the coarse mesh repeatedly to "define" a consistent mesh refinement pattern on the coarse mesh. The result of the logical refinement is stored in the data structure of the coarse mesh, which completely specifies whether and how each element in the coarse mesh should be refined. Our adaptive refinement scheme is based on "edge-marking" for both triangular and tetrahedral meshes. Starting from a predetermined subset of elements, the logical refinement scheme proceeds by marking (or logically refining) element edges wherever necessary, and the refinement pattern for each element is determined by the number of marked edges in that element. With information generated from the logical refinement step, the actual mesh refinement becomes conceptually simpler, since it is completely specified how each element should be refined. To make the physical refinement process simpler and efficient, low-level objects are refined before refining high-level objects. On a triangular mesh, it means edges are refined before refining elements.

  Figure 2. Refinement patterns for a triangular mesh.

A parallel logical adaptive refinement scheme, as shown in Figure 3 for triangular meshes, can be constructed based on a sequential scheme. In the parallel scheme, actions involving information exchanges with remote processors are shown in bold type. To perform a parallel logical adaptive refinement, we extend the serial scheme so that after traversing the local element set for edge-marking, each processor updates the status of edges on mesh partition boundaries by exchanging the edge status information (i.e. marked or not marked) with its neighboring processors. In step 2, the inner do-while loop stops when a processor can not find any local edges to mark in a particular sweep through its local elements, and the outer do-while stops when all processors can not find any edge to mark, which can be determined from the result of the global max operation. Using this strategy, the serial logical refinement scheme can

Step 1: mark all edges of each element with its refine_pattern = 4
Step 2: **update status for partition_boundary_edges**
    global_edges_refined = 1
    do while (global_edges_refined > 0)
        global_edges_refined = 0
        stop = false
        while (not stop)
            found = false
            for each element
                marked_edges = 0
                for each edge of the element
                    marked_edges++, if edge is marked
                end for
                if marked_edges > 1 and < 4
                    set refine_pattern to 4, mark all its edges
                    found = .true.
                    global_edges_refined = 1
                end if
            end for
            if(not found) stop = true
        end while
        global_max(global_edges_refined)
        if(global_edges_refined > 0)
            update status for partition_boundary_edges
        end if
    end while
Step 3: for each element

        if marked edge = 1, set refine_pattern to 2.
    end for

Figure 3. A parallel logical refinement scheme

be parallelized by inserting a few communication calls, with no change to Step 1 and 3 in the serial scheme.

The information generated from a parallel logical refinement, again, is sufficient for actual refinement on the local mesh in each processor, and the actual refinement in all processors can proceed independently of each other. The result of these local refinements is a distributed new mesh, but at that time the global new adaptive mesh is *not* completely defined yet because the global id for mesh objects (nodes, edges, elements) still need to be generated. The global id assignment can not be part of the local refinement process because clearly the global ids have to be consistent across the entire new mesh. We designed an efficient communication scheme that is used after the physical refinement to resolve global id for all mesh objects.

The logical refinement scheme we use for tetrahedral mesh is similar in structure to the one for triangular mesh, and the refine_.pattern for each element is still determined through an iterative edge_marking process. Refinement patterns for a fully refined element and transitional elements, as used in our implementation, are shown in Figure 4; the ratios in the figure indicate the number of child elements created from the parent element.

2.2 Mesh Quality Control

A problem associated with repeated AMR operations is the deterioration of mesh quality. Since elements in the mesh are not uniformly refined, transitional elements are created for consistency in the refined mesh. When repeated refinements are performed in some region of a mesh, the aspect-ratios of elements surrounding the region could degrade rapidly as adap-

Figure 4. Refinement patterns for tetrahedral elements: a fully refined element (left,

tive refinement proceeds; this mesh quality degradation becomes more serious for tetrahedral elements. Laplacian or optimization types of smoothing algorithms have been proposed [3,7] to improve elements shape either locally or globally. Most mesh smoothing schemes tend to change the structure of the input mesh to achieve the "smoothing effect" by rearranging nodes in the mesh. The changes made by a smoothing scheme, however, could modify the desired distribution of element density produced by the AMR procedure, and the cost of performing a global element smoothing could be very high. Given a finally refined mesh, applying a relatively efficient smoothing scheme over the entire mesh is probably a reasonable choice for mesh quality improvement, and it is going to part of our library as indicated in Figure 1. On the other hand, it is possible to prevent or slow down the degradation of element quality during a repeated adaptive refinement process. One idea [11] is to change the refinement pattern for an element in the previous refinement, when it was refined to produce transitional elements, if any of the children of that element need to be further refined in the current refinement. Figure 5 illustrates an example of this idea.

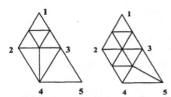

Figure 5. An example of mesh quality control. The refinement (left) on the coarse element 2-3-4 is modified (right) if any of the two elements in element 2-3-4 need to be further refined either due to their local errors or because their neighboring elements in 1-2-3 are to be

To integrate the mesh quality control feature into our adaptive refinement scheme for triangular mesh, we found that a relatively simple way to proceed is to define all possible refinement patterns for a pair of "twin" transitional element (child elements of element 2-3-4 in the left mesh in Figure 5), as shown in Figure 6. When trying to refine one of the twin elements in the logical refinement step, the modified logical refinement scheme checks all marked edges in the twin elements, and one of the refinement patterns in Figure 6 is selected as appropriate. To simplify the implementation of such a feature in a parallel adaptive refinement process, we require that the mesh partitioner not separate the twin elements onto two processors when partitioning the coarse mesh, so the actual refinement operation on the twin elements, performed after a mesh migration for load-balancing, will remain local in a processor. We find these ideas can be extended to tetrahedral mesh as well.

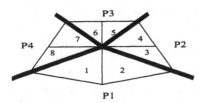

Figure 7. A triangular mesh is partitioned among four processors. The node at the center of the mesh needs all neighboring node data to assemble its row entry in the stiffness matrix..

Figure 6. Refinement patterns for a pair of "twin elements". The leftmost one is the twin elements to be further refined as a result of some of their neighboring elements are refined. The refinement pattern is determined by how the edges of the twin elements are refined.

2.3   A Plug-in Scheme for Parallel Finite Element Applications

In a typical finite element algorithm for a boundary-value problem, unknowns are defined on the interior nodes of a mesh, and a linear system of equations is solved to find these unknowns. To construct a row entry in the matrix (often referred to as a "stiffness matrix") for the linear system, where each row corresponds to an unknown, one needs to access data at the node as well as data at all its neighboring nodes. Since our mesh partitioning is based on partitioning mesh elements, all neighboring data of a given node at a partition boundary are not available in a single processor. Obtaining data from neighboring nodes residing on remote processors turns out to be far from trivial. Figure 7 gives an example to illustrate the problem, where the mesh is partitioned among four processors. To assemble a row entry for the node in the center of the mesh, the first question is which processor(s) will do it. A simple answer is to designate one processor sharing this node to do the assembling and have other sharing processors supply the data on neighboring nodes to that processor. If we assume processor P1 does the assembling, the question now is how would processor P3 send neighboring node data of the center node to P1; since in an element based mesh partitioning, P3 is not even aware that P1 is one of its "neighboring processors".

A simple (but not optimal in general) approach to making P1 and P3 realize that they are neighbors to each other is to perform a ring-style communication among all the processors, in which each processor sends either its own or received global ids of boundary nodes to its next neighbor processor in the ring. After the ring-style communication, each processor knows which processors share each of its boundary nodes, and subsequent node data exchanges, though still messy, can be restricted to communications among nearest neighboring processors. This scheme, except for the ring-style exchange in which a relatively small amount of information is communicated, is still mostly among neighboring processors, and thus is expected to be scalable.

2.4   Parallel Mesh Migration for Load Balancing

Adaptive mesh refinement generally introduces an imbalance of element distribution among processors. In order to reestablish a balanced element distribution, the mesh is repartitioned and mesh elements are redistributed among processors based on the new partition. In our parallel AMR library, ParMeTiS modules [8] are used to compute new partitioning. ParMeTis is a parallel graph-partitioning library with a collection of modules for partitioning or repartitioning graphs useful for AMR applications. To reduce mesh redistribution cost, ParMeTis is applied to the coarse mesh to be refined, and elements of the coarse mesh are migrated to reach a reasonably load-balanced distribution of the refined mesh. To use ParMetis in our library, a weighted dual-graph needs to be constructed for the coarse mesh, with each mesh element mapped to a weighted node in the dual-graph. Elements are weighted based on their refinement levels, and the ParMeTiS computes a new partitioning based on the weighted graph. In the context of an AMR process, it is possible and desirable to find a new mesh partition and a mapping of the new partition to processors that minimize both the movement of mesh elements and the interfaces among mesh components on different processors, since a smaller number of elements on partition boundaries imply less communication cost in the application code. ParMeTis offers a few functions that exploit the structure of an adaptively refined mesh to produce an optimized new partition and mapping.

The mesh is migrated in stages based on the component type for a clean and simple implementation. In two-dimensions, mesh elements, edges, nodes, and node coordinates are transported to new processors, in that order. Only information needed for rebuilding the mesh and for subsequent parallel adaptive refinement is migrated. The migrated mesh is reconstructed by a mesh_build module at new processors,

The parallel communication scheme for migration of mesh data is relatively straightforward. Processors first organize data that will remain local. Then, data that must be migrated is sent continually to processors that expect the data. While the sending is performed, processors probe for incoming messages, that are expected, and receive them immediately upon arrival (the probe is non-blocking). Probing has the added benefit that a processor can allocate storage for an incoming message before the message is actually received. When this process is completed, processors check to see if there are any remaining receives, processes them if necessary, and the migration completes. At this point, the mesh is reconstructed with the new data.

3. Fortran 90 Implementation

Fortran 90 modernizes traditional Fortran 77 scientific programming by adding many new features. These features allow programs to be designed and written at a higher level of abstraction, while increasing software clarity and safety without sacrificing performance. Fortran 90's capabilities encourage scientists to design new kinds of advanced data structures supporting complex applications. (Developing such applications, like P-AMR, might not have been considered by Fortran 77 programmers since development would have been too complex.) These capabilities extend beyond the well-known array syntax and dynamic memory management operations.

Derived-Types, for instance, support user-defined types created from intrinsic types and previously defined derived-types. Dynamic memory management in the form of pointers, and a variety of dynamic arrays, are also useful in data structure design. Many of the new Fortran 90 data types know information about themselves, such as their size, if they have been allocated, and if they refer to valid data. One of the most significant advances is the module, that supports abstraction modeling and modular code development. Modules allow routines to be associated with derived types defined within the module. Module components can be public and/or private leading to the design of clean interfaces throughout the program. This is very useful when multiple collaborators are contributing to the design and development of large projects. Other useful capabilities include function and operator overloading, generic procedures, optional arguments, strict type checking, and the ability to suppress the creation of implicit variables. Fortran 90 is also a subset of HPF and, while message passing programming using MPI is possible with Fortran 90, this link to HPF simplifies extensions to data parallel

```
module mesh_module
use mi_module; use heapsort_module
private
public :: mesh_refinement, mesh_migration ! and other public routines
! constant declarations
type node
      private
      integer :: id
      real :: coordinates(mesh_dim)
end type node
type edge
      private
      integer :: id, nodeix(nodes_), neighix(neigh_)
end type edge
type element
      private
      integer :: id, nodeix(nodes_), edgeix(edges_), neighix(neigh_)
end type element
type mesh
      private
      type(node), dimension(:), pointer :: nodes
      type(edge), dimension(:), pointer :: edges
      type(element), dimension(:), pointer :: elements
      type(r_indx), dimension(:), pointer :: boundary_elements
end type mesh

contains
      ! subroutine and function definitions placed here
end module mesh_module
```

Figure 8. Skeleton view of the mesh_module

programming. Both Fortran 90 and HPF are standards supported throughout industry which helps promote portability among compilers and machines.

One of the major benefits of Fortran 90 is that codes can be structured using the principles of object-oriented programming [6,9]. While Fortran 90 is not an object-oriented language, this methodology can be applied with its new features. This allows the development of a PAMR code where interfaces can be defined in terms of mesh components, yet the internal implementation details are hidden. These principles also simplify handling interlanguage communication, sometimes necessary when additional packages are interfaced to new codes. Using Fortran 90's abstraction techniques, for example, a mesh can be loaded into the system, distributed across the processors, the PAMR internal data structure can be created, and the mesh can be repartitioned and migrated to new processors (all in parallel) with a few simple statements.

A user could link in the routines that support parallel adaptive mesh refinement then, as long as the data format from the mesh generation package conforms to one of the pre-specified formats, the capabilities required for PAMR are available. We now describe the Fortran 90 implementation details that make this possible.

3.1 Fundamental Data Structures

Automated mesh generation systems typically describe a mesh by node coordinates and connectivity. This is insufficient for adaptive mesh refinement. Hierarchical information, such as the faces forming an element, the edges bounding each face, or elements incident on a common node is also useful. Additionally, large problems require the data to be organized and accessible across a distributed memory parallel computing system. These issues can be addressed by the creation of appropriate PAMR data structures.

The major data structure is the description of the mesh. While a variety of organizations are possible, where trade-offs between storage and efficiency of component access must be decided, most descriptions include elements, faces, edges, and nodes. These are related hierarchically where components generally contain references to other components that comprise its description. These references can be bi-directional, an edge may have references to its two node end points and references to the faces it helps form. However, some of these details can be omitted from the structure by using a combination of good data structure designs and efficient recomputation of the required information.

3.2 Basic Mesh Definition

Fortran 90 modules allow data types to be defined in combination with related routines. In our system the data structure for mesh is described as shown in Figure 8. In two-dimensions, the mesh is a Fortran 90 object containing nodes, edges, elements, and reference information about non-local boundary elements (r_indx). These components are dynamic, their size can be determined using Fortran 90 intrinsic operations. They are also private, meaning that the only way to manipulate the components of the mesh are by routines defined within the module. Incidentally, the remote index type r_indx (not shown) is another example of encapsulation. Objects of this type are defined so that they cannot be created outside of the module at all. A module can contain any number of derived types with various levels of protection, useful in our mesh data structure implementation strategy.

All module components are declared private, meaning that none of its components can be referenced or used outside the scope of the module. This encapsulation adds safety to the design since the internal implementation details are protected, but it is also very restrictive. Therefore, routines that must be available to module users are explicitly listed as public. This provides an interface to the module features available as the module is used in program units. The mesh_module uses other modules in its definition, such as the mpi_module and the heapsort_module. The mpi_module provides a Fortran 90 interface to MPI while the heapsort_module is used for efficient construction of the distributed mesh data structure. The routines defined within the contains statement, such as mesh_create_incore(), belong to the module. This means that routine interfaces, that perform type matching on arguments for correctness, are created automatically. (This is similar to function prototypes in other languages.)

3.3 Distributed Structure Organization

When the PAMR mesh data structure is constructed it is actually distributed across the processors of the parallel machine. This construction process consists of loading the mesh data, either from a single processor for parallel distribution (in_core) or from individual processors in parallel (out_of_core). A single mesh_build routine is responsible for constructing the mesh based on the data provided. Fortran 90 routine overloading and optional arguments allow multiple interfaces to the mesh_build routine, supporting code reuse. This is helpful because the same code that builds a distributed PAMR mesh data structure from the initial description can be applied to rebuilding the data structure after refinement and mesh migration. The mesh_build routine, and its interface, is hidden from public use. Heap sorting techniques are also applied in building the hierarchical structure so that reconstruction of a distributed mesh after refinement and mesh migration can be performed from scratch, but efficiently.

Figure 8 showed the major components of the mesh data structure, in two-dimensions. While Fortran 90 fully supports linked list structures using pointers, a common organization for PAMR codes, our system uses pointers to dynamically allocated arrays instead. There are a number of reasons why this organization is used. By using heap sorting methods during data structure construction, the array references for mesh components can be constructed very quickly. Pointers consume memory, and the memory references can become "unorganized", leading to poor cache utilization. While a pointer-based organization can be useful, we have ensured that our mesh reconstruction methods are fast enough so that the additional complexity of a pointer-based scheme can be avoided.

3.4 Interfacing Among Data Structure Components

The system is designed to make interfacing among components very easy. Usually, the only argument required to a PAMR public system call is the mesh itself. There are other interfaces that exist however, such as the internal interfaces of Fortran 90 objects with MPI and the ParMeTiS parallel partitioner [8] which written in the C programming language.

Since Fortran 90 is backward compatible with Fortran 77 it is possible to link to MPI for interlanguage communication, assuming that the interface declarations have been defined in the mpi.h header file properly. While certain array constructs have been useful, such as array syntax and subsections, MPI does not support Fortran 90 directly so array subsections cannot be (safely) used as parameters to the library routines. Our system uses the ParMeTiS graph partitioner to repartition the mesh for load balancing. In order to communicate with ParMeTiS our system internally converts the distributed mesh into a distributed graph. A single routine interface to C is created that passes the graph description from Fortran 90 by reference. Once the partitioning is complete, this same interface returns from C an array that describes the new partitioning to Fortran 90. This is then used in the parallel mesh migration stage to balance mesh components among the processors.

4. Performance

The numerical and computational performances of an overall parallel AMR process using the AMR library in an application, and the relative cost of components in the library are being studied. The scalability of communication intensive parts of the library, such as parallel adaptive refinement, mesh partition and migration, and the plug-in module for finite element applications are also being investigated in a few test cases.

Listed in each row of Table 1 is a breakdown of execution time of a parallel AMR process between load-balancing (including mesh partition and migration) and adaptive refinement (both logical and physical refinements). The input is a triangular mesh that contains about 650 elements in an electromagnetic waveguide filter region. Three successive adaptive refinements are performed on randomly selected subsets of mesh elements (roughly half of the total mesh elements each time), with each refinement followed by a load-balancing operation. Execution times were measured on a Cray T3E system on up to 32 processors. The quadratic speedup achieved for the parallel adaptive refinement part, as shown, may reflect the complexity of the (iterative) parallel adaptive refinement scheme as a function of the size of the submesh each processor. The relative cost of performing load-balancing becomes more significant as the number of processors increases, and it becomes dominant on 32 processors. The decision on whether to perform load-balancing, however, should be based on a comparison of the cost of performing load-balancing with the saving of the computational cost in an application from a load-balanced mesh distribution.

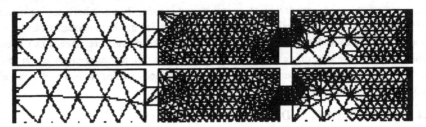

Figure 9. Illustration of mesh quality control during repeated adaptive refinement.

Processors	AMR time (sec.)	LB time (sec.)
2	56.0	14.3
4	14.4	2.6
8	2.6	0.7
16	0.41	0.31
32	0.15	0.30

Table 1: Initial performance analysis on the parallel AMR library

For many application problems, the mesh quality could degrade rapidly due to successive adaptive refinements, resulting a final adaptive mesh practically useless for those applications, as the effectiveness of the numerical schemes used depends on the (worst) aspect ratio of mesh elements. We have therefore implemented a technique, as discussed in Section 2.2, that, to some extent, controls the mesh quality from further deteriorating in a repeated adaptive refinement process. Figure 9 shows a comparison of repeated adaptive refinements on the filter mesh without and with mesh quality control. Three successive adaptive refinements were performed in two fixed subregions in the input mesh. The top mesh was produced without mesh quality control, and the bottom mesh is the result of mesh quality control. The difference in the number of elements in the final mesh for the two cases is about 3%.

5. Applications

We now present some results from applying parallel and serial AMR library modules in our library to a few test problems on triangular and tetrahedral meshes.destination processors. As a test case, as shown in Figure 10, on the robustness of our parallel AMR library, the complete parallel AMR procedure was applied to an input triangular mesh with 70 elements in the filter region, using 16 processors on a Cray T3E. Three successive parallel refinements were performed on half of the elements randomly selected in each processor. The final mesh displayed in Figure 10 shows a reasonably good quality and a roughly balanced mesh distribution across the processors.

Figure 10: Successive parallel adaptive refinements with mesh quality control and load-balancing on a Cray T3E using 16 processors.

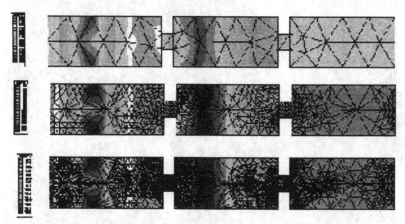

Figure 11 Adaptive finite-element solution in a waveguide filter. Adaptive refinement is guided by a local-error estimate procedure based on local residuals.

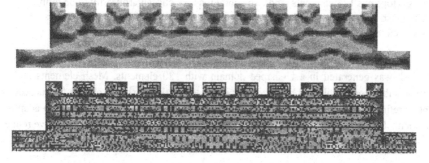

Figure 12. Adaptive finite-element simulation in a quantum well infrared photodetector for length infrared radiation. The computed magnetic field relative to an incident plane wave, an field on the adaptive mesh are shown respectively.

Our AMR module was also used in a finite element numerical simulation of electromagnetic wave scattering in the above waveguide filter [5]. The problem is to solve the Maxwell's equations for studying electromagnetic wave propagations in the filter domain. A local-error estimate procedure based on the Element Residue Method (ERM) [1] is used in combination with the AMR technique to adaptively construct an optimal mesh for an efficient final problem solution. Figure 11 shows a few snapshots of the mesh in a serial AMR solution process. The color and density distribution of mesh elements in the figure reflect the (estimated) error distribution in the computed fields. Another application using the serial AMR module is to an electromagnetic device simulation in a quantum well infrared photodetector (QWIP), as shown in Figure 12.

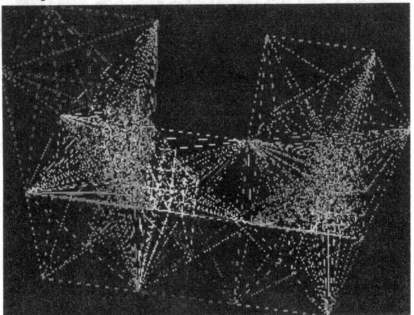

Figure 13. Adaptive refinement on a three-dimensional tetrahedral mesh. The initial mesh (top) has 128 elements, and two subregions are chosen arbitrarily to be refined. The mesh after adaptive refinements (bottom) has about 2,500 elements. Colors are assigned to meshe elements at various refinement levels.

A test on the AMR library was also made on a tetrahedral mesh. The initial tetrahedral mesh was generated in a U-shaped domain with 120 elements. Mesh elements in two spherical subregions, indicated by the circles in the top initial mesh, are chosen for successive adaptive refinements The radius of the refining spheres is reduced by 20% after each adaptive refinement. The image in Figure 13 is the resulting mesh after three successive adaptive refinements, which has about 2500 elements.

6. Concluding Remarks and Future Work

We have presented a complete framework for performing parallel adaptive mesh refinement in unstructured applications on multiprocessor computers. A robust parallel AMR scheme and its implementation in Fortran 90 and MPI, with mesh quality control, as well as a load-balancing strategy in the parallel AMR process, are discussed. A few test cases and application examples using our developed AMR modules are demonstrated. Initial measurements on the parallel performance of the AMR library on the Cray T3E were reported. Parallel implementation for the tetrahedral meshes are currently in process, and further performance analysis of the AMR library on a HP/Convex system and a Beowulf system (network of Linus PCs) will be performed in the near future.

7. References

[1] M. Ainsworth, J. T. Oden. A Procedure for a Posteriori error Estimation for *h-p* Finite Element Methods." Computer Methods in Applied Mechanics and Engineering, 101 (1972) 73-96.

[2] R. Biswas, L. Oliker, and A. Sohn. "Global Load-Balancing with Parallel Mesh Adaption on Distributed-Memory Systems." Proceedings of Supercomputing '96, Pittsburgh, PA, Nov. 1996.

[3] E. Boender. "Reliable Delaunay-Based Mesh Generation and Mesh Improvement." Communications in Numerical Methods in Engineering, Vol. 10, 773-783 (1994).

[4] Graham F. Carey, "Computational Grid Generation, Adaptation, and Solution Strategies". Series in Computational and PHysical Processes in Mechanics and Thermal Science. Taylor & Francis, 1997.

[5] T. Cwik, J. Z. Lou, and D. S. Katz, "Scalable Finite Element Analysis of Electromagnetic Scattering and Radiation." to appear in Advances in Engineering Software, V. 29 (2), March, 1998

[6] V. Decyk, C. Norton, and B. Szymanski. Expressing Object-Oriented Concepts in Fortran 90. ACM Fortran Forum, vol. 16, num. 1, April 1997.

[7] L. Freitag, M. Jones, and P. Plassmann. "An Efficient Parallel Algorithm for Mesh Smoothing." Tech. Report, Argonne National Laboratory.

[8] G. Karypis, K. Schloegel, and V. Kumar. "ParMeTiS: Parallel Graph Partitioning and Sparse Matrix Ordering Library Version 1.0". Tech. Rep., Dept. of Computer Science, U. Minnesota, 1997.

[9] C. Norton, V. Decyk, and B. Szymanski. High Performance Object-Oriented Scientific Programming in Fortran 90. Proc. Eighth SIAM Conf. on Parallel Processing for Sci. Comp., Mar. 1997 (CDROM).

[10] M. Shephard, J. Flaherty, C. Bottasso, H. de Cougny, C. Ozturan, and M. Simone. Parallel automatic adaptive analysis. Parallel Computing 23 (1997) pg. 1327-1347.

[11] R. Bank, Private communications, December, 1997

Quality Balancing for Parallel Adaptive FEM *

Ralf Diekmann[1], Frank Schlimbach[1], and Chris Walshaw[2]

[1] Department of Computer Science, University of Paderborn,
Fürstenallee 11, D-33102 Paderborn, Germany
{diek, schlimbo}@uni-paderborn.de
[2] School of Computing and Mathematical Sciences,
The University of Greenwich, London, UK.
C.Walshaw@gre.ac.uk

Abstract. We present a dynamic distributed load balancing algorithm for parallel, adaptive finite element simulations using preconditioned conjugate gradient solvers based on domain-decomposition. The load balancer is designed to maintain good partition aspect ratios. It can calculate a balancing flow using different versions of diffusion and a variant of breadth first search. Elements to be migrated are chosen according to a cost function aiming at the optimization of subdomain shapes. We show how to use information from the second step to guide the first. Experimental results using Bramble's preconditioner and comparisons to existing state-ot-the-art load balancers show the benefits of the construction.

1 Introduction

Finite Elements (or Finite Differences or Finite Volumes) can be used to numerically approximate the solutions of partial differential equations (PDEs) describing, for example, the flow of air around a wing or the distribution of temperature on a plate which is partially heated [2,34]. The domain on which the PDE is to be solved is discretized into a mesh of *finite elements* (triangles or rectangles in 2D, tetrahedra or hexahedra in 3D) and the PDE is transformed into a set of linear equations defined on these elements [34]. The coupling between equations is given by the neighborhood in the mesh. Usually, iterative methods like Conjugate Gradient (CG) or Multigrid (MG) are used to solve the systems of linear equations [2].

The quality of solutions obtained by such numerical approximation algorithms depends heavily on the discretization accuracy. In particular, in regions with steep solution gradients, the mesh has to be refined sufficiently, i.e. the elements should be small in order to allow a good approximation. Unfortunately the regions with large gradients are usually not known in advance. Thus, meshes either have to be refined regularly resulting in a large number of small elements

---

* This work is partly supported by the DFG-Sonderforschungsbereich 376 "Massive Parallelität: Algorithmen, Entwurfsmethoden, Anwendungen" and the EC ESPRIT Long Term Research Project 20244 (ALCOM-IT).

in regions where they are not necessary, or the refinement takes place during the calculation based on error estimates of the current solution [30]. Clearly, the second variant should be favored because it generates fine discretizations only in those regions where they are really needed. The resulting solutions have the same quality as if the mesh was refined regularly but the number of elements, and thus the time needed to determine the solution, is only a small fraction of the regular case.

The parallelization of numerical simulation algorithms usually follows the single-program multiple-data (SPMD) paradigm: The same code is executed on each processor but on different parts of the data. This means that the mesh is partitioned into P subdomains where P is the number of processors, and each subdomain is assigned to one processor [12]. Iterative solution algorithms mainly perform local operations, i.e. new values at certain parts of the mesh only depend on the values of nearby data (modulo some global dependencies in certain iteration scemes which we neglect here as they are independent of the way the mesh is partitioned). Thus, the parallelization requires communication mainly at the partition boundaries.

The efficiency of such a SPMD parallelization mainly depends on the equal distribution of data on the processors. As the individual iteration steps are synchronized by a data exchange at partition boundaries the time the processors need to perform one iteration step should be the same on each processor. This can be achieved by splitting the mesh into equal sized subdomains [12]. In the case of adaptive refinement, the distribution of data on the processors becomes unbalanced and the partition has to be altered in order to re-establish a balanced distribution.

Preconditioning is a powerful method to speed up linear solvers [2, 5, 34]. *Domain Decomposition* methods (DD) belong to the most efficient preconditioners. Unfortunately, their speed of convergence depends on the way the partition is generated. The shape (the Aspect Ratio) of subdomains determines the condition of the preconditioned system and has to be considered if partitions are changed due to load balancing.

PadFEM (**P**arallel **ad**aptive **FEM**) is an object-oriented environment supporting parallel adaptive numerical simulations [3, 24]. *PadFEM* includes graphical user interfaces, mesh generators (2D and 3D, sequential and parallel), mesh partitioning algorithms [12], triangular and tetrahedral element formulations for Poisson and Navier-Stokes problems [34], different solvers (especially DD-PCGs) [2, 5], error estimators [30], mesh refinement algorithms [21], and load balancers [12].

The load balancing module in *PadFEM* determines the amount of load to move between different subdomains in order to balance the distribution globally. It can calculate a *balancing flow* using different kinds of diffusive methods [4, 9], in particular first and second order diffusion iterations [13, 17], and a heuristic for convex non-linear min-cost flow [1]. In a second step, the elements to be moved are identified. When choosing these elements, the load balancer tries to optimize partition Aspect Ratios in addition to load balance. The elements at

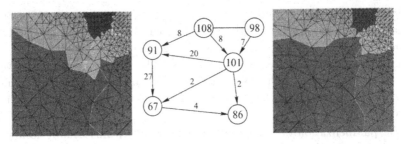

Fig. 1. (l.t.r.) unbalanced partition; corresponding quotient graph with balancing flow; resulting balanced partition after element migration.

partition boundaries are weighted by a cost function consisting of several parts. The migration chooses elements according to their weight and moves them to neighbors. The element weight functions are used to "guide" the balancing flow calculation. We consider the existing partition and define weights for the edges of the cluster graph expressing a kind of "cost" for moving elements over the corresponding borders.

The next section gives an introduction into the field of balancing adaptive meshes. Section 3 discusses some of the existing methods and Section 4 describes the element migration strategies used within *PadFEM*. This second step in load balancing is presented first, as parts of it are used in Section 5 where the balancing flow calculation is described. Section 6 finishes the paper with a number of results about Aspect Ratios and iteration numbers of the DD-PCG solver within *PadFEM*. Comparisons are made with *Jostle*, the Greenwich parallel partitioning tool [33].

2   Load Balancing and DD-Preconditioning

Balancing Adaptive Meshes: In most cases, the computational load of a finite element problem can be measured in terms of the numbers of elements. Equally balanced partitions can be generated using any of the existing graph partitioning algorithms (see e.g. [12] for an overview) or by generating the mesh in parallel [7]. If the mesh is refined adaptively, the existing partition will probably become unbalanced. Scalable load balancing algorithms can benefit from an existing partition by moving elements on the *quotient graph* defined by the adjacency between subdomains. The quotient graph contains one node for each partition. Edges denote common borders between the corresponding partitions. Figure 1 shows an unbalanced partition of a simple mesh (domain "Square") into four subdomains (left) and the resulting quotient graph (center). Any reasonable load balancing algorithm taking the existing distribution into account has to move data (elements) via edges of this graph. We add weights to nodes of the graph according to their load and determine a balancing flow on the edges. The flow is given as edge labels in Figure 1.

In a second step - the element migration phase - the load is actually moved. The input for this phase is the flow on the quotient graph and the task is to

Fig. 2. The three steps of the DD-PCG algorithm (left to right). The rightmost picture shows the solution after 4 iterations.

choose elements whose movements fulfill the flow demands. The choice of elements may consider additional cost criteria such as minimizing the boundary length or optimizing the shape of subdomains (which is actually done in *Pad-FEM*).

Preconditioning by Domain Decomposition: Due to space limitations, we only give a brief idea of the method. Please check [2] and [5] for a mathematical introduction, or to [14] for a more intuitive.

DD-methods take advantage of the partition of a mesh into subdomains. They impose artificial boundary conditions on the subdomain boundaries and solve the original problem on these subdomains (left of Fig. 2). The subdomain-solutions are independent of each other, and, thus, can be determined in parallel without any communication between processors. In a second step, a special problem is solved on the inner boundaries which depends on the jump of the subdomain solutions over the boundaries (mid-left of Fig. 2). This special problem gives new conditions on the inner boundaries for the next step of subdomain solution (mid-right of Fig. 2). Adding the results of the third step to the first gives the new conjugate search direction in the CG algorithm. The rightmost picture of Figure 2 shows the solution of the whole problem after four iterations.

The time needed by such a preconditioned CG solver is determined by two factors, the maximum time needed by any of the subdomain solutions and the number of iterations of the global CG. Both are at least partially determined by the shape of the subdomains. While the MG method as solver on the subdomains is relatively robust against shape, the number of global iterations are heavily influenced by the Aspect Ratio of subdomains [29]. Essentially, the subdomains can be viewed as elements of the "interface" problem [15], and just as with normal FEM, where the condition of the equational system is determined by the Aspect Ratio of elements, the condition of the preconditioning matrix is here dependent on the Aspect Ratio of subdomains.

3 Existing Approaches

Most existing implementations of adaptive grids perform load balancing by re-partitioning. This is sometimes the only possibility for parallelizing existing numerical codes whose data structures may support adaptivity but certainly not a movement of elements between subdomains of a distributed memory implementation. Attempts to minimize the amount of necessary data movement can be found in [23].

The tools *Metis* [25] and *Jostle* [33] are designed to support partitioning and load balancing of adaptive mesh calculations in parallel. Both use Hu&Blake's algorithm [19] for determining the balancing flow and both use a multi-level strategy for shifting elements, where the coarsening is done inside subdomains only. The tools differ in implementation details. *Jostle* uses the concept of *relative gain* optimization at partition boundaries and multilevel balancing and refinement at every graph level [31]. *Metis* uses multilevel diffusion on the coarsest graph levels, followed by greedy refinement. Both tools optimize according to *Cut*, Aspect Ratio is not directly considered.

The *SUMAA3D Project* at Argonne is especially designed to support largescale scientific computations in parallel. The mesh partitioning in *SUMAA3D* is based on *Unbalanced Recursive Bisection* (URB) which aims at generating better shaped partitions than standard recursive bisection [20]. By moving partition boundaries, the tool is also able to balance load in a parallel adaptive setting.

The group in Leeds has designed an environment supporting parallel adaptive mesh computations [27]. The load balancer uses a variant of multi-level Diffusion [18] which minimizes the amount of data movement.

Partitioners particularly designed to optimize the subdomain Aspect Ratio can be found in [11,15,29]. The tool PAR^2 was originally constructed as parallel partitioner, but it can also be used within an adaptive environment [11]. The method described in [15] is an iterative partitioner trying to improve the subdomain Aspect Ratios in a number of steps. In this work, DD-PCG solvers are used, but not within an adaptive environment. Other attempts try to optimize subdomain shapes by using Meta Heuristics such as Simulated Annealing or Tabu Search [29].

4 Element Migration

The basic idea of the element migration phase is to rate elements at partition boundaries according to a cost function and to greedily choose elements to be moved to neighboring processors based on this rating, until the necessary load has been moved. The migration phase starts with a valid balancing flow determined by the flow calculation phase which will be described in Section 5. For now, let us assume that the flow is represented as values x_{ij} on edges (i,j) of the quotient graph (meaning that x_{ij} elements have to be moved from partition i to partition j). The aim of the migration phase is to balance the load in such a way that the Aspect Ratio of subdomains is maintained, improved, or, at least, deteriorated as less as possible.

For a subdomain P_i, let A_i be its area[1] and B_i be the length of its boundary. We define the Aspect Ratio AR_i of P_i by $AR_i = B_i^2/(16A_i)$. This definition of the Aspect Ratio is quite different from the usual one found in the literature [11,

---

[1] Throughout this paper, we are dealing with 2D domains, only.

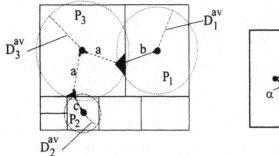 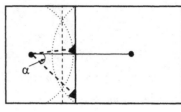

Fig. 3. Scaled Distances.   **Fig. 4.** Angle.

15] but in our experience we found that it expresses the desired tasks of shape optimization best [14].

The simplest way to rate elements at partition boundaries is to count the change in cut size (of the element graph) if each element were moved to another partition. For an element e, let $ed(e, i)$ be its number of edges adjacent to elements in P_i. Then, the change in cut when e is moved from P_i to P_j is given by $\Gamma_{cut}(e, i, j) = ed(e, i) - ed(e, j)$. For many FEM meshes constructed of triangles, the resulting element graph has a maximal degree of three and the change in Γ_{cut} (for nodes at the boundary) is at most 1. It is known [6] that local graph partitioning algorithms perform badly on degree-3 graphs. So it is no surprise that this element rating function does not produce good results. We will see comparisons later on. Nevertheless, it is used in several applications, and if it is combined with other sophisticated methods like multilevel schemes, it can still work very well [33].

The most promising rating functions consider the distances of elements from the center of their subdomain. If all boundary-elements have the same distance to this center, then the domain is a circle and has a good Aspect Ratio. Elements lying far away from the center might be good candidates to move. Let $dist(e, i)$ be the distance of element e to the center of subdomain P_i. The function $\Gamma_{rd}(e, i, j) = dist(e, i) - dist(e, j)$ is called the *relative distance* of element e. The idea behind this function is to prefer elements which are far away from the center of their own subdomain and, additionally, near to the center of the target partition. Such element rating functions can be found in some implementations [8, 11, 32]. The problem is that they do not account for the real physical size of a domain. A large partition might easily move elements to a small one, even if this does not improve the Aspect Ratio of any of them. We can define D_i as the average distance of the centers of all elements of subdomain P_i from its center. D_i can be used to normalize the distance function Γ_{rd}. $\Gamma_{srd}(e, i, j) = dist(e, i)/D_i - dist(e, j)/D_j$ is called the *scaled relative distance function*. Figure 3 shows the advantage of Γ_{srd} in contrast to Γ_{rd}. Without scaling, P_3 would prefer to move elements to P_2, whereas a move towards P_1 would improve the Aspect Ratios of P_1 and P_3.

If we consider the shape of a common border between two subdomains, the best would be to have a straight line perpendicular to the interconnection of

Fig. 5. Migration according to Γ_{srd}.   **Fig. 6.** Migration according to $\Gamma_{srd} + \Gamma_\alpha$.

the centers of both. If the border is circular in any direction, then one of the domains becomes concave which is not desirable. If elements are moved between subdomains which are very different in size, then the influence of the smaller one on Γ_{srd} becomes larger with increasing distance from the line connecting both centers. Thus, elements far away from this line are more likely to be moved than those lying on or nearby this line. The effect can be observed in Figure 5 where the domain "Cooler" is refined in 10 steps from an initial 749 elements to a final 7023. The main refinement takes place at the upper part of the domain[2] and the partitions lying in the corresponding region give up elements according to Γ_{srd}.

To avoid this effect, we superimpose Γ_{srd} by another function $\Gamma_\alpha(e,i,j) = \cos\alpha$, where α is the angle between the lines $C_i \leftrightarrow C_j$ and $C_i \leftrightarrow c_e$. Figure 4 shows the construction. The value of $\cos\alpha$ is large if the angle α is small, i.e. the superposition slightly increases the values of those elements which lie in the direction of the target partition. Figure 6 shows the effect on the same example of Fig 5. As desired, the borders between subdomains are much straighter and the target domains are less concave.

	AR: Max	AR: Avrg	Cut: Max	Cut: Avrg
Γ_{cut}	2.03	1.66	190	128
Γ_{rd}	1.53	1.33	142	92
Γ_{srd}	1.47	1.29	144	90

Table 1. AR and cut obtained with different rating functions.

A first comparison of the different rating functions can be found in Table 1. It shows the maximum and average Aspect Ratios as well as the Cuts if the domain "Square" is refined in 10 steps from an initial 115 elements to a final 10410. The values for Aspect Ratios given here and in the remaining of the paper are scaled to 1.0 being an "optimal" shape. Average values of ≤ 1.4 can be considered as acceptable, values of > 1.6 are unacceptable. It can be observed that Γ_{cut} is not only bad in optimizing AR but also in finding partitions with good cut. The explanation has already been given: As the graph has a maximal degree of 3, local methods fail. Γ_{rd} and Γ_{srd} are quite similar to each other because the difference in size of the subdomains is not very large in this example. The effects of Γ_α are not directly visible in terms of maximum or average AR. This kind of "cosmetic" operation is more visible in Figures 5 and 6.

---

[2] The figures show only the final partitions. Edges are omitted.

5 Balancing Flow Calculation

5.1 Possible Solutions

Let $G = (V, E, w)$ be the quotient graph of a load balancing problem (cf. Fig. 1). G has $n = |V|$ nodes, $m = |E|$ edges and there are load values w_i attached to each node $i \in V$. The task of the balancing problem is to find a flow x_{ij} of load on edges $(i, j) \in E$ of the graph such that after moving the desired numbers of elements, the load is globally balanced, i.e. $w_i = \bar{w} \; \forall i \in V$ (where $\bar{w} = \frac{1}{n} \sum w_i$ is the average load). As the flow is directed, we need to define (implicit) directions for edges. We may assume that edges are directed from nodes with smaller numbers to nodes with larger numbers. A flow of $x_{ij} < 0$ on edge $(i, j) \in E$ then means that $-x_{ij}$ load has to be moved from node j to i.

Let $A \in \{-1, 0, 1\}^{n \times m}$ be the *node-edge incidence matrix* of G. The edge directions are expressed by the signs (from $+1$ to -1). The matrix $L = AA^T$ is called the *Laplacian matrix* of G. If we define $x \in \mathbb{R}^m$ to be the vector of flow values, and if we would like to minimize the total amount of flow, then we are searching for an x minimizing $\sum |x_{ij}|$ such that $Ax = w - \bar{w}$. Such a problem is known as the *min-cost flow problem* [1].

The question of what are good cost criteria for such a flow is not easy to answer. It can be observed easily that any solution to the min-cost flow problem shifts the flow only via shortest paths in the network. Thus, it often uses only a small fraction of the available edges and this might not be desirable, especially when afterwards the shape of subdomains has to be considered.

A compromise between the conflicting goals of not shifting too much load but using all edges more or less equally might be the Euclidean norm of the flow defined as $\|x\|_2^2 = \sum x_{ij}^2$. This measure turns the min-cost flow into a non-linear, but convex optimization problem which is still solvable in polynomial time. For this special type of problem, solutions can be found easily. Hu and Blake, for example, propose the use of Lagrange multipliers for a solution of $Ax = w - \bar{w}$ [19]. The method requires to solve $L\lambda = w - \bar{w}$ and the flow is afterwards given by $x_{ij} = \lambda_i - \lambda_j$. $L\lambda = w - \bar{w}$ is easy to solve in parallel, because L directly corresponds to the quotient graph G (in fact, L is topologically equivalent to G).

Other local iterative methods determine x directly using $x^t = \alpha A^T w^t$ and $w^{t+1} = w^t - Ax^t$. This iteration is similar to $w^{t+1} = Mw^t$ with iteration matrix $M = (I - \alpha L)$ and parameter $\alpha \leq max\_deg(G)$. It is known as the *Diffusion Method* [4, 9]. Some schemes try to speed up the convergence of this local iterative method by using non-homogeneous iteration matrices M, optimal values of α [13], and over-relaxations like $w^{t+1} = \beta M w^t + (1 - \beta) w^{t-1}$ [17]. It can be shown that all these local iterative methods determine flows x which are minimal in the l_2-norm $\|x\|_2$ [10].

5.2 Average Weighted Flow

The *Average Weighted Flow* algorithm (*AWF*) proceeds in two steps. It first determines (source/ sink/amount) triples giving pairs of processors and the amount of load to move between them. Afterwards, these flows are routed via paths in the quotient graph. *AWF* is based on *edge qualities* ω_{uv} expressing the average quality of elements lying at the subdomain boundary corresponding to the edge (u, v). The element qualities are determined using any of the rating functions Γ.

The first step of *AWF* starts from each source node searching for sinks nearby which are reachable via "good" paths in the graph. For a path $p = v_1, \ldots, v_k$, we define the *path quality* $q(p) = \frac{1}{|p|} \sum_{i=1}^{k-1} \omega_{v_i v_{i+1}}$ as the average quality of edges on the path. As we are searching for paths with high quality, the task is to maximize $q(p)$. Unfortunately, this makes the problem intractable. If $q(p)$ were just the sum of the ω's, then we would search for longest paths without loops, an *NP*-complete problem [16]. Additionally, the averaging causes the loss of transitivity. The "best" paths from i to j via k need no longer include the best paths from i to k and from k to j.

We use a variance of breadth first search (BFS) as a heuristic to find approximately "best" paths. The algorithm starts from a source and searches - in normal BFS style - for paths to sinks. A sink node has a "distance" from the source according to the quality of the best path over which it has been reached during the BFS. If a node is found again (due to circle closing edges), its "distance" is updated. The algorithm visits each edge at most once and, thus, has a running time of $O(m)$. For each sink node, all the paths from the source found during the search are stored together with their quality. For building the (source/sink/amount) triples, the load $w_i - \bar{w}$ of a source i is distributed (logically) to its "best" sinks, each of them filling up to $\bar{w}$. The time for this phase is dominated by the search. As there are at most n sources and because we have to store all the paths, the total running time is $O(n^2 \cdot m)$.

The second phase routes the flow demands between source/sink pairs via paths in the network. The task is here to distribute the flow evenly over (a subset of) the possible paths from the source to the sink. The first phase already calculated possible paths. If a flow of x_{st} has to be shifted from a source s to a sink t and if there are two possible paths p_1 and p_2 between them, then the fraction of $\frac{q(p_1)}{q(p_1)+q(p_2)}$ is shifted via p_1, the rest via p_2. The case that there are more than 2 paths can be handled similarly.

	Square		Cooler		Smiley		Tower	
	Diff	AWF	Diff	AWF	Diff	AWF	Diff	AWF
Γ_u	1.351	1.340	**1.289**	**1.289**	1.387	1.411	1.341	1.320
Γ_{srd}	1.364	**1.320**	1.302	**1.289**	1.393	**1.340**	1.344	1.311
Γ_α	1.353	1.344	1.294	1.296	1.387	1.368	1.340	1.336
$\Gamma_{rd} + \Gamma_\alpha$	1.361	1.328	1.296	**1.289**	1.390	1.374	1.765	**1.300**

Table 2. AWF in comparison to Diffusion.

Table 2 shows a comparison of *AWF* and Diffusion. Again, several different element rating functions are chosen to guide the migration by weighting the edges of the quotient graph. The values in Table 2 are mean values of all results for a given problem and a given flow algorithm and edge weighing function. It

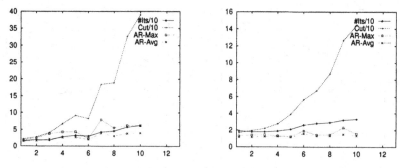

Fig. 7. Development of AR, Cut and #Iterations for domain "Square".

can be observed that an improvement in the achievable AR of around 5% is possible by weighing the edges properly (Γ_u is the non-weighted case). *AWF* behaves generally better than Diffusion, sometimes the improvements are large ($\approx 25\%$), in most cases they are around 5%. Γ_{srd} turns out to be a good edge weighing function, and also the use of Γ_α sometimes helps. More comprehensive evaluations of the different methods can be found in [26].

6 Results

Figure 7 shows the development of AR, Cut and number of iterations of the DD-PCG solver over a number of refinement levels of the Domain "Square" (shown in Fig. 1) refined in a number of steps from 736 to 15421 elements. Experiments using regular meshes and regular domains show an increase in the number of iterations by one when the number of elements are doubled. For irregular meshes such as those used here, one can not hope for such a good convergence behavior. The left picture shows the behavior if Γ_{cut} is used as element rating function, the right if $\Gamma_{srd} + \Gamma_\alpha$ is used (and note that the scaling on the y-axis is different for the two graphs). It can be observed that the number of iterations grows from 18 to 57 if Cut is the optimization goal, while if shape is optimized, the number of iterations for the final mesh is only 28.

Generally, AR expresses the development of the number of iterations much better than Cut. We conclude from the experiments that an optimization according to AR is reasonable.

We also compare the results of the AR-optimization in *PadFEM* to those achievable with a state-of-the-art load balancer for adaptive meshes. *Jostle* has been developed at the University of Greenwich and is a universal partitioning and load balancing tool for unstructured mesh computations [22,31,33]. It includes multi-level techniques for sequential partitioning as well as for parallel load balancing. The element migration decision uses Γ_{cut} as rating function. Together with a special border optimization based on KL and with sophisticated tie-breaking methods, *Jostle* is able to generate and maintain well shaped partitions, even if the Aspect Ratio is not directly considered.

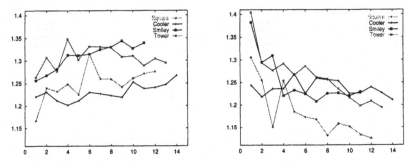

Fig. 8. The development of the AR for increasing levels of refinement.

Figure 8 shows the development of AR over typical runs of (solve/refine/balance...) for our four test examples. The left part shows the results if *Jostle* (without multi-level coarsening – the JOSTLE-D configuration described in [33]) is used as load balancer, the right gives the results if the AR-optimization of *PadFEM* is used (with best possible parameter combinations). It can be observed that the AR is decreasing with increasing numbers of elements if the *PadFEM*-balancer is used. The opposite is true if *Jostle* determines the elements to be moved. This shows that our strategy is able to operate as high-quality load balancer in adaptive environments maintaining well shaped subdomains over a long period of time. Standard load balancers like *Jostle* (but with *Metis*, it would be just the same) are not directly able to be used over a longer period without complete repartitionings from time to time - at least if the partition Aspect Ratio is a measure of importance. Numerical results show that *PadFEM* is able to improve the Aspect Ratio of mesh partitions by around 10% over a state-of-the-art general purpose load balancing tool.

Acknowledgements

We thank the *PadFEM*-team [24] in Paderborn for all their help. A. Frommer, B. Monien and S. Muthukrishnan provided ideas for parts of the load balancing.

References

1. R.K. Ahuja, T.L. Magnanti, J.B. Orlin. *Network Flows*. Prentice Hall, 1993.
2. S. Blazy, W. Borchers, U. Dralle. *Parallelization methods for a characteristic's pressure correction scheme*. Notes on Numerical Fluid Mechanics, 1995.
3. S. Blazy et al. *Parallel Adaptive PCG*. In: B.H.V. Topping (ed), *Advances in Computational Mechanics....* Civil-Comp Press, Edinburgh, 1998.
4. J.E. Boillat. Load Balancing and Poisson Equation in a Graph. *Concurrency - Practice and Experience* 2(4), 289-313, 1990.
5. J.H. Bramble, J.E. Pasciac, A.H. Schatz. The construction of preconditioners for elliptic problems by substructuring I.+II., *Math. Comp.*, 47+49, 1986+87.
6. T.N. Bui, S. Chaudhuri, F.T. Leighton, M. Sisper. Graph Bisection Algorithms with Good Average Case Behaviour. *Combinatorica* 7(2), 171-191, 1987.
7. M. Burghardt, L. Laemmer, U. Meissner. *Parallel adaptive Mesh Generation*. Euro-Conf. Par. Comp. in Computat. Mech., Civil-Comp Press, Edinburgh, 45-52, 1997.

8. N. Chrisochoides et al. *Automatic Load Balanced Partitioning Strategies for PDE Computations.* ACM Int. Conf. on Supercomputing, 99-107, 1989.
9. G. Cybenko. Load Balancing for Distributed Memory Multiprocessors. *J. of Parallel and Distributed Computing* (7), 279-301, 1989.
10. R. Diekmann, A. Frommer, B. Monien. *Nearest Neighbor Load Balancing on Graphs.* 6th Europ. Symp. on Algorithms (ESA), Springer LNCS, 1998.
11. R. Diekmann, D. Meyer, B. Monien. Parallel Decomposition of Unstructured FEM-Meshes. *Concurrency - Practice and Experience*, 10(1), 53-72, 1998.
12. R. Diekmann, B. Monien, R. Preis. *Load Balancing Strategies for Distributed Memory Machines.* Techn. Rep. tr-rsfb-96-050, CS-Dept., Univ. of Paderborn, 1997.
13. R. Diekmann, S. Muthukrishnan, M.V. Nayakkankuppam. *Engineering Diffusive Load Balancing Algorithms....* IRREGULAR, Springer LNCS 1253, 111-122, 1997.
14. R. Diekmann, R. Preis, F. Schlimbach, C. Walshaw. *Aspect Ratio for Mesh Partitioning.* Euro-Par'98, Springer LNCS, 1998.
15. C. Farhat, N. Maman, G. Brown. Mesh Partitioning for Implicit Computations via Iterative Domain Decomposition:.... *J. Numer. Meth. Engrg.*, 38:989-1000, 1995.
16. M.R. Garey, D.S. Johnson: *Computers and Intractability.* W.H. Freeman, 1979.
17. B. Ghosh, S. Muthukrishnan, M.H. Schultz. *First and Second Order Diffusive Methods for Rapid, Coarse, Distributed Load Balancing.* ACM-SPAA, 72-81, 1996.
18. G. Horton. A multi-level diffusion method... *Parallel Computing* 19:209-218, 1993.
19. Y.F. Hu, R.J. Blake. *An optimal dynamic load balancing algorithm.* Techn. Rep. DL-P-95-011, Daresbury Lab., UK, 1995 (to appear in CPE).
20. M.T. Jones, P.E. Plassmann. *Parallel Algorithms for the Adaptive Refinement and Partitioning of Unstructured Meshes.* Proc. IEEE HPCC, 478-485, 1994.
21. M.T. Jones, P.E. Plassmann. Parallel Algorithms for Adaptive Mesh Refinement. *SIAM J. Scientific Computing*, 18, 686-708, 1997.
22. *Jostle Documentation.* http://www.gre.ac.uk/~c.walshaw/jostle/
23. L. Oliker, R. Biswas. *Efficient Load Balancing and Data Remapping for Adaptive Grid Calculations.* Proc. 9th ACM SPAA, 33-42, 1997.
24. *PadFEM Documentation.* http://www.uni-paderborn.de/cs/PadFEM/
25. K. Schloegel, G. Karypis, and V. Kumar. Multilevel Diffusion Schemes for Repartitioning of Adaptive Meshes. *J. Par. Dist. Comput.*, 47(2):109–124, 1997.
26. F. Schlimbach. *Load Balancing Heuristics Optimizing Subdomain Aspect Ratios for Adaptive Finite Element Simulations.* MS-Thesis, Univ. Paderborn, 1998.
27. N. Touheed, P.K. Jimack. *Parallel Dynamic Load-Balancing for Adaptive Distributed Memory PDE Solvers.* 8th SIAM Conf. Par. Proc. for Sc. Computing, 1997.
28. D. Vanderstraeten, R. Keunings, C. Farhat. *Beyond Conventional Mesh Partitioning Algorithms....* 6th SIAM Conf. Par. Proc. for Sc. Computing, 611-614, 1995.
29. D. Vanderstraeten, C. Farhat, P.S. Chen, R. Keunings, O. Zone. A Retrofit Based Methodology for the Fast Generation and Optimization of Large-Scale Mesh Partitions:.... *Comput. Methods Appl. Mech. Engrg.*, 133:25-45, 1996.
30. R. Verfürth. *A Review of a posteriori Error Estimation and Adaptive Mesh-Refinement Techniques.* John Wiley & Sons, Chichester, 1996.
31. C. Walshaw, M. Cross. Mesh Partitioning: A Multilevel Balancing and Refinement Algorithm. Tech. Rep. 98/IM/35, University of Greenwich, London, 1998.
32. C. Walshaw, M. Cross, and M. Everett. A Localised Algorithm for Optimising Unstructured Mesh Partitions. *Int. J. Supercomputer Appl.*, 9(4):280–295, 1995.
33. C. Walshaw, M. Cross, and M. Everett. Parallel Dynamic Graph Partitioning for Adaptive Unstructured Meshes. *J. Par. Dist. Comput.*, 47(2):102–108, 1997.
34. O.C. Zienkiewicz. *The finite element method.* McGraw-Hill, 1989.

Parallelization of an Unstructured Grid, Hydrodynamic-Diffusion Code *

Aleksei I. Shestakov and Jose L. Milovich

Lawrence Livermore National Laboratory
Livermore CA 94550

Abstract. We describe the parallelization of a three dimensional, unstructured grid, finite element code which solves hyperbolic conservation laws for mass, momentum, and energy, and diffusion equations modeling heat conduction and radiation transport. Explicit temporal differencing advances the cell–based gasdynamic equations. Diffusion equations use fully implicit differencing of nodal variables which leads to large, sparse, symmetric, and positive definite matrices. Because of the unstructured grid, the off-diagonal non-zero elements appear in unpredictable locations. The linear systems are solved using parallelized conjugate gradients. The code is parallelized by domain decomposition of physical space into disjoint subdomains (SDs). Each processor receives its own SD plus a border of ghost cells. Results are presented on a problem coupling hydrodynamics to non-linear heat conduction.

1 Introduction

We describe the parallelization of ICF3D [1], a 3D, unstructured grid, finite element code written in C++. The ICF3D mesh consists of an arbitrary collection of hexahedra, prisms, pyramids, and/or tetrahedra. The only restriction is that cells share like-kind faces. We parallelize by first decomposing the physical domain into a collection of disjoint subdomains (SDs), one per processing element (PE). The decomposition tags each cell with the PE number which will "own" it. A collection of cells owned by a PE comprises the PE's SD. A cell owned by another PE and which shares at least one vertex with an owned cell is called a *ghost* cell. Each PE receives a terse description of only its SD plus a layer of ghost cells. The decomposition is especially suited to distributed memory architectures (DMP) such as the CRAY T3E. However, it may also be used on shared memory processors (SMP). ICF3D is portable; it runs on uniprocessors and massively parallel platforms (MPP). A single program multiple data (SPMD) model is adopted. In ICF3D, three levels of parallelization difficulties arise:

1. Embarrassingly parallel routines such as equation-of-state function calls which do not require any message passing since each cell is owned by only one PE.

---

* Work performed under the auspices of the U.S. Department of Energy by the Lawrence Livermore National Laboratory under contract number W-7405-ENG-48.

2. Straightforward parallelization of temporally explicit algorithms such as the hydrodynamic package.
3. Difficult problems requiring global communication, e.g., the solution of the linear systems which are the discretization of the diffusion equations.

The mesh consists of cell, face, and vertex *objects*. Since the input files assign PE ownership only to the cells, some faces and vertices lie on inter-PE "boundaries." Physics modules such as the one advancing the hydrodynamic equations, which update cell-centered data, also compute face-centered quantities, e.g., fluxes. If a face lies on an inter-PE boundary, the flux across it is computed by the two PEs which own the cells on either side. Fluxes are computed after the PEs exchange appropriate information to ensure that both obtain the same flux. Modules such as the diffusion solver update vertex-centered data. These equations are advanced by standard finite element (FE) techniques which lead to large, sparse, symmetric positive definite (SPD) linear systems that are solved using preconditioned conjugate gradient (CG) methods. The assembly of the linear systems requires integrating over cells, i.e., a cell-centered computation, and is done by each PE over only its owned cells. However, once the linear system has been completely assembled and properly distributed among the PEs the calculation is vertex-centered. A principal result of this paper shows how to assemble and solve such systems with the restriction that each PE sees only its SD and the surrounding ghost cells.

The PEs communicate using message passing function libraries. Two types are available, MPI and the native CRAY SHMEM libraries. The former is portable; it is available on both SMPs and DMPs.

In ICF3D two types of communication arise, global and point-to-point (PtP). An example of the former is the calculation of a new time step Δt. First, each PE loops through its cells or vertices and finds an acceptable value, then a *global reduction* function forms a single scalar and distributes it to all PEs. In PtP communication, PE[i] exchanges messages only with those PEs that own its ghost cells. For such exchanges, ICF3D relies on special "message passing objects" (MPO) which are constructed during the initialization of the run. The MPO *constructor* relies on mesh connectivity information that ICF3D computes as it builds the mesh objects. The actual calls to MPI (or SHMEM) functions are made by the MPO *member functions*.

In the next section, we discuss what is required for initialization. In Sect. 3 we describe the parallelization techniques required by the hydrodynamic module. Section 4 deals with analogous issues for the diffusion packages. Section 5 displays results on a problem which couples the explicit hydrodynamic and implicit diffusion schemes. Concluding remarks appear in Sect. 6.

2 Initialization

Before describing the procedures specific to MPP simulations, we discuss those required to initialize any run, even those for uniprocessors. The input and output files describe the mesh in the AVS UCD format [3] which uses two lists. The first, of length N_v, is the indexed list of vertices. Each vertex is specified by a 3-tuple – the three coordinates of the vertex.[1] The second list, N_c long, is the indexed list of cells. Each entry in the cell list contains a string denoting the cell's type, e.g., hex and a properly ordered list of indices into the vertex list. In addition, each vertex and cell has a global sequence number (GSN) and each cell has an assigned PE number. The GSN is unchanged during the run. The vertices, cells, and to-be-constructed faces also have a local sequence number, but these are relevant only to the code itself during the run.

After reading the input file, the mesh is created by constructing the cell, vertex, and face objects. The objects are accessed by pointers. The objects also have their own interconnecting pointers.

For MPP runs, as the cells are read in, the cell objects are constructed so that the cell pointers first list the owned cells, then the ghost cells. There is also a considerable amount of sorting of cells and vertices to facilitate the construction of the MPOs. However, this effort is a small overhead in the eventual problem running time because the mesh connectivity, and hence the logical data of the MPOs, do not change during the course of the run.

3 Hydrodynamics

Two types of message exchanges, face and vertex centered, arise in the hydro-dynamic scheme. The scheme [4] is conceptually straightforward to parallelize since it is compact and temporally explicit, although it is second order in both space and time. The method is an extension of the Godunov scheme in which all variables are cell-based. If applied to the equation, $\partial_t f + \nabla \cdot F = 0$, the scheme integrates over Δt and a cell to advance the average value f_j of the j-th cell:

$$V_j \left(f_j^{n+1} - f_j^n \right) + \Delta t \int dA \cdot F^n = 0 \,,$$

where V_j is the cell volume, the superscript denotes the time level, and the area integral is a sum over the cell's faces. The face fluxes are solutions to Riemann problems whose initial conditions are the cell-based f^n on either side of the face. In 1D, this yields the explicit dependence,

$$f_j^{n+1} = \mathcal{F}(f_{j-1}^n, f_j^n, f_{j+1}^n) \,.$$

Hence, if an inter PE boundary separates the j-th and (j+1)-st cells, if PE[i] owns the j-th cell, and if the latest value f_{j+1}^n has been passed and loaded into the proper ghost cell, then PE[i] will compute the correct, new cell average.

The responsibility for the message passing lies with an MPO, which in this case is face-cell-centered. That is, both send and receive MPOs run through the same faces, but the sending MPO reads and packs data into a buffer from owned cells, while the receiving MPO unpacks a buffer and loads data into ghost cells.

---

[1] ICF3D may be run in either 3D Cartesian, cylindrical, or spherical coordinate systems.

The second order aspect of the scheme complicates the above procedure. Temporal accuracy is obtained by a two-step Runge-Kutta scheme which makes two passes through the coding. This implies two sets of message exchanges per time cycle, but does not cause any other complications. However, the spatial accuracy does complicate matters since in each cell, the dependent variables are non-constant and may have different values at each of the vertices. Across each face, the initial data of the Riemann problem are the values of the variables on the vertices of the cell adjoining the face. This data is obtained by following the pointer of the face, to the cell, and then to the correct vertex values.

The second type of message exchange is vertex-centered and is required by the limiting procedure which removes local extrema from the cell's vertex values. For example, after the DFE pressures are calculated, they are restricted (limited) to a range obtained from the average values of the adjacent cells. The procedure is also explicit with compact support. If the vertex lies on an inter-PE boundary, by definition, there is at least one ghost cell attached. Special MPOs collect and distribute the extremal values to all PEs that own cells attached to the vertex.

4 Diffusion

In contrast to the hydrodynamic module in which calls to the parallelization functions appear in several places, e.g., before computing fluxes, in the diffusion modules, the parallelization occurs after the equation is discretized, a "local" linear system assembled, and the system solver called.

In ICF3D diffusion equations arise in simulating heat conduction, or in the two versions of the diffusion approximation for radiation transport. In all cases the equation is of the form

$$G\, \partial_t f = \nabla\cdot(D\nabla f) + S - Lf\ , \tag{1}$$

where G, D, S, $L \geq 0$. The unknown function is approximated as,

$$f(x, t^n) = \sum_j \phi_j(x) f_j^n\ ,$$

where, if x_i is a vertex, $\phi_j(x_i) = \delta_{ij}$ is the usual basis function. To advance Eq. (1), we use implicit temporal differencing and obtain,

$$(G' + L' - \nabla\cdot D\nabla)f^n = G' f^{n-1} + S'\ . \tag{2}$$

Note that Δt is absorbed into D', L', and S'. Next, Eq. (2) is multiplied by a basis function ϕ_i and integrated over the "domain." For MPP applications, the relevant domain is the SD of the PE, i.e., only its owned cells. Thus, the index i of the ϕ_i function ranges over the vertices of only the owned cells. Each multiplication by ϕ_i corresponds to one row of the linear system, $Af = b$ for the nodal unknowns f_j^n. The matrix is SPD, but the system is incomplete since equations corresponding to unknowns on the inter-PE boundary do not include integrals over ghost cells.

The MPP methodology is incorporated into the solver which, for MPP runs, first calls another routine that assembles the *distributed* linear system.

4.1   The MPP Distributed System

Since the linear systems are vertex-based, we extend the concept of PE ownership to the vertices. If all cells adjoining a vertex are owned by PE[i], we let PE[i] own the vertex. This procedure leaves ambiguous the ownership of vertices on inter-PE boundaries and those on the "exterior" of the ghost cells. Ownership of the former may be determined without requiring message passing. The simplest algorithm is for each PE to survey the ownership of all cells attached to the vertex and assign the vertex to the PE of lowest number. Assigning ownership of the exterior nodes requires message-passing since a PE does not have access to all cells that are attached to them. Since a PE unequivocally knows who owns the vertices attached to all its owned cells, during the initialization phase, each PE receives a message from the owners of its ghost cells regarding the ownership of this PE's exterior vertices. Although the coding for this procedure is complicated, it is called only once during initialization.

One essential operation within the CG iterations is MatVec, the multiplication of a matrix by a vector. On MPPs, after the system has been distributed, the matrices are rectangular. The number of rows equals the number of owned vertices and the columns correspond to the number of vertices linked to the owned vertices. The matrices are stored in compressed row form to avoid storing zeroes.

To facilitate the assembly of the distributed system, the vertices are sorted into six types: T_1 to T_6. Before describing them, we define $\mathcal{V}$ to be the set of vertices owned by the PE, $\mathcal{W}$ to be the set of vertices not owned by the PE, $\mathcal{S}(x)$ as the set of vertices connected to the vertex x by the stencil, and as *exterior* the vertices of ghost cells which do not lie on inter-PE boundaries. Thus,

- $T_1 = \{x \mid x \in \mathcal{V}. \ \forall y \in \mathcal{S}(x), \ y \in \mathcal{V}\}$
- $T_2 = \{x \mid x \in \mathcal{V}. \ \exists y \in \mathcal{S}(x), \ y \in \mathcal{W}\}$
- $T_3 = \{x \mid x \in \mathcal{V}. \ x \text{ is on inter-PE bdry.}\}$
- $T_4 = \{x \mid x \in \mathcal{W}. \ x \text{ is on inter-PE bdry.}\}$
- $T_5 = \{x \mid x \text{ is exterior. } \exists y \in T_3, \ x \in \mathcal{S}(y)\}$
- $T_6 = \{x \mid x \text{ is exterior. } \forall y \in \mathcal{S}(x), \ y \in \mathcal{W}\}$

The six types stem from the requirements for assembling the distributed system and for performing a MatVec. For the T_1 and T_2 vertices, the PE can compute the entire row of matrix coefficients without input from other PEs. However, on T_2 vertices, before completing a MatVec, the PE needs the latest value on some T_4 vertices. On T_3 vertices, the PE needs input from other PEs for completing the calculation of the row and for a MatVec operation. Linear equations which the PE computes on its T_4 vertices are sent to the PE which owns them. The T_6 vertices are not needed in the diffusion module.

Before calling the solver, each PE computes a partial linear system, $Ax = b$, after performing the required finite element integrations over the owned cells. If we index the matrix and vector elements by type, then

$$A = \begin{pmatrix} A_{11} & A_{12} & A_{13} & 0 \\ A_{12} & A_{22} & A_{23} & A_{24} \\ A_{13} & A_{23} & A_{33} & A_{34} \\ 0 & A_{24} & A_{34} & A_{44} \end{pmatrix}, \quad x = \begin{pmatrix} x_1 \\ x_2 \\ x_3 \\ x_4 \end{pmatrix}, \tag{3}$$

and similarly for b. In Eq. (3) A_{ij} denotes the, not necessarily square, matrix of coefficients of interactions between types T_i and T_j vertices and similarly x_i are the type T_i vector elements.

After computing the incomplete linear system whose matrix is given by Eq. (3), for MPP runs, the solver first calls certain MPOs which assemble the distributed system. Each row of T_4 vertices is sent to the PE which owns the vertex. In the context of how to compute a MatVec $y = Ax$, if the subscripts a and b divide the vertices into types $T_{1,2,3}$ and $T_{4,5}$ respectively, we have

$$y_a = A_a x_a + A_b x_b \tag{4}$$

where,

$$x_a^{\mathrm{T}} = (x_1, x_2, x_3), \quad x_b^{\mathrm{T}} = (x_4, x_5),$$

$$A_a = \begin{pmatrix} A_{11} & A_{12} & A_{13} \\ A_{12} & A_{22} & A_{23} \\ A_{13} & A_{23} & A_{33} + A'_{33} \end{pmatrix}, \quad \text{and } A_b = \begin{pmatrix} 0 & 0 \\ A_{24} & 0 \\ A_{34} + A'_{34} & A'_{35} \end{pmatrix}. \tag{5}$$

In Eq. (5) the primes denote matrix coefficients which have been computed by other PEs, i.e., on their T_4 vertices, and message-passed to this PE.

4.2   The Parallel Solver

The resulting linear equations are solved using preconditioned conjugate gradients (PCG). Two types of preconditioners are available: n-step Jacobi, and a parallel version of incomplete Cholesky (IC). Each iterative step of CG requires

- Three SAXPY, i.e., $\alpha x + y$ where α is scalar and x and y are vectors
- Two dot products
- One MatVec
- Solving the preconditioned system,

$$Pz = r . \tag{6}$$

The SAXPY operations do not require any message passing.

The dot products are calculated using a global reduction function.

The MatVec is computed according to the splitting defined in Eq. (5) except we interleave message passing and computation. The procedure is as follows. The PEs first call asynchronous receive functions and then halt at a barrier. Next, the PEs call the ready-to-send functions and, without waiting for the messages to arrive, calculate the first part of Eq. (4), i.e., $A_a x_a$. Afterwards, the PEs halt at another barrier before adding $A_b x_b$ to the result.

The key behind PCG is a preconditioning matrix P which closely resembles A and at the same time renders Eq. (6) easy to solve. Both the IC and n-step Jacobi preconditioners can be cast in the form of generalized polynomial preconditioners in which A is first factored into two parts,

$$A = M - N = M(I - G) , \quad G \doteq M^{-1}N .$$

If $||G|| < 1$, then the inverse may be approximated by

$$A^{-1} = (I - G)^{-1}M^{-1} \approx (I + G + \cdots + G^{i-1})M^{-1} .$$

We now define P as the inverse of the approximation, i.e.,

$$P \doteq M \left(\sum_{i=0}^{j-1} G^i \right)^{-1} \tag{7}$$

where $G^0 = I$.

For n-step Jacobi, $M = \mathrm{diag}(A)$ and we set $j = n$ in Eq. (7); 1-step is the easily parallelizable diagonal scaling preconditioner. If $n > 1$, the preconditioner requires $n - 1$ operations, each of which consists of a multiplication by the off-diagonal elements of A, then by M^{-1}.

Unfortunately, in our applications n steps are usually no better than simple diagonal scaling and can be significantly worse than 1 step – see ref. [1].

Our favored preconditioner is a parallel IC variant which we now describe. In the context of Eq. (7) we set $j = 1$ and let $P = LDL^T$, where L is lower triangular with unit diagonal, D is diagonal, and the sparsity pattern of L matches that of A except that we do not allow links to types T_4 and T_5 vertices. In other words, we form an incomplete factorization of matrix A_a in Eq. (5). Since A_a links only the owned nodes, the preconditioning step does not inhibit parallelization, nor does it require any message passing.

On each PE, our parallel IC preconditioner is equivalent to an incomplete decomposition of the underlying diffusion equation on the $T_{1,2,3}$ nodes with homogeneous Dirichlet data specified on the non-owned $T_{4,5}$ vertices. In that light, a possible improvement may be to replace the homogeneous data with "stale" values of the z vector on the $T_{4,5}$ vertices.

We end this section with a comparison of execution times of runs using two preconditioners, IC and 1-step Jacobi. Table 1 shows the effect of increasing the number of PEs while keeping the mesh size fixed. As expected, our parallel ICCG degrades as the SDs get smaller while Jacobi scales with the number of PEs. Although the problem size is small, the parallel IC preconditioner is superior to 1-step Jacobi.

5 Point Explosion

In this section, we present a calculation that uses the parallelization routines described above. Although the problem is spherically symmetric, we run in 3D and compare results to a 1D spherical calculation.

Table 1. Thermal Wave Problem. Execution time for different preconditioners and PE configurations, $32\times8\times8$ cells, 2337 vertices.

Preconditioner	4 PEs 512 cells/PE	16 PEs 128 cells/PE
ICCG	163.	82.
1-step Jacobi	404.	129.

The problem, suggested by Reinicke and Meyer-ter-Vehn [6] and later analyzed in ref. [7], consists of the sudden release of energy in the center of a cold, constant-density gas. The problem is a combination of the well-known hydrodynamic point explosion[8] and the spherically expanding thermal wave [9].

The equations of interest are the Euler equations with heat conduction:

$$\partial_t \rho + \nabla\cdot(\rho\mathbf{u}) = 0 \,,$$
$$\partial_t(\rho\mathbf{u}) + \nabla\cdot(\rho\mathbf{u}\mathbf{u} + p) = 0 \,,$$
$$\partial_t(\rho E) + \nabla\cdot(\mathbf{u}(\rho E + p)) = -\nabla\cdot H \,, \tag{8}$$

where the heat flux has the form,

$$H = -\chi\nabla T \,, \quad \text{where} \quad \chi = \chi_0 T^b \,,$$

and where χ_0, and b are constants. In Eqs. (8), $\mathbf{u}$, ρ, p, T, and E denote the velocity, density, pressure, temperature, and total specific energy respectively. The thermodynamic variables satisfy the ideal gas equation of state,

$$p = (\gamma - 1)\rho e = (\gamma - 1)\rho c_V T \,.$$

At $t = 0$, the energy is all internal and is concentrated at the origin,

$$\rho E|_{t=0} = \rho e|_{t=0} = \mathcal{E}_0\,\delta(r) \,,$$

while the quiescent gas is at constant density, $\rho|_{t=0} = 1$. We set,

$$\gamma = 7/5 \,, \quad b = 5/2 \,, \quad \chi_0 = c_V = 1 \,, \quad \text{and} \quad \mathcal{E}_0 = 0.76778 \,.$$

The numerical value for $\mathcal{E}_0$ was calculated a posteriori after we initialized the central T to a large value.

The solution may be estimated by separate analyses of the corresponding pure hydrodynamic and pure diffusion problems, both of which possess similarity solutions. However, in this case, the coupled problem is not self-similar.

The pure diffusion solution is characterized by a temperature front r_f which increases with time. The pure hydrodynamic problem is characterized by an infinite strength shock whose position r_s also grows with time. For the parameters chosen,

$$r_s \propto t^{2/5} \quad \text{and} \quad r_f \propto t^{2/19} \,.$$

Thus, in early times, diffusion dominates; in late times, hydrodynamics.

In ref. [7], the author computes an approximate time $t_\times$ and radius $r_\times$ for the two fronts to intersect. For the parameters chosen,

$$t_\times = 0.8945 \text{ and } r_\times = 0.9371 .$$

Figure 1 displays ρ and T at the late-time, hydrodynamically dominated regime, $t = 1.8$; at $r \approx 1.28$ we have a strong shock.

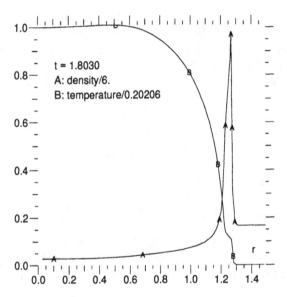

Fig. 1. Density and temperature vs. r at late time. T is normalized by $T(r = 0)$, ρ is normalized by the value across an infinite strength shock.

The results depicted in Fig. 1 were obtained by running ICF3D in "1D" mode, i.e., with only 1 cell in the transverse directions and 100 radial cells. We now display results of a 3D run in Cartesian coordinates on an unstructured grid on 64 PEs. The mesh consists of 11580 tetrahedra and 2053 nodes and is created by the LaGriT code [5] obtained from Los Alamos National Laboratory. The radial direction is discretized into initially 16 "spherical shells" of uniform width $\Delta_r = 0.125$. We partition the mesh using METIS [2]. Figure 2 displays the outside of the spherical domain. The innermost "sphere" consists of 20 tetrahedra all connected at the origin.

Figures 3 and 4 respectively display ρ and T across the $Z = 0$ plane at $t = 1.8$. The ρ result in Fig. 3 shows jaggedness in the transverse direction and an inability to reach the high compressions attained by the 1D result. Both errors are due to the coarseness of the grid and to the usage of tetrahedral cells. The result in Fig. 4 compares more favorably with Fig. 1. However, none of

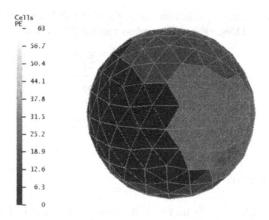

Fig. 2. Exterior of spherical domain. Colors represent PE numbers.

the discrepancies between the 3D and 1D results are due to the parallelization. There is no sign of the subdivision of the domain into 64 SDs.

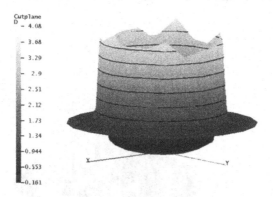

Fig. 3. Density at $t = 1.802$ across $Z = 0$ plane.

Lastly, we present a scalability study by varying the number of PEs while keeping the mesh size fixed. First, we define the parallelization efficiency E_N,

$$E_N = \frac{t_{N/2}}{2\,t_N}$$

where t_N is the execution time for N PEs. Table 2 lists E_N derived from the execution times of runs on an IBM SP2 for the above unstructured grid problem. Note that E_N remains close to 1 as N increases.

Table 2. Number of PEs N, execution time t_N, and parallelization efficiency E_N for a fixed size problem (11580 cells, 2053 nodes) run for 200 cycles.

N	t_N (sec)	E_N
8	1628.02	-
16	882.77	.922
32	489.78	.901
64	277.08	.884

6 Summary and Conclusion

We have presented a scalable method for parallelizing a 3D, unstructured grid, finite element code. The method is a SPMD model targeted for distributed memory architectures, but also works on shared memory computers. We decompose physical space into a collection of disjoint SDs, one per PE. The input files tag cells with a PE number. The code forms an analogous designation for the vertices. The SDs are surrounded by a layer of ghost cells which are used to store data that is owned, i.e., computed, by other PEs. Explicit calls to message passing functions communicate the data amongst the PEs. The calls are made by special MPOs which contain information that describes the PE network.

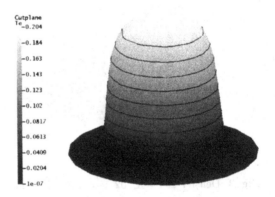

Fig. 4. Normalized T at $t = 1.802$ across $Z = 0$ plane.

We have demonstrated the method on a test problem requiring both temporally explicit and implicit schemes for partial differential equations. The concept of ghost cells leads to a straightforward parallelization of the explicit hydrodynamic scheme. The MPOs insure that the data computed by other PEs is placed in the expected locations.

Parallelization of the implicit discretization of the diffusion equations consists of a wrapper around the call to the linear system solver. For an MPP run, the linear system generated by the uniprocessor coding is incomplete since it is formed by integrating over only the owned cells. The parallelization completes the system by sending equations on non-owned vertices to the PE that owns them.

Since the linear systems are large, sparse, and SPD, they are solved using a parallelized PCG algorithm in which our noteworthy contribution is the parallel IC preconditioner. In a sense, it is even more incomplete than typical IC since our decomposition ignores links to non-owned vertices. Its utility depends on PEs having large SDs. In all cases studied, the number of CG iterations required using the parallel IC is less than that required by Jacobi. However, if the SDs are very small, and/or the matrix condition number is small, Jacobi is faster. In our applications, the parallel IC preconditioner has been very effective.

The important aspect of our parallelization strategy is its consistent reliance of domain decomposition of physical space. We are now parallelizing the laser deposition module along the same lines. Preliminary results are encouraging.

References

1. A. I. Shestakov, M. K. Prasad, J. L. Milovich, N. A. Gentile, J. F. Painter, G. Furnish, and P. F. Dubois, "The ICF3D Code," Lawrence Livermore National Laboratory, Livermore, CA, UCRL-JC-124448, (1997), submitted to *Comput. Methods Appl. Mech. Engin.*

2. G. Karypis and V. Kumar, "A fast and high quality multilevel scheme for partitioning irregular graphs," Technical Report TR 95-035, Department of Computer Science, Univ. Minn., 1995. To appear in *SIAM Journal on Scientific Computing 1997.* A short version appears in Intl. Conf. on Parallel Processing 1995. The METIS program is available on the web from: http://www-users.cs.umn.edu/ karypis/metis/metis/ main.html

3. *AVS Developer's Guide*, Advanced Visual Systems, Inc., Release 4, May 1992, p. E-1, 300 Fifth Ave., Waltham MA 02153.

4. D. S. Kershaw, M. K. Prasad, M. J. Shaw, and J. L. Milovich, *Comput. Methods Appl. Mech. Engin.,* **158** p. 81 (1998).

5. www.t12.lanl.gov/ lagrit.

6. P. Reinicke and J. Meyer-ter-Vehn, *Phys. Fluids* A **3** (7), p. 1807 (1991).

7. A. I. Shestakov, "Simulation of two Point Explosion-Heat Conduction Problems with a Hydrodynamic-Diffusion Code," Lawrence Livermore National Laboratory, Livermore, CA, UCRL-JC-130700, (1998), submitted to *Phys. Fluids* A.

8. L. D. Landau and E. M. Lifshitz, *Fluid Mechanics*, 2nd Ed., Pergamon Press, Oxford p. 404 (1987).

9. Ya. B. Zel'dovich and Yu. P. Raizer, *Physics of Shock Waves and High-Temperature Hydrodynamic Phenomena*, Vol. II, Academic Press, p. 668 (1966).

Acknowledgement

We gratefully acknowledge the help of D. George and A. Kuprat of Los Alamos National Laboratory for supplying the LaGriT mesh generation code.

Exchange of Messages of Different Sizes

A. Goldman[1], D. Trystram[1], and J. Peters[*][2]

[1] LMC-IMAG,   100, rue des Mathématiques
38041 Grenoble,   France   {goldman,trystram}@imag.fr
[2] School of Computing Science,   Simon Fraser University
Burnaby, BC V5A 1S6,   Canada   peters@cs.sfu.ca

Abstract. In this paper, we study the exchange of messages among a set of processors linked through an interconnection network. We focus on general, non-uniform versions of all-to-all (or complete) exchange problems in asynchronous systems with a linear cost model and messages of arbitrary sizes. We extend previous complexity results to show that the general asynchronous problems are NP-complete. We present several approximation algorithms and determine which heuristics are best suited to several parallel systems. We conclude with experimental results that show that our algorithms outperform the native all-to-all exchange algorithm on an IBM SP2 when the number of processors is odd.

1   Introduction

In this paper, we study the exchange of messages among a set of processors linked through an interconnection network. The goal is to minimize the time to exchange all messages. The motivation for this work comes from several concrete problems for which parallel or distributed algorithms need to alternate periods of computing with periods of data exchange among the processing nodes, and for which messages of arbitrary length need to be exchanged.

One example is the solution of partial differential equations in which computing and data exchange phases are repeated until a solution is found. The amount of data exchanged among the processors depends on the chosen initial problem partition, and is not necessarily uniform. Another example is the integration over time of N-body equations in molecular dynamics simulations [14]. The behavior exhibited in computations for problems like these can be modeled by the BSP model [19]. A good BSP implementation requires a good algorithm for message exchange and synchronization.

1.1   Related work

There are many papers devoted to message exchange and similar problems. One of the first was a study about scheduling file transfers in a complete network by Coffman, Garey, Johnson and Lapaugh [8]. They used an undirected weighted

---

[*] This work was done when the author was visiting LMC-IMAG in 1997.

multigraph model which did not allow either message forwarding or message splitting. They showed that computation of the optimal solution in the general case is NP-complete, and presented approximation algorithms, including some distributed algorithms, that guarantee near optimality.

Choi and Hakimi [7] investigated directed message transfer in complete networks using a weighted directed graph model in which the messages can be split in time, or among multiple arcs between two vertices. They described a polynomial algorithm to solve the problem when the start-up cost can be ignored. When the start-up cost is not negligible, the problem becomes NP-complete.

Bongiovanni, Coppersmith and Wong [4] proposed an algorithm for a similar problem. They presented an optimum time slot assignment algorithm for a satellite system with a variable number of transponders. Their model can be seen as a message exchange in a complete network in which each processor can send messages to itself, and the messages can be sent in several not necessarily consecutive parts. Other authors have studied several variations [5, 1].

Several papers consider the message exchange problem for specific interconnection structures such as meshes [18, 9, 15], and hypercubes [3]. Three algorithms for message exchange in meshes and hypercubes are proposed in [2] - one for short messages, another for long messages, and a hybrid algorithm. Most of these results concern messages of the same size.

Gonzalez was interested in multi-message multicasting [12, 13]. He worked with a fully connected network and proposed solutions with, and without message forwarding. His solutions assume that all messages have the same size.

A similar problem is TDM/WDM scheduling in passive star fiber optics networks [17]. A passive star with N nodes supports C different wavelengths (generally, $C < N$), and each node has a transmitter and a receiver. Each transmitter can be tuned, within time Δ, to any wavelength, but each receiver is assigned to just one wavelength. Traffic demand is given as an $N \times N$ matrix $D = [d_{i,j}]$ where $d_{i,j}$ is the number of packets to be sent from node i to node j. The goal is to find a valid schedule in which there is at most one transmitter using each wavelength at each instant. The problem of finding a minimum time, non-preemptive schedule was shown to be NP-hard in [11] (for $\Delta = 0$ and $C > 2$) and in [17] (for $\Delta \geq 0$ and $C \geq 2$).

In [16], the authors summarize several techniques for many-to-many personalized communication including a fixed pattern algorithm, a distributed random scheduling algorithm, and a two-stage algorithm.

We study the message exchange problem on an asynchronous model which corresponds closely to actual distributed memory machines. In the next section, we describe the problem and our model precisely and establish some lower bounds. In Section 3, we recall previous complexity results and show that the general asynchronous problem is NP-complete. In Section 4, we adapt some approximation algorithms that were developed for other problems to our all-to-all exchange problems. We analyze the algorithms, give suggestions about which algorithms are best suited to different parallel systems, and present experimental results.

2 Problem

The message exchange problem (MEP for short) can be described by a weighted digraph, $dG(V, E)$ in which the vertices represent the processors, and the arcs represent the messages to be transmitted. Each arc $e = (u, v)$ is labeled with an integer $w(e)$ that represents the size of the message to be sent from vertex u to vertex v. If several messages are to be sent from u to v, then $w(e)$ is the sum of the message sizes. The goal is to minimize the time to exchange all messages.

The port capacity of each processor is represented by two integers. For each $v \in V$, v can transmit $c_t(v)$ messages and receive $c_r(v)$ messages simultaneously. We will concentrate on the minimization problem with port capacity restricted to one for both sending and receiving, i.e. $c_t(v) = c_r(v) = 1$, for each $v \in V$. We refer to the graph $dG(V, E)$ of a message exchange problem as a *message exchange graph* (MEG for short). It is worth observing that there is not necessarily any relationship between the MEG and the topology of the interconnection network.

In addition to the labelled digraph representation described above, we will use two other representations of MEGs - a bipartite graph representaion, and a matrix of integers. Given a digraph $dG(V, E)$ we construct a bipartite graph $bG(V' = V_1 \bigcup V_2, E')$. For each vertex $v \in V$ we have two vertices, $v_1 \in V_1$ and $v_2 \in V_2$, in V'. For each arc $e = (u, v) \in E$ there is a corresponding edge $e' = \{u_1, v_2\} \in E'$ with weight $w(e') = w(e)$.

A problem instance $dG(V, E)$ can also be represented as a square matrix A of size $|V|$, with $V = \{v_1, \ldots, v_n\}$. The elements of A are the weights on the arcs of $dG(V, E)$. If there is a message to send from v_i to v_j, then $a_{i,j} = w(v_i, v_j)$, and $a_{i,j} = 0$ otherwise.

2.1 Model

In this section we present the communication, transmission, and synchronization models. We assume that the network is fully connected. The models that we have chosen can be used to approximately model most current parallel machines with distributed memory.

Communication Model: The links between the processors are bidirectional, and each processor can use at most one link to transmit and one link to receive data at any given time.

Transmission Model: The time to send a message of size L from a processor u to another processor v, is $\beta + L\gamma$, where β is the start-up time and $\frac{1}{\gamma}$ is the bandwidth.

We will study two different message transmission properties: message splitting, and message forwarding. If message splitting is permitted, then a message can be partitioned and sent as several (not necessarily consecutive) transmissions. If message splitting is not permitted, then once the message transmission begins, it continues without interruption until completion. There are also two possibilities for message forwarding: either files must be transmitted directly between the source and the destination, or files can be forwarding along a path of intermediate processors.

Synchronization Model: The given algorithms assume an asynchronous model in which transmissions from different processors do not have to start at the same time. However, all processors start the algorithm at same time, that is, there is a synchronization before the start of the algorithm. The total time for an algorithm is the difference between the initial time and the completion time of the last processor.

2.2 Notation and lower bounds

In this section, we present theoretical bounds on the minimum number of communications and on the bandwidth for the general message exchange problem. The number of communications is determined by the number of start-ups, and the bandwidth depends on the message weights. We consider two cases - either message forwarding is permitted or it is not permitted.

Given a MEG $dG(V, E)$, we denote the in-degree of each vertex $v \in V$ by $\Delta_r(v)$, and the out-degree by $\Delta_s(v)$. Let $\Delta_r = \max_{v \in V}\{\Delta_r(v)\}$ and $\Delta_s = \max_{v \in V}\{\Delta_s(v)\}$.

Since our model does not assume any additional overhead to provide synchronization, we can not focus the study on steps as on a BSP-like model [19]. Instead, we consider the number of start-ups to be the number of start-ups that must be disjoint in time. If message forwarding is not allowed, we can compute the following straightforward bound on the number of start-ups.

Claim. To solve a message exchange problem on a digraph $dG(V, E)$ without message forwarding, the number of start-ups is at least $\max\{\Delta_s, \Delta_r\}$.

When message forwarding is allowed, a message $e = (u, v)$ can be routed along a directed path from u to v. In this case, a vertex can send one or more messages on the same link, using the same startup time, so we can derive a smaller lower bound as follows:

Claim. The number of time-disjoint start-ups needed to complete a personalized all-to-all exchange among n processors with message forwarding is at least $\lceil \log_2 n \rceil$.

Proof: This is easy to verify by observing the lower bound for broadcasting. As we assume a 1-port model, the number of informed nodes can at most double during each start-up time. $\square$

Given a MEG $dG(V, E)$, the bandwidth bounds are determined by two obvious bottlenecks for each vertex - the time to send the messages and the time to receive the messages. Each vertex v has to send the messages $S_v = \{e \in E | e = (v, u), u \in V\}$ and this takes time at least $t_s = \max_{v \in V} \sum_{e \in S_v} w(e)\gamma$. The time for v to receive the messages $R_v = \{e \in E | e = (u, v), u \in V\}$ is at least $t_r = \max_{v \in V} \sum_{e \in R_v} w(e)\gamma$. The next two claims are straightforward.

Claim. If message forwarding is permitted, then the time to complete a personalized all-to-all exchange is at least $\max\{t_s, t_r\}$.

Claim. If message forwarding is not allowed, the time to complete a personalized all-to-all exchange is at least $\max\{\Delta_r, \Delta_s\}\beta + \max\{t_s, t_r\}\gamma$.

3 Complexity

In this section, we will present some complexity results about the message exchange problem. To give a clear mathematical meaning to our goal of minimizing the time, we will formally define the scheduling function.

Two different possibilities corresponding to our classification of problems will be studied - the case with no preemptive message transmissions, and the case where the messages can be split and sent as sub-messages. Both analyses assume that message forwarding is not allowed.

3.1 Indivisible messages

Given a MEG $dG(V, E)$, a schedule is defined to be a function $s : E \to [0, \infty)$ that assigns to each arc e, a starting time $s(e)$, in such a way that at each instant, for every vertex v, at most one of the arcs of v is used for transmissions and at most one arc is used for receptions. An arc e is used from the starting time $s(e)$ to the finishing time $s(e) + \beta + \gamma w(e)$.

The length of a schedule s (also called *makespan*) is the latest finishing time, that is, $\mathrm{mak}(s) = \max_{e \in E} s(e) + \beta + \gamma w(e)$. The goal is to find the minimum schedule length $t_{\mathrm{opt}} = \min_{s \in S} \mathrm{mak}(s)$, where S is the set of all feasible schedules.

Property 1. Given a schedule s, and an instant t, the active arcs e such that $s(e) \leq t < s(e) + (\beta + \gamma w(e))$ form a subgraph of dG, where for each vertex v, the in-degree and out-degree are each at most one.

It is worth observing that if all the arc weights are the same, the scheduling problem can be viewed as finding a set of subgraphs $G_1, \ldots, G_k$ of dG, such that each arc $e \in E$ belongs to only one subgraph G_i.

In [8], the authors studied the file transfer problem. Their model uses a graph $G(V, E)$ where the edge weights represent the amount of time to transfer a file between the endpoints of the edges, and the start-up time is not considered. They do not use the concept of send or receive, and each vertex $v \in V$, has an integer port constraint $p(v)$. Their goal was to solve the problem in minimum time using a scheduling function similar to ours, but without any distinction between transmission and reception.

Finding the minimum makespan for file transfer problems is NP-complete [8]. The problem is NP-complete even when all the edge weights are equal to one $(w(e) = 1, e \in E)$ and all port capacities are equal to one $(p(v) = 1, v \in V)$ (by a reduction from the chromatic index problem). The problem is also NP-complete when there are only two vertices in a multi-graph and $p(v) \geq 2$ for every $v \in V$ (by a reduction from a multiprocessor scheduling problem).

Claim. Given an instance of the message exchange problem, there is an equivalent instance of the file transfer problem.

Proof: The main idea is that when preemption is not allowed, the startup time is not meaningful. Using the bipartite representation $bG(V, E)$ of the MEG, an edge with weight $w(e)$ in the MEG will correspond to an edge with weight $\beta + w(e)\gamma$ in the file transfer graph $G(V, E')$. $\square$

The following two theorems from [8] concern the difficulty of finding a minimum makespan schedule for the file transfer problem. Given a file transfer problem instance, $G(V, E)$:

Theorem 1. *An optimal schedule for the message exchange problem can be found in polynomial time when all arc weights are the same.*

Theorem 2. *If $G(V, E)$ is a tree of single port vertices, finding the minimum makespan for the file transfer problem is an NP-hard problem.*

Using the previous claim we can state the following corollary:

Corollary 1. *Optimal scheduling for the message exchange problem is an NP-hard problem.*

3.2 Divisible messages

When the messages can be partitioned and sent in several transmissions, the previous scheduling formulation cannot be used. Instead, a scheduling formulation similar to the one presented in [7] will be used. First, we present some notation.

- Given a graph g and a positive number t we define $t.g$ to be the weighted graph obtained from g by assigning weight t to each edge of g.
- Given two graphs $g_1(V, E_1)$ and $g_2(V, E_2)$ with the same set of vertices, and two positive integers t_1, t_2, we define $t_1.g_1 + t_2.g_2$ to be the graph $g(V, E)$ with edge set $E = \{(u, v)|(u, v) \in E_1 \bigcup E_2\}$. The edge $e = (u, v) \in E$ has weight equal to the sum of the weights of all edges in $t_1.g_1$ and $t_2.g_2$ that join u and v.
- Given two weighted graphs g_1, g_2 with the same vertex set, we use $g_1 \leq g_2$ to mean that g_1 is a subgraph of g_2, and the weight of each edge in g_1 is no larger than the weight of the corresponding edge in g_2.

Given an instance of the message exchange problem, $dG(V, E)$ we will consider the bipartite graph representation, $bG(V', E')$. A schedule of $bG(V', E')$ is a sequence of subgraphs $g_1, \ldots g_n$ and a sequence of positive integers $t_1, \ldots, t_n$ such that $\sum_{i=1}^{n} t_i.g_i \geq bG$. The scheduling time is $n\beta + \sum_{i=1}^{n} t_i$. The first two of the following three theorems are from [7].

Theorem 3. *When $\beta = 0$, an optimal schedule for $bG(V', E')$ can be found in polynomial time.*

For a synchronous model, we have the following theorem [7].

Theorem 4. *When $\beta > 0$, the problem of deciding if there is a schedule for $bG(V', E')$ with completion time at most M is NP-complete.*

The problem is also NP-complete in an asynchronous model.

Theorem 5. *When $\beta > 0$, the problem of deciding if there is a schedule for $bG(V', E')$ with completion time at most M is NP-complete for the asynchronous model.*

Proof: The proof uses a reduction from the 3-partition problem [10].

4 Algorithms

All of the algorithms presented here have been studied in different settings. We analyze them here for the message exchange problem. We describe an algorithm as a graph $dG(V, E)$, in which we label the $n = |V|$ vertices (processors) as $p_0, \ldots, p_{n-1}$, and let $m_{i,j}$ denote the weight of the arc from p_i to p_j. Let M be the largest message size ($M = \max_{e \in E} w(e)$), and S be the smallest message size ($S = \min_{e \in E} w(e)$). We use k to denote the ratio between the sizes of the largest and the smallest messages ($k = \frac{M}{S}$).

In our model, each processor knows the messages to be sent and their sizes, but does not know the sizes of messages to be received. In our algorithm descriptions, we will ignore the extra communications needed to determine the sizes of messages to be received.

4.1 Hypercube-like all-to-all algorithm

We present this algorithm for n a power of 2. Let $n = 2^p$ and let $b_i = (b_{i_1}, \ldots, b_{i_p})$ be the binary representation of p_i. We use b_i^j to denote $(b_{i_1}, \ldots, \bar{b}_{i_j}, \ldots, b_{i_p})$.

```
for t = 1 to p do
    do in parallel for all j (0 ≤ j < n)
        bᵢ exchange messages with bᵢʲ
```

Analysis: In this algorithm, each processor executes $\log n$ phases. In each phase, each processor exchanges its message, and all the previously received messages with one neighbor. Notice that this algorithm does message forwarding. We do not consider the times for copying and rearrangement of messages in each processor.

For the first exchange, the time is bounded by $\beta + (n-1)M\gamma$ ($\geq \beta + t_s\gamma$). For the second exchange, the time is bounded by $\beta + 2(n-1)M\gamma$. In general, at step i the time for the exchange is bounded by $\beta + 2^{i-1}(n-1)M\gamma$. The total time of this algorithm is at most $\log_2 n\beta + (n-1)^2 M\gamma$.

Worst case: Each message m is transmitted by one processor in the first phase, by two processors in the second phase, by four processors in the third phase, and so on. After phase i the message m is known by 2^i processors, and all processors know m at the end. As message m has only one destination, there is an overhead of $n - 2$ transmissions for each message. This overhead can be ignored in the case of a large start-up time with small messages. On the other hand, with large messages the overhead will be the predominant factor.

This algorithm can be improved with tighter control on which messages should be forwarded. Even with this improvement the algorithm will be $O(n)$ times slower than the bandwidth lower bound when message forwarding is not allowed. There is a clear trade-off between the β term and the γ term. To obtain an algorithm with only $O(\log n)$ phases, we have to use message forwarding; it is impossible using only direct transmissions.

4.2 Fixed pattern algorithm

```
for t = 1 to n − 1 do
    do in parallel for all j (0 ≤ j < n)
        p_j sends the message addressed to p_{j+t mod n}
        p_j receives the message from p_{j−t mod n}
```

Analysis: Each processor executes $n - 1$ steps in this algorithm. In each step t, processor p_i sends (non-blocking send) its message to the processor $p_{i+t \bmod n}$. There are no conflicts because each processor sends to only one other processor and receives from one other processor at any given time. The time to complete the algorithm is $(n - 1)\beta + \sum_{t=1}^{n-1} \max_i \{m_{i,i+t \bmod n}\}\gamma \le (n - 1)\beta + (n - 1)M\gamma$.

Worst case: Suppose that there are only two message sizes M and S, and the algorithm behaves synchronously. If there are $n - 1$ large messages such that each one will be sent in a different step, then the time will be $(n - 1)\beta + (n - 1)M\gamma$. In contrast, the lower bound on the bandwidth is $((n - 2)S + M)\gamma$ (when each processor has to receive at most one message). Since $M/S = k$, the algorithm will be $\frac{k(n-1)}{(n-2)+k}$ times slower than the bandwidth lower bound. On the other hand, when all the messages have the same size, this algorithm produces an optimal schedule (see Theorem 1).

4.3 Centralized pattern algorithms

The general structure of these algorithms is the following.

```
all-to-all exchange of the message sizes
best strategy decision
strategy execution
```

Analysis: These algorithms have three different phases. The purpose of the first phase is to provide each processor with global knowledge of the message sizes. In the second phase, a sequential algorithm is used to find a communication pattern. The third phase is the execution of the communication pattern found. To improve the efficiency of the algorithm we overlap the calculation in the second step with the communication in the third step.

In the following algorithms we look for a sequence of matchings, where a matching is a graph in which each node has in-degree at most one and out-degree at most one. We look for the first matching, M_1, in the graph $dG(V, E)$. We look for the second matching in the graph $dG(V, E) - M_1$[1], and so on. We stop when all the communications are in the sequence of matchings.

Max-Min strategy. In this strategy, we look for a sequence of maximal matchings in which the minimum weight is maximized. The algorithm for finding such matchings is polynomial. Given a problem instance, this strategy finds a solution that solves the problem in at most $n - 1$ phases.

Max-Weight strategy. In this strategy, we find matchings in which the sum of the weights is maximal. The algorithm for finding such matchings is polynomial.

Uniform strategy. In this strategy, we look for a sequence of matchings in which the weights of all edges are the same. To do this, we have to do a preprocessing step to obtain a special matrix.

Given the graph $dG(V, E)$, let A be the square matrix $(m_{i,j})$. We generate a quasi-doubly stochastic matrix (a matrix with all line sums being equal) Q. Let r_i be the sum of row i and c_j be the sum of column j. Let $S = \max_{i=1}^{n} \{r_i, c_i\}$, that is $S = \max\{t_s, t_r\}$. We use the following algorithm to obtain Q.

```
Q ← A
for i = 1 to n do
    for j = 1 to n do
        if i ≠ j then q_ij ← q_ij + min{S − r_i, S − c_j}
```

At the end, all lines (rows and columns) of Q will have sum S. Let $dQ(V, E)$ be the corresponding graph. In the graph $dQ(V, E)$, all processors have to transmit (and also receive) messages with weights that sum to S.

Find a sequence of maximal matchings, and then normalize each matching so that all edges in a matching have the same (minimum) weight. Since we are working with an auxiliary graph, there will be some *dummy* traffic, but there is no need to do the corresponding dummy transmissions. In this algorithm we suppose that the messages can be partitioned. The number of phases will almost always be greater than $n - 1$ (in general $O(n^2)$). The generated schedule is optimal when there is no start-up time.

Worst case: For all these centralized pattern algorithms, when all the message sizes are small the first two phases can be as time-consuming as the strategy execution. For the uniform strategy, the start-up times are important $(O(n^2))$.

Recommendations The choice of the right algorithm depends on the parameters of the linear model (β and γ) and the average message length L.

$\beta >> L\gamma$. The hypercube-like algorithm minimizes the number of start-ups.

$L\gamma >> \beta$. The bandwidth should be used in the best way, so the centralized algorithm with the uniform strategy is the right choice.

$L\gamma \sim \beta$. If the messages have approximately the same size, the fixed pattern is a good choice. With messages of different sizes the Max-Min or Max-Weight strategies for the centralized pattern algorithm perform better.

---

[1] This operation also removes arcs with weight zero.

4.4 Experimental results

In this section we analyze the behavior of the algorithms on a real machine. We implemented three of the algorithms on an IBM SP2 with 32 processing nodes, interconnected by a high performance switch (HiPS). The algorithms were coded in MPICH, the IBM version of MPI. The implemented algorithms were:

Step Fixed pattern algorithm

MaxMin Centralized pattern algorithm with Max-Min strategy

MaxWeight Centralized pattern algorithm with Max-Weight strategy

These algorithms were implemented using the send and receive primitives from MPI, without any special optimization. As a basis of comparison, we utilized the native `MPI_Alltoall` procedure. Since we did not have access to the source code of MPICH, where we expect that some low-level primitives are used, we consider performances close to the performance of `MPI_Alltoall` to be good results.

In the algorithms, there is always an initial phase in which the message sizes are exchanged. This phase is needed to allocate the receive buffers. In the first algorithm, each processor receives only the size of the messages that it will receive. In contrast, in the last two algorithms (MaxMin and MaxWeight) there is a total exchange of all message sizes. For the initial phases of all algorithms we used the fixed pattern algorithm (small messages of equal size).

We measured the times of the message exchange algorithms with the number of processors ranging from 3 to 14. For each experiment we generated a matrix composed of random message sizes following a Gaussian distribution with average 800k and deviation 400k. To avoid experimental errors, we calculated the average of 30 runs for each algorithm on each message exchange matrix. For the whole experiment, 30 different matrices were generated for each number of processors. We present below the results for 10 and 11 processors.

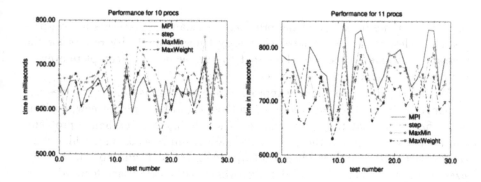

Analysis Table 1 compares the average performance of our algorithms to the performance of the MPICH algorithm. The numbers in the table are the aver-

# of procs	3	4	5	6	7	8	9	10	11	12	13	14
Step (%)	20	-6	14	-3	6	-4	3	-5	4	-5	0	-6
MaxMin (%)	19	-6	15	-3	8	-2	5	-2	4	-4	1	-6
MaxWeight (%)	19	-5	14	-2	10	1	9	2	8	0	6	-1

Table 1. Average gain compared to MPI_Alltoall

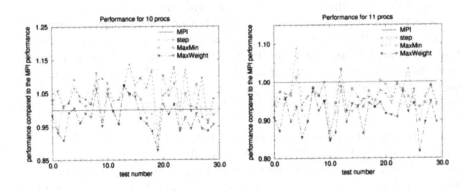

age percentage gains of our algorithms compared to the MPICH performances. We can see that the best performances were obtained when the number of processors was odd, and that the MaxWeight algorithm has the best performance of our three algorithms. The MaxWeight algorithm clearly outperforms the MPI_Alltoall primitive when the number of processors is odd.

5   Conclusions

In this paper, we investigated the problem of exchanging messages among processing nodes. We have extended previously known complexity results for the message exchange problem to show that the problem is NP-complete even in the more realistic asynchronous model.

We presented five polynomial approximation algorithms to solve the message exchange problem. The heuristics used in these algorithms are not new, but this is the first time that they have been used for the message exchange problem. Our analyses showed that the most suitable algorithm depends on machine characteristics (startup time and bandwidth), and on the uniformity of the message sizes. We implemented three of our algorithms, without message forwarding or message preemption, on an actual parallel machine. Our experiments showed that the centralized knowledge algorithms are the best ones when the message sizes are not uniform. Our algorithms outperform the native MPI routine of the machine when the number of processors is odd. We are currently implementing an all-to-all primitive for Athapascan-0 [6], a library for parallel processing with threads, that was developed in our research group.

References

1. P. Barcaccia, M. Bonuccelli, and M. Ianni. Minimun length scheduling of precedence constrained messages in distributed systems. In *EUROPAR'96*, volume LNCS 1123, pages 594–601, 1996.

2. M. Barnett, L. Shuler, R. van de Geijn, S. Gupta, D.G. Payne, and J. Watts. Interprocessor collective communication library. In *Proceedings of the Scalable High Performance Computing Conference*, pages 357–364, 1994.

3. S.H. Bokhari. Multiphase complete exchange on a circuit switched hypercube. In *Proceedings of the 1991 International Conference on Parallel Procesing*, volume I, pages 525–529, 1991.

4. G. Bongiovanni, D. Coppersmith, and C. Wong. An optimum time slot assignment algorithm for an ss/tdma system with variable number of transponders. *IEEE Transactions on Communications*, 29:721–726, 1981.

5. M.A. Bonuccelli. A polynomial time optimal algorithm for satellite-switched time-division multiple access satellite communications with general switching modes. *SIAM J. Disc. Math.*, 4:28–35, February 1991.

6. J. Briat, I. Ginzburg, and M. Pasin. Athapascan-0b reference manual. Technical Report, Apache, LMC - IMAG, Grenoble, France, 1997.

7. H. Choi and S.L. Hakimi. Data transfer in networks. *Algorithmica*, 3:223–245, 1988.

8. E.G. Coffman, M.R. Garey, D.S. Johnson, and A.S. Lapaugh. Scheduling file transfers. *SIAM J. Comput.*, 14(3):744–780, August 1985.

9. P. Fraigniaud and J.G. Peters. Structured communication in cut-through routed torus networks. Technical Report TR 97-05, School of Computing Science, Simon Fraser Univ, 1997.

10. A. Goldman, J. Peters, and D. Trystram. Exchange of messages of different sizes. manuscript, www-apache.imag.fr/apache/.

11. T. Gonzales and S. Sahni. Open shop scheduling to minimize finish time. *Journal of the Association for Computing Machinery*, 4(23):665–679, October 1976.

12. T.F. Gonzales. Multimessage multicasting with forwarding. Technical Report, UCSD Department of Computer Science, TRCS-96-16.

13. T.F. Gonzales. Multi-message multicasting. In *IRREGULAR'96*, volume LNCS 1117, pages 217–228, 1996.

14. Y.S. Hwang, R. Das, J. Saltz, M. Hodoscek, and B. Brooks. Parallelizing molecular dynamics programs for distributed memory machines. *IEEE Computational Science & Engineering*, 2(2):18–29, 1995.

15. J.G. Peters and C.C. Spencer. Global communication on circuit-switched toroidal meshes. Technical Report TR 97-02, School of Computing Science, Simon Fraser Univ., 1997. to appear in PPL.

16. S. Ranka, R.V Shankar, and K.A. Alsabti. Many-to-many personalized communication with bounded traffic. In *The Fifth Symposium on the Frontiers of Massively Parallel Computation*, pages 20–27, February 1995.

17. G.N Rouskas and V. Sivaraman. On the design of optimal tdm schedules for broadcast wdm networks with arbitrary transceiver tuning latencies. In *INFOCOM'96*, pages 1217–1224, 1996.

18. R. Tahkur and A. Choudhary. All-to-all communication on meshes with wormhole routing. In *IPPS'94*, pages 561–565, 1994.

19. L.G. Valiant. A bridging model for parallel computation. *Communications of the ACM*, 33(8):103–111, August 1990.

The Distributed Object-Oriented Threads System DOTS

Wolfgang Blochinger, Wolfgang Küchlin, and Andreas Weber*

Wilhelm-Schickard-Institut für Informatik
Universität Tübingen, 72076 Tübingen, Germany
{blochinger,kuechlin,weber}@informatik.uni-tuebingen.de
http://www-sr.informatik.uni-tuebingen.de/

Abstract. We describe the design and implementation of the *Distributed Object-Oriented Threads System* (DOTS). This system is a complete redesign of the *Distributed Threads System* (DTS) using the object-oriented paradigm both in its internal implementation and in the programming paradigm it supports. DOTS extends the support for *fork/join* parallel programming from shared memory threads to a distributed environment. It is currently implemented on top of the *Adaptive Communication Environment (ACE)*. A heterogeneous network of Windows NT PC's and of UNIX workstations is transformed by DOTS into a homogeneous pool of anonymous compute servers. DOTS has been used recently in applications from computer graphics and computational number theory. We also discuss the performance characteristics of DOTS for a workstation cluster running under Solaris and a PC network using Windows NT, as they were obtained from a prototypical example.

1 Introduction

The Distributed Object-Oriented Threads System (DOTS) is a programming environment for the parallelization of irregular and highly data-dependent algorithms. It extends support for *fork/join* parallel programming from shared memory threads to a distributed memory environment. DOTS works on the principle that a forked thread can be executed on a remote machine provided its input and output parameters can be copied over the network and it does not engage in shared memory communication. This programming paradigm can be characterized as SPMD (*single program, multiple data*) with asynchronous remote procedure calls.

In DOTS, each network node is a multi-threaded shared-memory multiprocessor (possibly consisting of a single processor). Under DOTS, the network again forms a symmetric multiprocessor. Together they form a hierarchical multiprocessor. The idea is to support a uniform parallelization paradigm on this machine. Large grained threads are to be forked over the network, small grained ones over the processors in a shared memory multiprocessor. Since each network node is to be oversaturated with threads, communication latency is hidden implicitly by running another thread.

DOTS uses an object-oriented approach both in its internal implementation and in the programming paradigm it supports. It provides a convenient way for the object serialization of the parameter and result objects that are involved in the distributed part of the computation process.

---
* Supported by *Deutsche Forschungsgemeinschaft* under grant Ku 966/4-1.

2 Designing Distributed Applications with DOTS

2.1 Overview of DOTS

DOTS consists of two main components: The DOTS run-time class library and the so called DOTS Commander.

The DOTS Run-Time Class Library is a C++ class library that contains classes and functions which are necessary for coding a DOTS application. This library is to be linked with the user program.

The DOTS Commander is a program which controls the distributed computation. It starts and terminates all processes on the different computers that take part in the distributed computation. It also provides the user interface.

In order to use DOTS for a distributed computation the following main steps have to be carried out.

— Code parts suitable for distributed computations have to be identified and have to be rewritten as static functions of the appropriate classes. The distributed computation of these functions has to be performed by forking and joining DOTS threads.
— Code for the serialization of parameter objects has to be provided. DOTS provides a good infrastructure to make this a straight-forward programming task. Basically, all members of the parameter classes have to be listed in a statement involving a DOTS_Archive class and operators for packing ($<<$) and unpacking ($>>$). Using the template and operator overloading mechanisms of C++, recursion through the member classes and picking the appropriate objects for functions to be forked and joined is done automatically by the DOTS library.

2.2 Object-Serialization

The execution of a DOTS thread implies the transmission of argument and result objects over a network connection. This process requires the transformation of the involved objects into serial data structures before transferring them, and their reconstruction after the transmission from this serial representation. The system class DOTS_Archive implements a suitable serial data structure and contains predefined operators for packing ($<<$) and unpacking ($>>$) all simple C++ data types into and from this structure. The class DOTS_Archive supports computations in heterogeneous networks by automatically converting the data into a standard representation within the serialization process (homogenizing e. g. big-endian and little-endian architectures).

DOTS offers two ways for suppling code for (de-)serialization using the DOTS_Archive class:

Implicit Object-Serialization. This technique is based on defining a (de-)serialization operator for each argument and result class. Consider the example given in Fig. 1. The function dots_func will be forked over the network. The main advantage of this technique lies in reusing code in nested classes which leads to compact and clearly structured programs.

```
class X {
public:
  int a;
  int b;

  /* The operators << and >> have to be defined for X.
   * The corresponding ones for the simple datatypes
   * can be used within the definition.
   */
  friend DOTS_Archive& operator<<(DOTS_Archive& arch, X& x)
    {return arch << x.a << x.b;}
  friend DOTS_Archive& operator>>(DOTS_Archive& arch, X& x)
    {return arch >> x.a >> x.b;}
};

class Y {
public:
  int c;
  X x;

  /* The operators << and >> have to be defined for Y.
   * Notice that now the ones defined for X can be used
   * in the definition as can the corresponding ones
   * for the simple datatypes.
   */
  friend DOTS_Archive& operator<<(DOTS_Archive& arch, Y& y)
    {return arch << y.c << y.x;}
  friend DOTS_Archive& operator>>(DOTS_Archive& arch, Y& y)
    {return arch >> y.c >> y.x;}
};

Y* dots_func(X* arg);
```

Fig. 1. Implicit Object-Serialization

Explicit Object-Serialization. This method is based on redefinition of the system's standard methods for (de-)serialization with user supplied code. The main reason for us to provide this option is to facilitate the integration of marshaling code written in C in DOTS. Since for new applications the implicit serialization is the method of choice, we will not give further details here.

2.3 Forking DOTS Threads

Forking a function over the network can be done via the DOTS library function dots_fork. This function takes three parameters: the function to be forked, its argument object, and the *thread group* to which the forked thread is assigned to. We also provide a variant of the fork function, called dots_lfork. This variant takes the same parameters but allows a more efficient execution of the generated DOTS thread: the argument object will not be serialized immediately. It might happen that the DOTS thread can be executed locally; in this case no serialization and deserialization is necessary resulting in improved performance compared to the dots_fork function, which will always serialize and deserialize its argument object. However, the argument object of a dots_lfork must not be changed until a corresponding dots_join is called. Thus using dots_lfork requires more care on the side of the programmer than using dots_fork. The differences in the internal execution between these variants are explained in more detail in Sec. 3.3.

2.4 Joining DOTS Threads

The functional result of a DOTS thread can be obtained via a call to dots_join. This function takes a thread group as input parameter and returns the result of an arbitrary thread in the group that has already terminated or of the thread that will be the first one to terminate if none has terminated so far. Moreover, a status flag indicating an error will also be returned. If no thread in the group has already terminated, dots_join will block until one thread in the group will be finished with its computations. Thus the semantics of dots_join is the one of a *join any* construct.

2.5 Cancellation of DOTS Threads

With subsequent calls to dots_join the results of all forked threads in a group can be retrieved. However, in many applications—e. g. in search problems—it is not necessary to obtain the results of all forked threads in the group, but one or some of the returned results are sufficient to allow the computation to continue.

In such a case all remaining threads in a thread group, which are still queued (cf. Sec. 3), can be terminated by a call to the dots_cancel function, which takes a thread group as a parameter. Thus subsequent computations do not have to compete for computational resources with others that are now known to be unnecessary.

By the combination of the *join any* semantics of dots_join and the functionality of dots_cancel DOTS is a powerful tool for the distributed computation of irregularly structured search problems.

3 Design and Implementation of DOTS

3.1 System Overview

Fig. 2 shows the internal structure of the DOTS Commander. Its main component is the Global Manager, which—in conjunction with the Scheduler—conducts the execution process. The Logging Manager controls the distributed logging system of DOTS. DOTS follows the SPMD model, which means that all involved hosts run the same program. The task of the Session Manager is to manage the node processes running the same program on all hosts. Fig. 3 shows the components of a node process. Together with the Global Manager the Local Manager controls the execution of DOTS threads. The Logger processes the logging messages of the node process. All the described functional components of the DOTS Commander and of the node processes are executing in parallel.

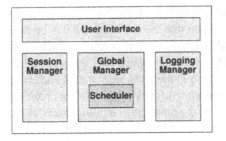

Fig. 2. The DOTS Commander

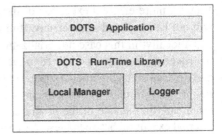

Fig. 3. A DOTS node process

3.2 General Execution Process of DOTS Threads

We first look at the general execution process for a DOTS thread, beginning with its creation by dots_fork and finishing up by delivering its result through dots_join. Fig. 4 shows the example of three hosts, one running the DOTS Commander and two running a node process. The execution process consists of the following five steps (cf. Fig. 4):

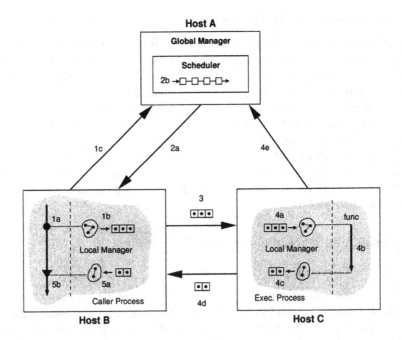

Fig. 4. General Execution Process

Step 1: The process begins with a call to dots_fork in the user program (1a). The argument object is serialized and the corresponding DOTS_Archive object is stored for later use (1b). In order to register the new DOTS thread in the system, a message containing a thread information block is sent to the Global Manager (1c). In DOTS thread information blocks are used to store data associated with the execution of DOTS threads, like a thread ID or the ID of the node process chosen to execute the thread. Thread information blocks are passed from one execution stage to the next and updated during a step when necessary. When all described actions are completed, dots_fork returns to the user program.

Step 2: The Global Manager extracts the thread information block from the message and passes it to the Scheduler. The task of the Scheduler is to determine a node process for the execution of the DOTS thread according to its scheduling strategy. The default scheduling strategy is to assign a DOTS thread to the node which currently executes the smallest number of DOTS threads. In order to prevent the hosts from being overloaded, for each host the maximal number of DOTS threads to be executed at a time can be individually determined by the user. The best bounds depend on the application as well as on the number of available CPUs and speeds of the CPUs. The scheduling strategy can easily be changed by the programmer by subclassing the scheduler interface class. If a node process is determined, its ID is stored in the thread information block, which is sent back to the caller process (2a). If no node process is currently available (because all hosts execute the maximal number of threads) the thread information block is queued (2b). In this case the execution process of the DOTS thread will be continued when a node process completes the execution of another DOTS thread (see step 4).

Step 3: The Local Manager of the caller process sends a message containing the thread information block and the archive with the serialized argument object to the executor node process.

Step 4: The Local Manager of the executor process produces a copy of the argument object from the archive contained in the message (4a). The thread function is now executed with the argument object in a new system thread (4b). If the execution has completed the result object is serialized (4c) and sent in a message to the caller process (4d). The Global Manager is now informed by a message that a new DOTS thread can be executed on this host (4e).

Step 5: After receiving the message from the executor process, the Local Manager of the caller process extracts the DOTS_Archive from the message and produces a copy of the result object (5a). Now the result is ready to be obtained via dots_join. If a call to dots_join has already occurred, the caller is now unblocked.

3.3 Optimizations of the Execution Process

In two special cases the described execution process is changed in order to minimize the communication overhead of DOTS threads. The optimizations presented below will have a maximal effect, if a DOTS thread is created with dots_lfork. In this case the serialization of the argument object is deferred to Step 3 in the normal execution process and thus can be completely eliminated in the situations described below.

Local execution:
This optimization takes effect when the Scheduler determines that the caller process is

to be identical with the executor process. In this case Step 3 to Step 5 are combined into one single step, which is entirely executed in the caller process. Almost all of the communication overhead can be avoided, leading to an efficient local execution of the DOTS thread fork.

Execution as local procedure call:
When a call of dots_join appears and the affected DOTS thread is still waiting in the queue of the Scheduler for its execution, the DOTS thread is removed from the queue and executed as local procedure call. Instead of blocking and waiting for the remote execution of the DOTS thread, the caller thread of dots_join is continuing its execution with a function call for the function with the argument object. (Cf. Fig. 5).

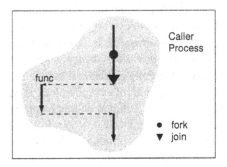

Fig. 5. Execution as Local Procedure Call

Using dots_fork instead of dots_lfork requires in both cases that the serialized argument object must first be retrieved from the stored archive. For this reason dots_lfork should be used when ever possible.

3.4   Implementation of DOTS

The implementation of DOTS is based on the *Adaptive Communication Environment* (ACE) toolkit [12]. ACE is a system-independent, object-oriented platform for developing communication software. There are implementations of ACE for all major UNIX systems (e.g. Solaris 2.x, AIX, SGI IRIX), for Windows NT (Win32) and for MVS OpenEdition. Fig. 6 shows the overall structure of DOTS and those parts of ACE which were used for its implementation.

Using ACE as the implementation foundation for DOTS provides the following advantages:

– The platform independence of ACE makes DOTS available on a wide range of operating systems and hardware platforms. Currently we are using implementations of DOTS on Solaris 2.x, IRIX 6.x and on Windows NT. It is to be expected that DOTS can be easily ported to all other platforms that are supported by ACE.

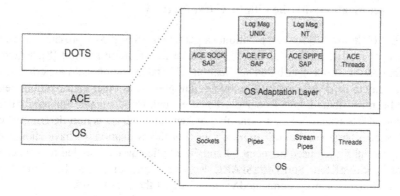

Fig. 6. Layered Structure of DOTS and ACE

- ACE's class encapsulation of the C based APIs of the different operating systems substantially improves the type-safety of the implementation. This can prevent the occurrence of run time errors and therefore improves the correctness and software quality.
- Performance measurements presented in [13] show that the additional abstraction layers of ACE do not cause any significant performance loss compared with using operating system calls directly.

4 Performance Measurements and Example Applications

For all performance measurements the wall clock time of several program runs was quantified. All performance results which are presented below are based on the arithmetic mean of the measured times. The running time of the sequential versions of the described application is denoted as $time_1$; while $time_n$ is the time elapsed with $n > 1$ hosts involved.

4.1 Volume Rendering on a PC Network

While requiring enormous capacities of computational power, the fast rendering of 3D volumetric data becomes more and more crucial for many applications in the field of medical science and in other scientific domains. As is described in full detail in [10], DOTS was successfully used to considerably speed up a volume rendering application on inexpensive and widely used Windows NT PCs. The pool consisted of up to five Intel PentiumPro and up to four Intel PentiumII PCs running at 200 Mhz and at 300 MHz respectively. The PCs were connected via a 10 Mbps Ethernet network. The process of casting 256^2 rays into a 128^3 data set with oversampling in ray direction could be accelerated almost linearly by the number of involved hosts. Detailed performance results are given in [10].

4.2 Factorization with the Elliptic Curve Method

The LiDIA library [9] for computational algebraic number theory provides an efficient sequential implementation of the elliptic curve factorization method [8], which was the foundation of our distributed application. This factorization method is of a too fine granularity to be distributed by a simple e-mail method as other factorization methods have been distributed, but is coarse grained enough to be suitable for parallelization via our DOTS system. Since the number of generated curves is non-deterministic, we performed a large number of program runs. For the evaluation we have chosen those program runs which used between 60 and 70 elliptic curves. This led to a speedup of 3.5 on a pool of four Sun UltraSPARC 1 workstations running at 143 Mhz with 32 MByte of main memory, connected via a 10 Mbps Ethernet network.

4.3 Comparing the Performance of DOTS on Solaris 2.x and Windows NT

For the realization of a performance comparison between DOTS running on Solaris and DOTS running on Windows NT we used a program for the distributed computation of the Mandelbrot set. The selected complex area is divided into a number of horizontal stripes, and each stripe is treated by a separate DOTS thread. Since this simple application does not depend on particular properties of one of the two different operating systems, we were able to use exactly the same source code for both platforms making the comparison more accurately. In spite of the simplicity of the Mandelbrot application, we were able to simulate essential properties of more sophisticated distributed applications by varying the following semi-independent program parameters:

— The computation time *(the granularity)* of a DOTS thread depends on the maximal number of allowed iterations per point.
— The amount of network traffic per DOTS thread *(the weight of a DOTS thread)* can be varied by choosing different sizes of the area to be calculated.
— The relative running time of the individual DOTS threads used in the computation *(the irregularity)* depends on the distribution of points of convergence in the chosen area. This distribution can vary considerably leading to non uniformly sized tasks.

For all subsequently presented measurements we have formed DOTS threads with a granularity of about 5 seconds and a network weight of about 8 KByte on both platforms. This setting represents a rather worst case approximation (low granularity, high weight) of the properties to be found in typical application domains of DOTS.

Performance Measurements under Solaris. Table 1 shows the results of the performance measurements in a Solaris based network. The pool of hosts consisted of 3 Sun UltraSPARC 1 workstations (each running at 143 MHz with 32 MByte of main memory) and 3 four-processor Sun HyperSPARC 10 workstations (running at 90 Mhz with 160–512 MByte main memory). All hosts were connected by a 10 Mbps Ethernet network. One host was used to execute the DOTS Commander, up to five other hosts were used to execute the application program.

Performance Measurements under Windows NT. For performance measurements under Windows NT we used a pool of 6 PCs with Intel Pentium processors (150 Mhz, 64 MByte main memory) connected by a 10 Mbps Ethernet network. One host was used to execute the DOTS Commander, up to five other hosts were used to execute the application program. Table 2 shows the results.

n = #hosts	speedup$_n$ = $\frac{time_1}{time_n}$	efficiency$_n$ = $\frac{speedup_n}{n}$
1	1.0	1.0
2	1.9	1.0
3	2.7	0.9
4	3.2	0.8
5	4.0	0.8

Table 1. Performance Measurements on a Solaris Pool

n = #hosts	speedup$_n$ = $\frac{time_1}{time_n}$	efficiency$_n$ = $\frac{speedup_n}{n}$
1	1.0	1.0
2	1.6	0.8
3	2.3	0.9
4	2.6	0.7
5	3.5	0.7

Table 2. Performance Measurements on a Windows NT Pool

Discussion. Although with both configurations substantial speedups of the computation could be achieved, DOTS running in a network of workstations under Solaris shows better results for speedup and efficiency compared to the pool of Windows NT PCs. One possible reason for this behavior might be the comparatively heavy design of the Windows NT system threads, which leads to longer response times of the completely multi-threaded DOTS Local Manager. Since Windows NT and PC Hardware nowadays is getting increasing attention in computer science—mainly in the domain of computer graphics—this issue will be one focus in our future research activities.

5 Related Work

5.1 DTS

DOTS is a complete object-oriented redesign of the PVM [6] based Distributed Threads System (DTS) [2]. DTS has been used successfully for a variety of projects whose aim was to speed up the computation of scientific code by parallelizing it on a network of workstations—e. g. theorem provers [3], complex root finders [11], or symbolic solvers [2].

In spite of its success the DTS system has some drawbacks, which motivated the new design. The central idea of DTS—providing an extension from *fork/join* parallel programming from shared memory threads to network threads—was kept in DOTS. However, the following major advantages could be achieved by the redesign:

- The overhead of using the PVM-system could be avoided. The main problem of the use of PVM on the DTS system has been that all of the software bulk and the instabilities and non-portabilities of PVM resulted in a corresponding problem of DTS.

- The DTS system is programmed in C and supports distributed programming based on C. It gives no support for an object-oriented language. Parameter marshaling was achieved via user supplied copying functions, which were thus not encapsulated. DOTS supports the object-oriented concept of object serialization to communicate parameters. The way in which the DTS copying functions had to be provided by the user was felt to be somewhat user unfriendly by many developers outside of computer science who used DTS to distribute their scientific simulation computations. This concern led to the development of a precompiler [7] which automatically generates the copy functions out of simple pre-processor directives. However, the additional level of source code that has to be precompiled has several disadvantages, e. g., it might conflict with development tools that assume that the source code is syntactically correct C or C++ code (such as an object-oriented CASE tool). In our newly designed DOTS system, archive classes are provided for the serialization and deserialization of objects. The serialization of objects that will be distributed over the network is encapsulated in the class definition, and is thus compatible with the class abstractions provided by C++.
- In the redesigned DOTS system, also PC's running under Windows NT can be integrated into the computation environment.

5.2 Nexus

Nexus [4, 5] provides an integration of multithreading and communication by disjoining the specification of the communication's destination and the specification of the thread of control that should respond to that communication. Dynamically created global pointers are used to represent communication endpoints. Remote service requests are used for data transfer and to asynchronously issue a remote execution of specified handler functions. Thus in Nexus threads are created as a result of communication.

In contrast to DOTS, Nexus provides no explicit join construct to retrieve return values from a remote thread. Nexus is primarily designed as foundation for high level communication libraries and as target for parallel programming language compilers, whereas DOTS is intended to be used directly by the application programmer.

5.3 Cilk-NOW

Cilk is a parallel multithreaded extension of the C language. Cilk-NOW [1] provides a runtime-system for a functional subset of Cilk, that enables the user to run Cilk programs on networks of UNIX workstations.

Cilk-NOW supports multithreading in a continuation passing style, whereas DOTS uses the fork/join paradigm for coding distributed multithreaded programs. Cilk-NOW implements a work-steeling scheduling technique. In DOTS the Global Manager can be configured with application dependent schedulers by subclassing the system's scheduler interface.

Acknowledgment: We are grateful to C. Ludwig for implementing the distributed elliptic curve factorization method.

Availability of DOTS: If you are interested to obtain a copy of DOTS , write an e-mail to one of the authors.

References

1. BLUMOFE, R. D., AND LISIECKI, P. A. Adaptive and reliable parallel computing on networks of workstations. In *USENIX 1997 Annual Technical Symposium* (January 1997).
2. BUBECK, T., HILLER, M., KÜCHLIN, W., AND ROSENSTIEL, W. Distributed symbolic computation with DTS. In *Parallel Algorithms for Irregularly Structured Problems, 2nd Intl. Workshop, IRREGULAR'95* (Lyon, France, Sept. 1995), A. Ferreira and J. Rolim, Eds., vol. 980 of *Lecture Notes in Computer Science*, pp. 231–248.
3. BÜNDGEN, R., GÖBEL, M., AND KÜCHLIN, W. A master-slave approach to parallel term rewriting on a hierachical multiprocessor. In *Design and Implementation of Symbolic Computation Systems — International Symposium DISCO '96* (Karlsruhe, Germany, Sept. 1996), J. Calmet and C. Limongelli, Eds., vol. 1128 of *Lecture Notes in Computer Science*, Springer-Verlag, pp. 184–194.
4. FOSTER, I., KESSELMAN, C., AND TUECKE, S. The Nexus task-parallel runtime system. In *1st Intl. Workshop on Parallel Processing* (Tata, 1994), McGrawHill.
5. FOSTER, I., KESSELMAN, C., AND TUECKE, S. The Nexus approach to integrating multithreading and communication. *Journal of Parallel and Distributed Computing 37* (1996), 70–82.
6. GEIST, G. A., AND SUNDERAM, V. S. The PVM system: Supercomputer level concurrent computation on a heterogeneous network of workstations. In *Sixth Annual Distributed-Memory Computer Conference* (Portland, Oregon, May 1991), IEEE, pp. 258–261.
7. GRUNDMANN, T., BUBECK, T., AND ROSENSTIEL, W. Automatisches Generieren von DTS-Kopierfunktionen für C. Tech. Rep. 48, Universität Tübingen, SFB 382, July 1996.
8. LENSTRA, A. K., AND LENSTRA, JR, H. W. Algorithms in number theory. In *Algorithms and Complexity*, J. van Leeuwen, Ed., vol. A of *Handbook of Theoretical Computer Science*. Elsevier, Amsterdam, 1990, chapter 12, pp. 673–715.
9. LIDIA GROUP. *LiDIA Manual Version 1.3—A library for computational number theory*. TH Darmstadt, Alexanderstr. 10, 64283 Darmstadt, Germany, 1997. Available at http://www.informatik.th-darmstadt.de/TI/LiDIA/Welcome.html.
10. MEISSNER, M., HÜTTNER, T., BLOCHINGER, W., AND WEBER, A. Parallel direct volume rendering on PC networks. In *The proceedings of the International Conference on Parallel and Distributed Processing Techniques and Applications (PDPTA'98)* (Las Vegas, NV, U.S.A., July 1998). To appear.
11. SCHAEFER, M. J., AND BUBECK, T. A parallel complex zero finder. *Journal of Reliable Computing 1*, 3 (1995), 317–323.
12. SCHMIDT, D. C. The ADAPTIVE Communication Environment: An object-oriented network programming toolkit for developing communication software., 1993. http://www.cs.wustl.edu/~schmidt/ ACE-papers.html.
13. SCHMIDT, D. C. Object-oriented components for high-speed network programming, 1995. http://www.cs.wustl.edu/~schmidt/ ACE-papers.html.

Graph Partitioning and Parallel Solvers: Has the Emperor No Clothes?*
(Extended Abstract)

Bruce Hendrickson

Sandia National Labs, Albuquerque, NM 87185–1110

Abstract. Sparse matrix-vector multiplication is the kernel for many scientific computations. Parallelizing this operation requires the matrix to be divided among processors. This division is commonly phrased in terms of graph partitioning. Although this abstraction has proved to be very useful, it has significant flaws and limitations. The cost model implicit in this abstraction is only a weak approximation to the true cost of the parallel matrix-vector multiplication. And the graph model is unnecessarily restrictive. This paper will detail the shortcomings of the current paradigm and suggest directions for improvement and further research.

1 Introduction

Over the past several years, a comfortable consensus has settled upon the community concerning the applicability of graph partitioning to the parallelization of explicit methods and iterative solvers. This consensus is largely the result of two happy occurrences. First is the availability of good algorithms and software for partitioning graphs, motivated by parallel computing applications (eg. Chaco [2], METIS [5], JOSTLE [11], PARTY [10] and SCOTCH [9]). Second is the excellent parallel efficiencies which can be obtained when solving differential equations. Clearly, the latter is a consequence of the former, right?

Yes, and no. Although the Emperor may not be completely naked, he is wearing little more than his underwear. The graph partitioning abstraction is seriously flawed, and its uncritical adoption by the community has limited investigation of alternatives. The standard partitioning approach suffers from two kinds of shortcomings. First, it optimizes the wrong metric. And second, it is unnecessarily limiting. These shortcomings will be discussed in detail below. The purpose of this paper is to elucidate these failings, and to suggest some improvements and directions for further research. More generally, I hope to stimulate renewed interest in a set of problems that has been erroneously considered to be solved.

---

* This work was funded by the Applied Mathematical Sciences program, U.S. Department of Energy, Office of Energy Research and performed at Sandia National Laboratories, operated for the U.S. DOE under contract number DE-AC04-76DP00789.

2 Matrices, Graphs and Grids

Before discussing the problems with graph partitioning in detail, it will be helpful to review the relationship between grids, matrices and graphs. A grid or mesh is the scaffolding upon which a function is decomposed into simple pieces. Standard techniques for solving differential equations represent their solutions as a union of simple functions on the pieces the grid. The grid determines the nonzero structure of the corresponding sparse matrix, but the relationship between the grid and the matrix can be complex. The solution values can be associated with grid points, or with grid regions. Multiple unknowns per grid point lead to block structure in the matrix. A nonzero value in the matrix might occur for each pair of grid points which are connected by an edge. Alternatively, a nonzero might occur if two grid points share a region. More complicated relationships are also possible. But an important point is that the matrices associated with computational grids are usually structurally symmetric. That is, if element $[i, j]$ is nonzero, then so is element $[j, i]$.

The standard graph of a symmetric $n \times n$ matrix has n vertices, and an edge connects vertices i and j if matrix element $[i, j]$ is nonzero. Note that this graph is generally not identical to the grid, although the two are related. Also note that every symmetric matrix has a graph, whether or not it has a corresponding grid.

For parallelizing iterative solvers, three graphs are commonly used.

1. The standard graph of the (symmetric) matrix.
2. The grid.
3. The dual of the grid. That is, each region becomes a vertex and two vertices are connected if their corresponding regions are adjacent to each other

Each of these graphs has its advantages and disadvantages. The graph of the matrix is unambiguous and always available for symmetric matrices. But if the matrix is never explicitly formed, the graph may be difficult to construct. And if there are multiple unknowns per grid point, then the graph of the matrix may be unnecessarily large.

The grid does not suffer from the problem of multiple unknowns, and so can be a smaller representation of the basic structure of the matrix. However, as discussed above, it only approximates the structure of the matrix. Also, if the matrix doesn't come from a grid-based differential equation, then there may be no grid to use.

The dual of the grid has the same advantages and disadvantages as the grid, and its construction requires some nontrivial effort on the part of the user. However, as will be discussed in §4, when the standard graph partitioning approach is used, the dual is often a better model of communication requirements than the grid itself.

3 Parallel Matrix-Vector Multiplication

For sparse matrices, the standard approach to parallelizing matrix-vector multiplication involves dividing the rows of the matrix among the processors. Each processor is responsible for computing the contributions from the matrix values it owns. Consider forming the product $y = Ax$ on p processors. The rows and columns of A are divided into p sets, and the elements of x and y are divided in the same way. As we will discuss below, graph partitioning is only applicable to structurally symmetric matrices, so for the remainder of this section we will assume that A is structurally symmetric. Conceptually, the partitioning can be thought of as a symmetric reordering of the rows and columns of A, and also of x and y, followed by an organization of A into block structure as illustrated below.

$$
A = \begin{bmatrix} A_{11} & A_{12} & \cdots & A_{1p} \\ A_{21} & A_{22} & \cdots & A_{2p} \\ \vdots & \vdots & \ddots & \vdots \\ A_{p1} & A_{p2} & \cdots & A_{pp} \end{bmatrix},
\tag{1}
$$

Processor i begins with segment i of x and computes segment i of y. It has all the necessary elements of x to compute the contribution from its diagonal block A_{ii}. To compute the contribution of off-diagonal blocks, processor i needs some x values from other processors. Specifically, if $A(k, l) \in A_{ij}$ is nonzero, then processor i needs $x(l)$ from processor j. This exchange of elements of x is all the communication required to form the matrix-vector product.

Several important observations are in order. First, the block structure of A can be interpreted in terms of the standard graph of the matrix of A. Since each row of A is a vertex in the graph, the block structure induces a partitioning of the vertices into p sets. Second, the total number of nonzero values in A_{ij} is equal to the number of edges in the graph which connect vertices in partition i to those in partition j. Third, the number of x values that processor j must send to processor i is equal to the number of vertices in partition j which have neighbors in partition i. We will call vertices which have neighbors on other processors *boundary vertices*.

Explicit solvers have exactly the same communication pattern as matrix-vector multiplication. Consequently, all the issues discussed here are relevant to them as well.

It is worth noting that the same row-wise partitioning of the matrix can also be used to efficiently compute $z = A^T w$. This is useful for nonsymmetric matrices since many iterative methods for such matrices require both matrix-vector and matrix-transpose-vector products. Details can be found in [3].

4 Problems with Graph Partitioning

The standard graph partitioning problem addressed by codes like Chaco [2] and METIS [5] is the following. Divide the vertices of the graph into equally sized sets in such a way that the number of edges crossing between sets is minimized. (This formulation can be generalized to consider weights on the vertices and edges, but details are beyond the scope of this paper.) The construction of an appropriate graph is the responsibility of the user, and typically one of the three graphs described in §2 is used. By limiting the number of edges crossing between sets, the off-diagonal blocks of the matrix have few nonzeros. Thus, in matrix terms, graph partitioning methods try to put most of the nonzero values in the diagonal block. This is clearly advantageous for limiting the communication, but unfortunately, the number of cut edges is not the most important thing to minimize. The inappropriateness of this metric are discussed in §4.1, while more general limitations of the graph partitioning model are described in §4.2.

4.1 Problems With Edge Cuts

Minimizing edge cuts has three major flaws.

A. Although widely assumed to be the case, edge cuts are not proportional to the total communication volume. Obviously, this depends on the graph presented to the partitioner. If the graph is the standard graph of the matrix then, as discussed in 3, the communication volume is instead proportional to the number of boundary vertices. It is for this reason that the dual of the grid is often used as the graph to partition. Edges in the dual roughly correspond to pairs of boundary vertices in the grid. But the dual may be difficult to construct, and is not defined for general matrices. It would be better to have a partitioner that works directly with the structure of the matrix and tries to minimize an accurate model of the communication cost.

B. Sending a message on a parallel computer requires some startup time (or latency) and some time proportional to the length of the message. Graph partitioning approaches try to (approximately) minimize the total volume, but not the total number of messages. Unfortunately, in all but the largest problems, latencies can be more important than volume.

C. The performance of a parallel application is limited by the slowest processor. Even if all the computational work is well balanced, the communication effort might not be. Graph partitioning algorithms try to minimize the total communication cost. However, optimal performance will be obtained by instead minimizing the maximum communication cost among all processors. In some applications it is best to minimize the sum of the computation and communication time, so imbalance in one can be offset in the other. The standard graph partitioning approaches don't address these issues.

With these problems, why has the standard graph partitioning approach proved so successful for the parallel solution of differential equations? There are two reasons. First, grid points generally have only a small number of neighbors. So cutting a minimal number of edges ensures that the number of boundary vertices is within a small multiple of the optimal. This isn't true of more general matrices. Also, well structured grids have good partitions. If the number of processors is kept fixed while the problem size increases, latencies become unimportant and the computational work for a matrix-vector multiplication should grow linearly with n, while the communication volume should only grow as $n^{2/3}$ in 3D and $n^{1/2}$ in 2D. Thus, very large problems should exhibit excellent scalability, even if the partition is nonoptimal. For more general matrices, this inherent scalability doesn't hold true, and the quality of the partition matters much more.

4.2 Limitations of Graph Partitioning

Besides minimizing the wrong objective function, the standard graph partitioning approach suffers from several limitations, all reflective of a lack of expressibility of the model.

A. **Only square, symmetric matrices.** In the standard graph model, a single vertex represents both a row and a column of a matrix. If the matrix is not square, then the model doesn't apply. Also, in the standard model the single edge between i and j can't distinguish a nonzero in location $(i.j)$ from one in location (j, i). So even square matrices aren't handled well if they are not symmetric.

B. **Only symmetric partitions.** Since a single vertex represents both a row and a column, the partition of the rows is forced to be identical to the partition of the columns. This is equivalent to forcing the partition of y to be identical to the partition of x in $y = Ax$. If the matrix is symmetric, then this restriction simplifies the operation of an iterative solver. However, for nonsymmetric solvers, this restriction is unnecessary.

C. **Ignores the application of the preconditioner.** Preconditioning can dramatically improve the performance and robustness of iterative solvers. In a preconditioned iterative method, the matrix-vector product is only one part of a larger operation. For some preconditioners (eg. Jacobi) optimal performance of $y = Ax$ will lead to good overall performance. But for more complex preconditioners it is better to consider the full calculation instead of just the matrix-vector product. Unfortunately, the standard graph model has no capacity to consider the larger problem.

5 A Better Combinatorial Model

Some of the limitations discussed in §4.2 can be addressed by a more expressive model proposed by Kolda and Hendrickson [3, 4, 8]. The model uses a *bipartite* graph to represent the matrix. In this graph, there is a vertex for each row and another vertex for each column. An edge connects row vertex i to column vertex j if there is a nonzero in matrix location $[i, j]$. The structure of nonsymmetric and nonsquare matrices can be unambiguously described in this way, which addresses limitation (**A**).

The row vertices and column vertices can now each be partitioned into p sets, which corresponds to finding a $p \times p$ block structure of the matrix. An edge between a row vertex in partition i and a column vertex in partition j corresponds to a nonzero value in off-diagonal block A_{ij}. By minimizing such edges, most of the nonzero values will be in the diagonal blocks. The same matrix-vector multiplication algorithm sketched above can now be applied. The partition of x will correspond to the column partition of A, while the partition of y reflects A's row partition. The freedom to have different row and column partitions removes limitation (**B**).

The freedom to decouple row and column partitions has an additional advantage. Consider the case where two matrices are involved, so $y = BAx$. In [3], Hendrickson and Kolda show that the row partition can be used to get good performance in the application of A, while the column partition can optimize the application of B. So if a preconditioner is explicitly formed as a matrix (eg. approximate inverse preconditioners), then the bipartite partitioning problem can model the full iteration. This is a partial resolution of limitation (**C**).

Although the bipartite model is more expressive than the standard model, the algorithms in [3] still optimize the flawed metric of edge cuts.

6 Conclusions and Directions for Further Research

Despite the success of the standard graph partitioning approach to many parallel computing applications, the methodology is flawed. Graph partitioning minimizes the wrong metric, and the graph model of matrices is unnecessarily limiting. A bipartite graph representation resolves some of the limitations, but a number of important open questions remain.

1. **More accurate metric.** The principle shortcoming of standard graph partitioning is that the quantity being minimized does not directly reflect the communication cost in the application. Unfortunately, minimizing a more appropriate metric is challenging. Boundary vertices, the correct measure of communication volume, are harder to optimize than edge cuts. A good refinement algorithm which works on this quantity would be a significant advance. More generally, the true communication cost in a parallel matrix-vector multiplication operation is more complex than just total volume. The number of messages can be at least as important as the volume, but graph partitioning algorithms seem ill suited to addressing this quantity directly. Similarly, since the processors tend to stay synchronized in an iterative solver, the maximum communication load over processors is more important than the total. These issues have received insufficient attention.

2. **Matrix and preconditioner.** As discussed in §4.2, the matrix-vector multiplication is generally only a piece of a larger operation. Methodologies that optimize a full step of the iterative solver would be better. The bipartite model described in §5 is an improvement, but much work remains to be done.

3. **Multicriteria partitioning.** In many parallel applications there are several computational kernels with different load balancing needs. In many of these

situations, the individual kernels are synchronized so for peak performance each kernel must be load balanced. A simple example is the application of boundary conditions which only occurs on the surface of a mesh. A good partition for such a problem will divide the problem in such a way that each kernel is balanced. In the boundary condition example, each processor should own a portion of the surface of the mesh, and also of the full volumetric mesh. This important problem has not yet received sufficient attention, but some researchers are beginning to address it [7].

4. **Partitioning for domain decomposition.** Domain decomposition is a numerical technique in which a large grid is broken into smaller pieces. The solver works on individual subdomains first, and then couples them together. The properties of a good decomposition are not entirely clear, and they depend upon the details of the solution technique. But they are almost certainly not identical to the criteria for matrix-vector multiplication. For instance, Farhat, et al. [1] argue that the domains must have good aspect ratios (eg. not be long and skinny). It can also be important that subdomains are connected, even though the best partitions for matrix-vector multiplication needn't be. For the most part, practitioners of domain decomposition have made due with partitioning algorithms developed for other purposes, with perhaps some minor perturbations at the end. But a concerted effort to devise schemes which meet the need of this community could lead to significant advances.

5. **Parallel partitioning.** Most of the work on parallel partitioning has been done in the context of dynamic load balancing. But several trends are increasing the need for this capability. First is the interest in very large meshes, which won't easily fit on a sequential machine and so must be partitioned in parallel. Second, for a more subtle reason, is the growing interest in heterogeneous parallel architectures. Generally, partitioning is performed as a preprocessing step in which the user specifies the number of processors the problem will run on. With heterogeneous parallel machines, the number of processors is insufficient – the partitioner should also know their relative speeds and memory sizes. A user will want to run on whatever processors happen to be idle when the job is ready, so it is impossible to provide this information to a partitioner in advance. A better solution is to partition on the parallel machine when the job is initiated. A number of parallel partitioners have been implemented including Jostle [11] and ParMETIS [6]. This is an active area of research

Despite the general feeling that partitioning is a mature area, there are a number of open problems and many opportunities for significant advances in the state of the art.

Acknowledgements

The ideas in this paper have been influenced by many discussions with Rob Leland and Tammy Kolda. I have also benefited from conversations with Ray Tuminaro, David Day, Alex Pothen and Michele Benzi.

References

1. C. FARHAT, N. MAMAN AND G. BROWN, *Mesh partitioning for implicit computation via domain decomposition: impact and optimization of the subdomain aspect ratio*, Int. J. Num. Meth. Engrg. 38 (1995), pp. 989–1000.

2. B. HENDRICKSON AND R. LELAND, *The Chaco user's guide, version 2.0*, Tech. Rep. SAND95-2344, Sandia Natl. Lab., Albuquerque, NM, 87185, 1995.

3. B. HENDRICKSON AND T. G. KOLDA, *Parallel algorithms for nonsymmetric iterative solvers*. In preparation.

4. ———, *Partitioning Sparse Rectangular Matrices for Parallel Computation of Ax and $A^T v$*. In *Lecture Notes in Computer Science*, Springer-Verlag, 1998. Proc. PARA'98. To appear.

5. G. KARYPIS AND V. KUMAR, *A fast and high quality multilevel scheme for partitioning irregular graphs*, Tech. Rep. 95-035, Dept. Computer Science, Univ. Minnesota, Minneapolis, MN 55455, 1995.

6. ———, *Parallel multilevel graph partitioning*, Tech. Rep. 95-036, Dept. Computer Science, Univ. Minnesota, Minneapolis, MN 55455, 1995.

7. ———, *Multilevel algorithms for multi-constraint graph partitioning*, Tech. Rep. 98-019, Dept. Computer Science, Univ. Minnesota, Minneapolis, MN 55455, 1998.

8. T. G. KOLDA, *Partitioning sparse rectangular matrices for parallel processing*. In *Proc. Irregular'98*, 1998.

9. F. PELLEGRINI, *SCOTCH 3.1 user's guide*, Tech. Rep. 1137-96, Laboratoire Bordelais de Recherche en Informatique, Universite Bordeaux, France, 1996.

10. R. PREIS, R. DIEKMANN, *The PARTY Partitioning-Library, User Guide - Version 1.1*, Tech. Rep. tr-rsfb-96-024, University of Paderborn, Paderborn, Germany, 1996.

11. C. WALSHAW, M. CROSS, AND M. EVERETT, *Mesh partitioning and load-balancing for distributed memory parallel systems*, in Proc. Parallel & Distributed Computing for Computational Mechanics, Lochinver, Scotland, 1997, B. Topping, ed., 1998.

Parallel Simulation of Particulate Flows[*]

Matthew G. Knepley, Vivek Sarin, and Ahmed H. Sameh

Department of Computer Sciences,
Purdue University,
West Lafayette, IN 47907
{knepley,sarin,sameh}@cs.purdue.edu
Phone: (765)494-7816, FAX: (765)494-0739

Abstract. Simulation of particles in fluids requires the solution of non-linear Navier-Stokes equations for fluids coupled with Newton's equations for particle dynamics, in which the most time consuming part is the solution of nonsymmetric and indefinite sparse linear systems. In this paper, we present a comprehensive algorithm for the simulation of particulate flows in two dimensional domains. A backward Euler method is used for time evolution, and a variant of Newton's method is used to solve the nonlinear systems. The linear systems are solved efficiently by a novel multilevel algorithm that generates discrete divergence-free space for the incompressible fluid. Unlike incomplete factorization preconditioners, our technique has the desirable properties of robust and effective precondi-tioning along with efficient implementation on parallel computers. We present experiments on the SGI Origin2000 that demonstrate the par-allel performance of our algorithm, and discuss various aspects of the simulation package and the associated software design.

1 Introduction

This work has grown out of a Grand Challenge Project aimed at developing highly efficient methods for simulating the motion of large number of particles in fluids under the action of the hydrodynamic forces and torques exerted by the suspending fluid. Such large scale simulations will then be used to elucidate the fundamental dynamics of particulate flows and solve problems of engineer-ing interest. The goal is to develop a high-performance, state-of-the-art software package called *Particle Movers* that is capable of simulating the motion of thou-sands of particles in two and three dimensions. The project aims to develop computational models for simulations in Newtonian fluids that are governed by the Navier-Stokes equations as well as in several popular models of viscoelastic fluids.

Particulate flow simulations are extremely computationally intensive, and re-quire the efficient use of parallel computers. It has been observed that the most time consuming part of the computation is the solution of large sparse linear

---

[*] This work is supported, in part, by the NSF through HPCC Grand Challenge grant ESC-95-27123. See http://www.aem.umn.edu/Solid-Liquid_Flows for more details.

systems. These systems are nonsymmetric and indefinite, causing severe difficulties for commonly used preconditioners based on incomplete factorizations. It is therefore imperative to develop novel algorithms for preconditioning that are robust and efficiently parallelizable. Our experiments demonstrate that the techniques developed in this paper have been successful in addressing these issues for systems arising in two dimensional simulations. We expect our method to apply equally effectively to three dimensional simulations as well.

The purpose of these simulations is used to study the microstructural (pair interaction) effects that produce clusters and anisotropic structures in particulate flows, to produce statistical analyses of particulate flows (mean values, fluctuation levels and spectral properties), to derive engineering correlations of the kind usually obtained from experiments, and to provide clues and closure data for the development of two-phase flow models and a standard against which to judge the performance of such models. In addition, they will be used to solve practical problems of industrial interest such as sedimentation, fluidization and slurry transport of solid particles in Newtonian and viscoelastic fluids.

The numerical simulations are validated against experimental data available in literature, experiments done in Prof. Daniel Joseph's laboratory at the University of Minnesota, or field data from industry, especially from our industrial sponsors. The project serves as a platform for the advancement of the science of solid-liquid flows using all the available tools: theory, experiments, and numerical simulation.

The paper is organized as follows: Section 2 outlines the equations governing the solid-liquid flows. In Section 3 we present the algorithm for simulation of particulate flows along with a description of the discretization techniques and fluid-particle interactions. Section 4 discusses implementation issues pertaining to algorithms, parallel processing, and software design. Conclusions and future work are presented in Section 6.

2 Problem Formulation

Consider an incompressible fluid flowing in a two dimensional channel of length L and width W, in which a pressure gradient is set against gravity (Fig. 1) and N_p solid particles are moving freely. The Reynolds number is large enough that inertia cannot be neglected. We are interested in determining the motion of both fluid and individual particles. The continuity and momentum equations for the fluid are given by:

$$\nabla \cdot \mathbf{u} = 0, \tag{1}$$

$$\rho \frac{\partial \mathbf{u}}{\partial t} + \rho \mathbf{u} \cdot \nabla \mathbf{u} = \rho \mathbf{g} - \nabla p + \nabla \cdot \tau, \tag{2}$$

where ρ is the fluid density, $\mathbf{g}$ is gravity, p is the pressure, and $\mathbf{u}$ is the fluid velocity. The tensor τ represents the extra-stress tensor; for a Newtonian fluid, it is given by

$$\tau = \mu \left(\nabla \mathbf{u} + \nabla \mathbf{u}^T \right). \tag{3}$$

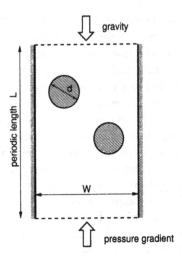

Fig. 1. Two solid particles moving in a periodic channel.

The particles obey Newton's law:

$$M\frac{dU}{dt} = F,\tag{4}$$

$$\frac{dX}{dt} = U,\tag{5}$$

where X and U are the generalized position and velocity vectors of the particles, respectively. The unknown vectors X and U incorporate both translational and angular components of the particle velocities. The matrix M denotes the generalized mass matrix and the vector F comprises of the forces and torques acting on the particles by the fluid as well as external fields such as gravity. Imposing a no-slip condition for the fluid on the surface of each particle, we obtain the following expression for fluid velocity at a point on the particle:

$$u = U + r \times \Omega.\tag{6}$$

Here, r is the vector from the center of the particle to the point on its surface, and Ω is the angular velocity of the particle.

Following Hu [6], the set of equations (1)–(6) is solved by a finite element technique in which the Galerkin formulation is generalized to incorporate both the fluid and particle degrees of freedom. As the domain occupied by the fluid changes with time, a moving mesh method is used, formulated in an arbitrary Lagrange-Euler (ALE) referential. More precisely, the momentum equation is rewritten as:

$$\rho\frac{\delta u}{\delta t} + \rho(u - u_m)\cdot\nabla u = \rho g - \nabla p + \nabla\cdot\tau,\tag{7}$$

where u_m is the mesh velocity field, and the time derivative is computed in the new reference frame.

3   Solution Methodology

The physical system is evolved from an initial state by an implicit time stepping scheme. At each time step, a nonlinear system of equations is solved that requires the solution of a series of linear systems involving the Jacobian of the nonlinear equations. The differential equations are approximated using the finite elements method on a nonuniform mesh. In this section, we outline the various algorithms used for the particulate flow.

3.1   Basic Algorithms

The first order accurate backward Euler method is used for time evolution. The choice of the time step is severely constrained by the requirement that particle dynamics be resolved accurately. For instance, no particle is allowed to translate farther than its radius, or rotate more than 180° in one time step. With such constraints, a first order scheme appears adequate. This choice has not impacted the stability or accuracy of the simulations. The change in particle velocity was also constrained, but this played no role in the simulations presented in this paper.

In our ALE scheme, the positions of the particles as well as the mesh are updated using an explicit formula based on the computed velocities. Furthermore, the velocity of nodes in the mesh must be computed at every nonlinear iteration prior to computing the Jacobian. In order to ensure a smooth distribution of the mesh nodes over the domain, the mesh velocities are constrained to satisfy the Laplace equation with an inhomogeneous weighting function $j(\mathbf{x})$:

$$\nabla \cdot (j(\mathbf{x})\nabla \mathbf{u}_m) = 0. \tag{8}$$

The weighting function $j(\mathbf{x})$ is chosen to be inversely proportional to the element area so that smaller elements are stiffer, i.e., less easily distorted. We also must maintain a well-shaped mesh; therefore, when particle motion deforms the elements to an unacceptable degree, the mesh is regenerated and the fields are projected from the old mesh.

At the time step t_n, the discretized coupled equations (1)–(8) lead to a system of nonlinear algebraic equations. Our discretization gives rise to a large sparse set of equations for the particle and fluid velocities, which is solved using a variant of the Newton's method. At each Newton step, a linear system with the Jacobian as the coefficient matrix must be solved. In this case, the Jacobian is a saddle-point problem with a nonsymmetric $(1,1)$ block. Moreover, this block becomes real positive for a sufficiently small Reynolds number. We choose to solve the full coupled problem with a preconditioned Krylov method. The general algorithm for simulation is presented below.

Algorithm *Particulate_Flow*

1. Generate the mesh for the domain.
2. Solve the nonlinear finite element problem (1)–(8) at time t_n.
3. Move the particle positions.
4. Move the nodes of the mesh.
5. If *mesh is distorted* then
    5.1. Generate a new mesh for the domain.
    5.2. Interpolate the fields onto the new mesh.
6. Set $t_{n+1} = t_n + \Delta t$; if $t_{n+1} < t_{max}$, go to 2, else terminate.

3.2 Finite Element Discretization

Our finite element formulation extends to both the fluid and particle equations of motion. Due to the boundary condition on the surface of the particles, the forces acting on the particle due to the fluid are canceled by the integral of the traction over the surface. Thus, hydrodynamic forces need not be calculated explicitly [5]. Spatial coordinates are discretized on a triangular mesh by a mixed finite element formulation using P2/P1 pair of elements. In other words, we use linear elements for pressure and quadratic elements for the velocity fields. The quadratic velocity elements are necessary to capture the behavior of closely spaced particles.

3.3 Fluid-Particle Interactions

In constructing the system of nonlinear equations, it is convenient to utilize building blocks from elementary computations. This simplifies greatly the subtle issues of parallelism, as well as, promoting code reuse and modular programming. We begin by assuming that discretization of the relevant fluid equations on our domain is a primitive operation. In this case, we refer to the discretized Jacobian, i.e., differential of the evolution operator. Our intention is to accommodate the presence of particles using projection operators that describe the constraints imposed on the system matrix by the physical boundary conditions on the particle surfaces [8]. Formally, we have

$$\tilde{J} = \bar{P}^T J \bar{P}, \tag{9}$$

where $\tilde{J}$ is the Jacobian of the system, J is the Jacobian of the *decoupled* fluid-particle system, and $\bar{P}$ is the projection matrix enforcing the fluid-particle coupling. In particular, J takes the form:

$$J = \begin{pmatrix} A & B & 0 \\ B^T & 0 & 0 \\ 0 & 0 & A_p \end{pmatrix},$$

where A is discrete operator for velocity in (2), B is the discrete gradient operator for pressure in (2), B^T is the negative of the discrete divergence operator in

(1), and A_p is the discrete evolution operator for particles obeying Newton's equations of motion.

To clarify this further, let us divide the velocity unknowns into two categories, u_I for interior velocity unknowns and u_Γ for velocity unknowns on the surface of the particles. The Jacobian can now be expressed as

$$
J = \begin{pmatrix}
A_I & A_{I\Gamma} & B_I & 0 \\
A_{\Gamma I} & A_\Gamma & B_\Gamma & 0 \\
B_I^T & B_\Gamma^T & 0 & 0 \\
0 & 0 & 0 & A_p
\end{pmatrix}.
$$

The no-slip condition on the surface of the particles demands that the fluid velocity at the surface equal the velocity of the surface itself. To enforce this condition, the projection matrix $\bar{P}$ is expressed as

$$
\bar{P} = \begin{pmatrix}
I & 0 & 0 \\
0 & P & 0 \\
0 & 0 & I \\
0 & I & 0
\end{pmatrix},
$$

where P is the projector from the space of surface velocity unknowns onto particle unknowns. Due to the simple form of the no-slip condition, P can be stored and applied in time proportional to the number of particles. Finally, the Jacobian in (9) takes the form:

$$
\tilde{J} = \begin{pmatrix}
A_I & A_{I\Gamma}P & B_I \\
P^T A_{\Gamma I} & P^T A_\Gamma P + A_p & P^T B_\Gamma \\
B_I^T & B_\Gamma^T P & 0
\end{pmatrix},
\tag{10}
$$

giving rise to a saddle-point problem.

4 Implementation

4.1 Algorithm Design

The nonlinear system at each time step was solved using an inexact Newton method [1] with a cubic line search algorithm [2]. However, a fixed tolerance on the inner linear system proved to be expensive; therefore, an algorithm for adaptively adjusting these tolerances proposed by Eisenstat and Walker [3] was also tested.

The incompressibility condition imposed on the fluid constrains the fluid velocity to be discretely divergence-free. The multilevel algorithm developed by Sarin and Sameh [11, 10] was used to obtain a well-conditioned basis, $N_{\tilde{B}^T}$, for the divergence-free velocity space. In particular, $N_{\tilde{B}^T}$ is a full column rank matrix corresponding to the null space of $\tilde{B}^T$, where $\tilde{B}^T = (B_I^T, B_\Gamma^T P)$. The discrete divergence-free velocity is expressed as $u = N_{\tilde{B}^T} z$, where z is an arbitrary vector. This approach eliminates pressure as an unknown, resulting in a smaller system.

The nonlinear equations for the fluid velocity can now be solved in divergence-free velocity space specified by $N_{\tilde{B}^T}$. This effectively constrains the Jacobian further, restricting the intermediate velocity fields in the Newton's method to this divergence-free space. The resulting linear system is real positive, which is a distinct advantage over the (indefinite) saddle-point problem (10). In addition, our formulation eliminated the variables constrained by Dirichlet boundary conditions from the system. This had the desirable effect of making the discrete gradient in the momentum equation (2) the transpose of the discrete divergence operator in the continuity equation (1), thereby allowing reuse of the multilevel factorization of the divergence-free velocity.

The resulting reduced system was solved using full GMRES [9]. The multilevel algorithm for divergence-free basis preconditions the linear system *implicitly*, and therefore, no additional preconditioning was necessary. Encouraging preliminary results have also been obtained using BiCGSTAB and alternative orthogonalization routines.

The saddle-point problem to be solved at each Newton iteration can be quite ill-conditioned, and we observed that relaxation preconditioners were ineffective in accelerating convergence. Note that the modified gradient operator $P^T B$ also has a greater connectivity, as rows corresponding to particle unknowns couple all the fluid unknowns on the particle surface. The use of projection matrices for the no-slip constraint restricts the use of incomplete factorization to the submatrix of the interior unknowns.

Our approach is based on early experience with systems in which the constraints on the surface of the particles were added explicitly to the system. In that case, the linear system did not involve the projector P, and ILU preconditioners could be used for the entire matrix. This strategy was moderately successful for uniprocessor implementations, but proved to be a severe bottleneck for parallel implementations. Moreover, attempts to increase parallelism by resorting to block ILU preconditioning [14] showed a sharp drop in the preconditioner's effectiveness for large problem instances.

4.2 Parallelism

The computational profile of our simulations suggests that the most time consuming sections of the algorithm are the following:

- generation of the Krylov subspace, and
- application of divergence-free basis.

While the matrix-vector products are efficiently parallelizable, the main limitation to parallelism is posed by the orthogonalization step in GMRES. As a result, the maximum achievable speedup is limited by the synchronization costs of dot products of the modified Gram-Schmidt algorithm. On the other hand, the multilevel algorithm for the divergence-free basis is very scalable and achieves speedups similar to parallel matrix-vector product computations.

The remainder of the computation consists of linear system solves for mesh velocity using conjugate gradient acceleration, and line search in the Newton algorithm which relies on evaluations of the nonlinear function. Both of these computations are highly parallel and scale very well.

4.3 Software Issues

The prototype simulation code was developed using Petsc [13] from Argonne National Laboratory. This provided a solid framework and structure for the rapid development of a robust and efficient code, a large portion of which is expected to be incorporated into future versions of Petsc. MPI [4] is employed for problem specific communication routines that aren't already supported by Petsc. As a result of these design choices, the code is very portable, and has been successfully executed on a variety of parallel architectures including the SGI Origin2000, Cray T3E, and a network of Sun workstations.

The modular design of our code has derived maximum benefit from quality existing software, which was seamlessly integrated into our code. Our mesh is generated using Triangle [12] from Shewchuk and partitioned using Parallel METIS [7] from Kumar and Karypis.

5 Numerical Experiments

In this section, we describe a set of benchmark problems that have been devised to compare simulation results with actual experiments conducted in Prof. Daniel Joseph's laboratory at the University of Minnesota.

5.1 Simulation Benchmarks

In the benchmark problems, we assume that the particles are circular disks in two dimensions and spheres in three dimensions. Our simulations were conducted for two dimensional domains with several disks. These disks are confined by closed side walls to move in two dimensions although the flow is actually three dimensional. A disk has diameter 0.25 inches, specific gravity is 1.14, and is fluidized in water. The channel has a solid impenetrable bottom and is 3.20 inches wide.

5.2 Sedimentation of a Single Particle

This problem simulates the sedimentation of a single disk from rest at a position in which the disk center is 30 inches above the bottom of the channel and half way between the left wall and the centerline. The most interesting aspect of the simulation is the trajectory of the particle with particular attention paid to the final stages when it approaches the bottom, at which point lubrication forces act with differing results depending on the collision model used. It is also useful to compare the equilibrium position in the channel with the experiments. *However, for the purpose of analyzing the parallel and numerical performance of the algorithm, we concentrate on the initial stages of the computation.*

At the first time step, the computational mesh for this problem has 2461 elements and 1347 nodes, generating 9418 unknowns in the unconstrained problem. Figure 2 shows the initial mesh and the associated partitioning into 8 subdomains. The simulation was run for five time steps ($\Delta t = 0.01s$) starting with the particle and fluid at rest. Table 5.2 presents parallel performance of the algo-

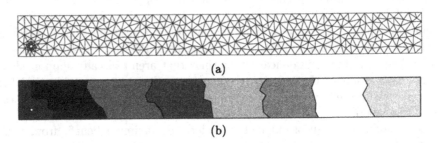

(a)

(b)

Fig. 2. Sedimentation of a single particle: (a) mesh with 2461 elements and 1347 nodes, and (b) partitioning into 8 domains. The domain has been rotated by 90° such that gravitational force pulls the particles towards the right.

rithm on the SGI Origin2000 at NCSA utilizing up to 8 processors. Clearly, the small size of this problem precludes efficiency on a large number of processors. The superlinear behavior observed for two processors is due to improved cache performance when the size of data on individual processors shrinks.

Table 1. Simulation of a single sedimenting particle on SGI Origin2000.

Processors	Time (sec.)	Speedup	Efficiency
1	1819	1.0	1.00
2	822	2.2	1.11
4	502	3.6	0.91
8	334	5.3	0.66

5.3 Sedimentation of Multiple Particles

The next benchmark simulates the sedimentation of 240 disks, starting from the initial configuration of a *crystal*. This crystal consists of an array of 240 particles arranged in 20 rows with 12 in each row. The lines between particle centers are vertical and horizontal, so that the array is not staggered. The centers of adjacent particles are $0.20/13 \approx 0.06154$ inches apart. The centers of particles along the walls are also at the same distance from the wall. The particles will sediment from a crystal whose top is initially 30 inches from the bottom of the channel.

At the first time step, the computational mesh for this problem has 8689 elements and 6849 nodes, generating 43,408 unknowns in the unconstrained the algorithm on the SGI Origin2000 utilizing up to 16 processors.

problem. Figure 3 shows the initial mesh and the associated partitioning into 8 subdomains. The simulation was run for five time steps ($\Delta t = 0.01s$) starting with the particles and fluid at rest. Table 5.3 presents parallel performance of

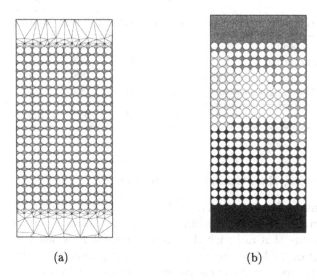

(a)   (b)

Fig. 3. Sedimentation of multiple particles: (a) mesh with 8689 elements and 6849 nodes, and (b) partitioning into 8 domains. The domain has been truncated to highlight the region of interest.

Table 2. Simulation of multiple sedimenting particle on SGI Origin2000.

Processors	Time (sec.)	Speedup	Efficiency
1	3066	1.0	1.00
2	1767	1.7	0.85
4	990	3.0	0.75
8	570	5.3	0.66

Table 5.3 presents the computational cost of critical steps. The nonlinear system solution time consists of the calculation of the Jacobian (*matrix assembly*), application of the nonlinear operator, formation of the multilevel preconditioner, and the solution of the linear system. The linear system solution time is dominated by *matrix-vector multiplication*, application of the multilevel preconditioner (*preconditioning*), and *orthogonalization* for the Krylov subspace vectors. The nonlinear solve takes most of the time and its parallelization is fairly well achieved.

Table 3. Parallel performance of critical steps for multiple sedimenting particles on SGI Origin2000.

Simulation step	$P = 1$		$P = 8$		
	Time (s)	Percent	Time (s)	Percent	Speedup
Matrix assembly	224	11	33	10	6.8
Preconditioning	1010	49	143	41	7.1
Matrix-vector multiplication	452	22	86	25	5.3
GMRES orthogonalization	360	18	83	24	4.3
Total	2046	100	345	100	5.9

5.4 Analyzing the Parallel Performance

The experiments described in Tables 5.2, 5.3, and 5.3 define *speedup* as the improvement in speed over the *best* implementation of the algorithm on a uniprocessor. This implies that although the parallel algorithm demonstrates good *speed improvement* on multiple processors, the speedup may be modest. We believe that this is the most realistic metric for any parallel algorithm.

The experiments indicate that the critical steps in the nonlinear solver are scalable and efficient even on a moderate sized problem. The performance may be further improved by using a less expensive alternative to GMRES in the solution of Newton's equation, or an orthogonalization routine with a smaller serial component. Another factor in the deterioration of efficiency is the slight increase in iterations due to degradation of the parallel preconditioner. In principle, one can use the *same* preconditioner on multiple processors without increase in iterations, but with marginal decrease in efficiency. This requires more sophisticated implementation, and will be focus of the next phase of the project. (The reader is referred to [10] for a scalable implementation of the preconditioner.) The decrease in parallel performance is also an artifact of the partitioning strategy which aims at good load balance at the expense of preconditioning. We anticipate improved parallel performance for the entire code after optimizing portions responsible for mesh movement and projection of fields onto new meshes.

6 Conclusions and Future Work

We have produced a portable, scalable, extensible, and robust simulation of the fully dynamic interaction between particles and fluid, capable of generating results that may be compared with the most recent experiments in the field. We have described a comprehensive algorithm for the simulation of particulate flows. The linear systems, which are the most time consuming part of these simulations, are solved by a novel multilevel algorithm. Our experiments demonstrate that this approach is a viable alternative for the parallel solution of the Navier-Stokes equations, as it exhibits reasonable scalability and robust preconditioning of the linear systems involved. Ongoing work on this project extends our simulation algorithm to three dimensions with more particles.

Acknowledgements

The authors wish to express their gratitude for helpful comments and suggestions from the grand challenge team: Todd Hesla, Howard Hu, Peter Huang, Betrand Maury, Tsorng-Whay Pan, Denis Vanderstraeten, and Andy Wathen; and the Petsc team: Barry Smith, Bill Gropp, Lois McInnes, and Satish Balay. The authors are especially grateful to Daniel Joseph for insightful discussions and support of this work. The first author thanks ANL for the opportunity to develop a large part of the software with the Petsc team.

References

1. R. S. Dembo, S. C. Eisenstat, and T. Steihaug. Inexact Newton methods. *SIAM Journal on Numerical Analysis*, 19(2):400–408, 1982.
2. J. E. Dennis and R. B. Schnabel. *Numerical Methods for Unconstrained Optimization and Nonlinear Equations*. Prentice-Hall, 1983.
3. S. C. Eisenstat and H. F. Walker. Choosing the forcing terms in an inexact Newton method. *SIAM Journal on Scientific Computing*, 17(1):16–32, 1996.
4. W. D. Gropp, E. Lusk, and A. Skjellum. *Using MPI: Portable Parallel Programming with the Message Passing Interface*. MIT Press, 1994.
5. T. Hesla. A combined fluid–particle formulation. Presented at a Grand Challenge group meeting.
6. H. Hu. Direct simulation of flows of solid–liquid mixtures. *International Journal of Multiphase Flow*, 22, 1996.
7. G. Karypis and V. Kumar. Parallel multilevel k-way partitioning scheme for irregular graphs. Technical Report 96–036, University of Minnesota, 1996.
8. B. Maury and R. Glowinski. Fluid–particle flow: A symmetric formulation. *C. R. Acad. Sci. Paris*, 324(I):1079–1084, 1997.
9. Y. Saad and M. H. Schultz. GMRES: A generalized minimum residual algorithm for solving nonsymmetric linear systems. *SIAM Journal on Scientific and Statistical Computing*, 7:856–869, 1986.
10. V. Sarin. *Efficient Iterative Methods for Saddle Point Problems*. PhD thesis, University of Illinois, Urbana-Champaign, 1197.
11. V. Sarin and A. H. Sameh. An efficient iterative method for the generalized Stokes problem. *SIAM Journal on Scientific Computing*, 19(1):206–226, 1998.
12. J. R. Shewchuk. Triangle: Engineering a 2d quality mesh generator and delaunay triangulator. In *First Workshop on Applied Computational Geometry*, pages 124–133. ACM, Philadelphia, PA, 1996.
13. B. F. Smith, W. D. Gropp, L. C. McInnes, and S. Balay. Petsc 2.0 Users Manual. Technical Report TR ANL-95/11, Argonne National Laboratory, 1995.
14. D. Vanderstraeten and M. G. Knepley. Parallel building blocks for finite element simulations: Application to solid–liquid mixture flows. In *Proceedings of the Parallel Computational Fluid Dynamics Conference*, Manchester, England, 1997.

Parallel Vertex-To-Vertex Radiosity on a Distributed Shared Memory System*

Adi Bar-Lev   Ayal Itzkovitz   Alon Raviv   Assaf Schuster
Department of Computer Science
Technion - Israel Institute of Technology

Abstract. In this paper we describe the parallel implementation of the Vertex-To-Vertex Radiosity method on a cluster of PC hosts with a Distributed Shared Memory interface (DSM). We first explain how we use stochastic rays to compute the Form-Factor. We then proceed to describe the implementation of this method on top of the MILLIPEDE system, a virtual parallel machine that runs on top of available distributed environments. We discuss a step-by-step process for exploiting MILLIPEDE's optimization mechanisms. Despite the relatively slow communication medium, the optimization process leads from initial slowdown to high speedups.

1 Introduction

Fast generation of photorealistic rendering of a scene using computer graphics techniques is one of the most challenging tasks facing researchers today. In the real world, a constant-in-time amount of energy that is emitted by a light source establishes an energy transfer balance between all the objects in the scene in a matter of just a few nano-seconds. It is this state of energy balance we wish to simulate and display as our final stage in the generation of a synthetic scene.

One popular and relatively efficient method for generating realistic scenes is the *Ray-Tracing* method. Though effective in scenes containing shiny and highly reflective (specular) objects, the Ray-Tracing technique will produce unrealistic effects for closed space scenes where most of the effects result from the diffusion of light from the walls and other objects. More on this subject can be found in [7].

The Radiosity method, adapted from the field of Heat Transfer (Sparrow & Cess [14]), attempts to render the scene by simulating this diffusion process. Diffusive materials distribute the unabsorbed energy from light sources in a uniform manner in all directions. The Radiosity method simulates the resulting energy transfer by discretizing the environment, representing it in terms of polygons, and finding an approximation of the energy each polygon receives. The computation eventually assigns to each polygon in the scene a fraction of the total energy which was distributed by the light sources. The energy is then translated into terms of color and intensity, thus providing the shade and lighting of the scene.

---

[1] Preliminary results of this work were presented at the Workshop on Algorithm Engineering, Venice, September 1997.

Most of the shading which we see in every day life is in fact a result of the diffusive property of substances, which causes the light to appear "soft", and to shade areas which are not in line of sight of its source, such as the area near the corners of a room or under a table. Thus, although radiosity does not work well for specular materials, it is highly effective in the rendering of such closed scenes, which are composed mostly of objects which diffuse light.

Radiosity also has the advantage of being *view independent*: once the scene has been generated, a viewer may be placed at any point, and since the energy distribution has already been computed, it will take a short time to generate the appearance of the scene from this specific point of view (POV). Ray-Tracing, in contrast, is *view dependent*; the full computation must be repeated for every new POV. Radiosity also has the ability to simulate not only lights from light sources, but also the shade caused by heat sources such as furnaces in a plant, thus enabling the simulation scenes far more realistic than those generated by other methods.

The solution for the Radiosity method is, however, known to be highly exhaustive in terms of both memory and computation, and can be as much as two orders of magnitude slower than the Ray-Tracing technique. In this work we present a parallel implementation for the Radiosity method, using an algorithm based on iterating energy casting. Note that spreading the energy between the elements of the scene typically involves an "all-to-all" type communication which may prevent speedups with naive parallel implementations. In particular, we developed in this work a parallel version of Radiosity on a network of PCs for which the communication media is relatively inefficient (when compared to other parallel computing platforms). Furthermore, we used the MILLIPEDE system which implements (among its other features, see below) Distributed Shared Memory, sometimes creating implicit bursts of communication due to false or true sharing of variables. In order to obtain speedups in this environment we had to design the data structures carefully, to use optimization techniques that are provided by the MILLIPEDE system, and to fine-tune the involved parameters.

Related works include [10, 6, 3, 16], see the full version for details.

1.1 The Millipede System

MILLIPEDE is a work-frame for adaptive distributed computation on non-dedicated networks of personal computers, [5, 9, 13, 8].[1] MILLIPEDE implements distributed shared memory (DSM) and load sharing mechanisms (MGS); it provides a convenient environment to distributively execute multithreaded shared-memory applications. MILLIPEDE provides several techniques for reducing false-sharing and other communication which is created implicitly when using the distributed shared memory. These include a strong support for reduced consistency of the shared-memory pages (thus allowing for duplicate local copies on different hosts),

---

[1] MILLIPEDE is currently implemented in user-level using the Windows-NT operating system, see also www.cs.technion.ac.il/Labs/Millipede.

with the consistency type assigned on a per-variable basis, a mechanism for allocating different variables on different pages of memory (thus avoiding ping-pong and false-sharing situations), and instructions to copy-in-and-lock pages of memory (to stabilize the system and optimize resource usage).

1.2 Radiosity - An Overview

Given a geometric representation environment, we can divide it into sets of polygons denoted as *patches*. The light computation in such an environment involves transmission of energy between the patches until an energy balance is reached.

Based on the theory of energy transfer between objects of diffused material, the relative amount of energy transfered between two patches is dependent on the area which is not occluded by other patches in the scene, and on their geometric relation (angles of planes, distance, size and shape). This geometric relation, called the *Form-Factor*, represents the relative amount of visibility between a point on a patch and the area of another patch in the scene.

After discretizing the scene into polygons, we can view the problem as a set of N equations with N variables. The variables in these equations are the total amount of energy received by each patch from the environment after reaching energy balance, and the coefficients of the equations are the Form-Factors between the given patch and the other patches in the scene.

Prior to the discretization, the differential equation is as follows (in the equations a $\cdot$ denotes a scalar multiplication, a $\bullet$ denotes the inner product with the normal of the plane, and a $\times$ denotes the cross-product of two vectors): $B_i \bullet dA_i = E_i \bullet dA_i + \rho_i \cdot \int_{A_j} (B_j \cdot F_{j,i} \bullet dA_j)$, where dA_i, dA_j denote the differential area of points i and j in the direction of the plane, respectively, B_i, B_j denote the energy gathered in dA_i and dA_j, respectively, E_i denotes the amount of energy emitted from dA_i, ρ_i denotes the reflectivity coefficient of dA_i, and $F_{j,i}$ denotes the Form-Factor from dA_j to dA_i.

After the discretization of the surfaces into polygons we get: $B_i \bullet A_i = E_i \bullet A_i + \rho_i \cdot \Sigma_j (B_j \cdot F_{j,i} \bullet A_j)$. Also, by using geometric relations we get: $B_i = E_i + \rho_i \cdot \Sigma_j (B_j \bullet F_{i,j})$.

This set of equations can be solved by using *Gaussian elimination* or the *Kramer method* for solving sets of equations with time complexity $O(N^3)$, where N is the number of patches in the scene. *Cohen and Greenberg* [4] used the *Gauss-Seidel technique* to develop the *Progressive-Refinement Radiosity* method of complexity $O(N^2)$. In this approach, for each iteration, each patch in the scene computes the amount of energy that it casts over the rest of the patches in the scene (rather than the amount of energy it receives from each of the other patches). Although it does not alter the solution, this approach enables the algorithm to pick the patch with the largest amount of energy for distribution, thus speeding up the approximation process by focusing on the equations of highest weight in the set. In addition, the iteration process can be stopped at any point and the outcome will still approximate the solution in a uniform way, as the patch with the largest amount of energy has less energy to distribute than a certain threshold.

2 The Radiosity Algorithm

We now present the Radiosity algorithm which is based on the vertex-to-vertex Form-Factor calculation. Our initial experiences with it were reported in [2]. Additional approaches using similar methodology were also introduced by others [11, 1].

2.1 Adaptive Area Subdivision

In this work we refer to a scene as a set of objects comprised of convex polygons (or that can be easily polygonized as such). In most scenes the shape and size of the polygons do not match the "shading lines" that are created in nature; hence, one needs to decompose a given scene in order to achieve best results. Thus the scene should first be split in the best possible way, and only then should the eventual energy of the patches that were created during the process be decided upon. According to the equations, the time complexity increases with the level to which the devision is refined, which implies a tradeoff between the time complexity and the accuracy of the result.

Before dividing the scene into new polygons (i.e., new patches), it is important to notice that the major energy contributors according to the equations are the light sources which also define the shadow lines in the scene. Although there may be patches which subsequently emit a lot of energy, they do not appear as point light-sources, and thus do not form sharp shadow borders. Thus, the heuristic choice of which polygons to split takes into account the energy differences between polygon vertices due to the energy casted by the primary light-sources only. Once the energy difference between two adjacent vertices exceeds a certain threshold, the polygon is split in order to match the shadow borderline.

The technique we chose for splitting the polygons is intuitive, relying on the notion of *Polygon Center of Gravity* (PCG). When a polygon P is to be divided, we construct its PCG as the "middle point" of its vertices, i.e., the point with the axis values that are the average of the corresponding values of all of the polygon's vertices. Let $V_{E_{i,j}}$ denote the point in the center of the edge $(V_i, V_j) \in P$. For each $V_i \in P$ with neighbors j and k we construct a new polygon with four vertices: V_i, $V_{E_{i,j}}$, $V_{E_{i,k}}$, and the PCG of P. As a result of this process, each divided polygon is split into several new polygons, one for each of its original vertices.

Assume that the energy of each patch can be measured at its vertices. We now denote: dA_i – the differential area around a given vertex V_i. In the discretization process, this area is represented by the area of the polygons surrounding V_i, and is thus refined at each step of their subdivision. $\Delta(B_i, B_j)$ – denotes the difference between the energy of two adjacent vertices V_i, V_j.

The recursive division algorithm is as follows: For each polygon P in the scene: **(1)** Construct the PCG of P and dA_i for each vertex V_i. **(2)** Receive energy from each energy source in the scene. **(3)** For each vertex pair V_i, V_j such that their energy difference $\Delta(B_i, B_j)$ is larger than a certain threshold subdivide the polygon P and repeat 1.

2.2 Form-Factor Computation

Computing the Form-Factors is the major computational bottleneck of the Radiosity method. In our approach, all the energy in the scene is divided between the patches, and the energy of each patch is subdivided between its own vertices. Thus, to calculate Form-Factors, we compute them between each vertex in the (subdivided) scene and every visible patch. For simplicity's sake, we refer to the vertices henceforth as *energy sources*. This approach helps us deal with simple light-sources as well, since we can use the light sources in exactly the same way that we use vertices. Given a vertex V_i, and a patch P_j, we compute the corresponding Form-Factor $F_{i,j}$ in two stages. The first stage computes the *Initial Form-Factor* $\widehat{F_{i,j}}$ which reflects the relation between V_i and P_j, but does not represent any occlusions which might have been caused by other patches in the scene.

Let $\widehat{F_{i,j}}$ denote the initial Form-Factor from V_i to P_j, v denote the number of vertices in P_j, R_k denote the vector beginning at V_i and ending at the kth vertex of P_j, β_k denote the angle between the two vectors R_k and R_{k+1} which connect V_i to two adjacent vertices of P_j, N_i denote the plane normal of P_i. Now, $\widehat{F_{i,j}}$ can be computed using the following equation for convex polygons: $\widehat{F_{i,j}} = 1/(2\pi) \cdot \Sigma_k \{\beta_k \cdot N_i \bullet \overline{(R_k \times R_{k+1})}\}$.

Given an energy-source V_i and a patch P_j, we compute $\widehat{F_{i,j}}$. Next we need to calculate the occlusions for each source of energy. We achieve an approximation of this occlusion by casting stochastic rays. Each ray is cast from a random point in the area near V_i (dA_i), to a random point on patch P_j. Both the start point and the end point of the ray are selected by using a uniform probability distribution. If N_j rays are cast towards P_j with Form-Factor $\widehat{F_{i,j}}$, $F_{i,j}$ is approximated by calculating the number of hits that actually occured on P_j, i.e., the integral over the non-occluded area of P_j from V_i's POV. The error function in this case can be computed by using the Poisson probability distribution.

Note that N_j is directly related to the energy of the source V_i, and to the size of $\widehat{F_{i,j}}$. Let n denote the number of rays cast over the whole hemisphere with an acceptable error function. Let τ denote the ratio between the energy of V_i and some pre-specified value, i.e., this ratio limits the number of rays cast for a single energy source as a function of its energy. Now, the number of rays N_j to be cast over P_j is calculated as follows: $N_j = \tau \cdot \widehat{F_{i,j}} \cdot n$.

2.3 The Progressive Refinement Algorithm

We now describe the global algorithm which uses the *Progressive Refinement* approach. Let $F_{l,i}$ denote the Form-Factor between a given light-source l and dA_i, $B_{l,i}$ denote the energy transfered from a light-source l to V_i (which depends on $F_{l,i}$), ΔB_i denote the energy that V_i received since the last time it casted energy, and B_i denote the total energy V_i received from the beginning of the process. The algorithm is as follows:

1. Compute the Form-Factors from each light-source over all polygons in the scene. Cast energy at the polygons accordingly.
2. Subdivide the scene into patches and repeat step 1 until a good division is obtained.
3. Having the final division, cast the energy of the light-sources towards the patches in the scene.
4. $\forall V_i$ assign: $B_i \leftarrow B_{l,i}$, $\Delta B_i \leftarrow B_{l,i}$.
5. Repeat the following operations: For each vertex V_i whose ΔB_i is larger than the threshold: **a.** for every patch in the scene P_j compute $F_{i,j}$ (first iteration only). **b.** Distribute ΔB_i between all patches; $B_i \leftarrow B_i + \Delta B_i$. **c.** $\Delta B_i \leftarrow 0$. **d.** Insert V_i at the end of the vertex list.

The algorithm uses two different notations for the energy. The first is the global energy B_i which the patch does not absorb, and which gives the patch its shade when the scene is rendered. The second is the energy received by V_i from other vertices in the scene during their energy distribution phase. The latter is added to the energy gathered by V_i (denoted B_i), and is distributed as ΔB_i to all the patches in the scene when V_i reaches the top of the list. In the described algorithm, each vertex is inserted into a sorted list of vertices, where the sorting is done by the ΔB of each vertex. A vertex is picked for energy casting from the head of the list, i.e., the most significant energy contributor at that current time.

3 Parallelizing the Algorithm

The Radiosity algorithm which was initially designed as a sequential application, was modified to work efficiently in parallel. This was done in two main steps. The first step was to analyze the potential maximal level of parallelism of the problem, i.e., to locate places in the code where work could potentially be carried in parallel. We put in minimal code modifications to make the application run concurrently on the MILLIPEDE system and verified the correctness of the outcome. The second step was to improve the performance by handling three factors: *true sharing*, *false sharing* and the *scheduling policy*. The results show that the second step was vital for improving the performance and obtaining speedups in a loosely-coupled distributed shared memory environment.

3.1 The Naive Parallelization

The Radiosity algorithm is based on iterating energy casting as described earlier, where each step is in fact independent of the other steps. Parallelism can be found mainly in three places. The first is the subdivision stage where the scene is split among the hosts so that each of them computes the subdivision on the polygons it receives. The second is the main stage of the algorithm where each vertex is assigned to a different host, which calculates the Form-Factor and distributes the energy difference to the other vertices. The last stage is the real-time display of the scene.

Parallelizing the subdivision stage is straightforward and we proceed to describe the second and third stages. The implementation of the second stage maintains for each vertex V_i, a data structure which contains the following information: its current energy B_i; its current delta energy ΔB_i (which the vertex can further distribute to the other vertices); and the new delta energy $\widehat{\Delta B_i}$ (received by V_i when some other vertex V_k distributes its ΔB_k towards V_i). Calculating the Form-Factor and distributing the energy can therefore be done in parallel by using following **Parallel Step**: for vertex V_i with non-zero $\widehat{\Delta B_i}$ do: **1.** Lock V_i and $\widehat{\Delta B_i}$. **2.** $\Delta B_i \leftarrow \widehat{\Delta B_i}$, $B_i \leftarrow B_i + \widehat{\Delta B_i}$. **3.** Unlock V_i and $\widehat{\Delta B_i}$. **4.** If the array $F_{i,j}$ does not exist then compute it. **5.** Distribute the ΔB_i according to the Form-Factor array.

Mutual exclusion should be enforced in two places. The first appears when a vertex V_i takes the $\widehat{\Delta B_i}$ and adds it to its B_i and its current ΔB_i before distributing its energy to the scene (Parallel Step, stages 1 through 3). No concurrent update of the $\widehat{\Delta B_i}$ is allowed at that time. The second place is when the $\widehat{\Delta B_k}$s of the other vertices is modified, an operation that should be taken exclusively by a single vertex at a time (Parallel Step, stage 5).

Using Millipede MILLIPEDE is designed to support multithreaded parallel programming on top of clusters of PCs. Therefore, the code changes for parallelizing the sequential Radiosity algorithm were minimal. Basically each vertex could execute the Parallel Step independently of all other vertices, thus making it a parallel execution task (a *job* in MILLIPEDE terminology). Memory allocation and deallocation operators **new** and **delete** were overloaded to use MILLIPEDE interface API for allocating memory from the shared memory. MILLIPEDE atomic operation **faa** (*Fetch And Add*) was used to preserve mutual exclusion when the new delta energy of the current job vertex was updated, or when this value was updated for the other job vertices.

Starting a new job for each vertex turned out to cause thrashing and calculation of redundant data. Thus we created execution jobs for a limited number of vertices that were picked from the beginning of the vertex array after the sorting phase, i.e., only those vertices with the highest energy to be cast are handled exclusively by a dedicated job.

Scheduling is done by a special job, the *job manager*, which is in charge of scheduling new jobs, as follows: while(not finished): **1.** Sort the vertices according to their energy. **2.** If the number of working threads is below a minimal number, then for each vertex whose energy is above the threshold spawn a new job running the Parallel Step. **3.** Sleep Δ milliseconds (let the working threads cast some more energy before the next round).

In Figure 1 we see that the naive implementation performs poorly. When the number of machines grows, the computation slows down and the parallel algorithm has no advantage over the sequential one.

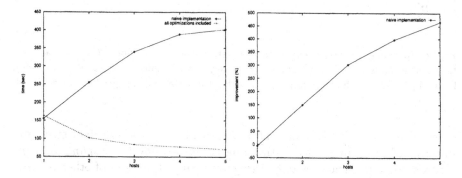

Fig. 1. Two versions of the parallel Radiosity algorithm. In the left graph we see that the naive implementation performs poorly when the number of machines involved in the computation increases. In the right graph we see that performance can improve by more than 450 percent over performance of the naive for the fully optimized implementation, when the number of machines grows to 5.

3.2 The Optimizations

The parallel implementation, as described above, suffers from several inefficiencies which greatly influence the performance. We attempted to increase performance by focusing on three factors: the sharing of the data structures, the false sharing in the algorithm implementation, and the scheduling policy. We show that each change greatly improves the execution time, while the combination of the three results in good speedups for various scenes, as can be seen in Figure 2.

True Sharing Calculating the Form-Factor of a vertex is a time consuming operation. The calculation involves reading the characteristics of other vertices, performing the mathematical operations and finally storing the Form-Factor in the private data structure of each vertex. In terms of DSM operations, calculating the Form-Factor involves *reading* from other vertices data structures and *writing* to that of its own vertex.

On the other hand, distributing the vertex energy (called *shooting Radiosity*) is a very short operation, and requires merely updating the data structures of

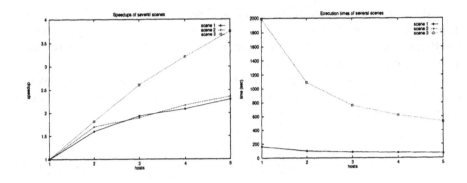

Fig. 2. Speedups for three scenes, two relatively small ones (numbers 1 and 2, each less than 200 seconds sequentially), and one relatively large (number 3, about 2000 seconds). For problems with high computational demands (like scene 3) good speedups are achieved.

other vertices according to current values of the vertex's own data structure. In terms of DSM operations, distributing the energy involves *reading* from the vertex's own data structure and *writing* to the data structures of other vertices.

In the naive parallel algorithm a job does both operations, namely, calculating the Form-Factor (if required) and distributing the energy to other vertices. Thus, in the naive parallelization, a true sharing of the vertices data structures occurs. The sharing of the data structures by several jobs leads to massive page shuttling and long delays in execution. Note, however, that the Form-Factor calculation takes the majority of the computation time and has a high computation-to-communication ratio (vertices' data structures are only being read). Therefore, in our solution, only Form-Factor calculations are carried out in parallel. All energy distribution is done on one of the machines, while calculations of the Form-Factors are sent to other machines to be computed concurrently. As a result of this change, access to shared data is reduced significantly.

The Parallel Step therefore changes. New jobs are spawned, executing Parallel Step 1 (see below) in the case that the Form-Factor has not yet been calculated, and Parallel Step 2 (see below) otherwise.

Parallel Step 1 for a vertex V_i: calculate its Form-Factors towards all patches. **Parallel Step 2** for vertex V_i with non-zero new delta energy: **1.** Copy the $\widehat{\Delta B_i}$ to the current B_i and the current ΔB_i. **2.** Distribute the current ΔB_i of V_i according to the Form-Factor to the $\widehat{\Delta B_k}$s of the vertices that are visible to/from V_i.

Figure 3 shows the performance boost of the improved algorithm when the parallel steps are separated into local and remote execution. The improvement is the result of a dramatic reduction in the sharing of the vertices' data structures. In fact, for the sake of locality of memory reference, the potential level of parallelism is *reduced* such that distributing the energy (shooting Radiosity) is *not* carried out concurrently.

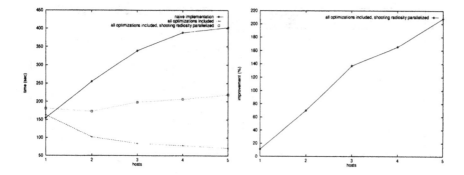

Fig. 3. Maximizing the level of parallelism does not give optimal execution times. When the creation of remote concurrent jobs is restricted to those calculating the Form-Factor, then an improvement of 60-200 percent is achieved due to less sharing and increased locality of memory references.

False Sharing It turned out that the naive parallel Radiosity implementation contained several falsely shared variables in the vertex structure, which led to significant performance degradation. Among others, this structure contains the values that reflect the energy in the node (B, ΔB and $\widehat{\Delta B}$). This data is referenced only at the stage of energy distribution, and is not required during the Form-Factor calculation. Thus, the false sharing on the vertices' data structure can be avoided by extracting the energy fields to a stand-alone allocation. As a result, jobs which calculate the Form-Factor do not (falsely) share data with shooting Radiosity operations, and thus would not be disturbed by energy distribution carried out concurrently by other vertices.

Figure 4 shows that a performance improvement of 30-50 percent is obtained when the energy variables are excluded from the vertex data structure to a stand-alone allocation. There were several additional falsely shared variables in the naive implementation, and we used the same technique for them.

Scheduling Policies The previously described naive scheduler wakes up once in every Δ milliseconds. It sorts the array of vertices according to their energy level, and then spawns new jobs for handling the most significant vertices in the scene at this point. It turned out that a fine tuning of the value of Δ is very important. When Δ is too small, too many jobs may end up working concurrently on different vertices, making redundant calculations which may be insignificant in terms of energy dispersal. Also, choosing too small a value for Δ causes frequent reads of energy information by the scheduling job (for its sorting task), which may cause data races with those jobs updating the energy values in the vertices. At the other extreme, choosing too large a value for Δ might cause idle machines to wait too long for the scheduler to wake up and spawn new jobs.

Figure 5 shows how performance is dependent on Δ. Further optimization

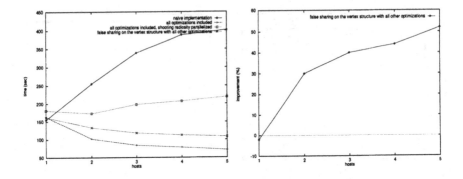

Fig. 4. False sharing causes a slowdown of 30-50 percent in the potential optimal execution time of the parallel Radiosity algorithm. The graph shows the effect of solving it by separating the related allocations. The right graph shows some performance degradation on a single machine, which is the result of an additional level of indirection in accessing some variables. In a distributed system this overhead can be neglected.

may adjust the value of Δ dynamically during execution. This requires further study.

We also tried another approach in which the scheduler is set to be event triggered. We used the integrated mechanism for message passing in MILLIPEDE (named MJEC) to implement the scheduler in the following way. When the Parallel Step ends, the job sends a FINISHED message to the scheduler, which triggers the scheduler to spawn new tasks. Upon a FINISHED message: **1.** Sort the array of vertices according to their energy. **2.** For each of the first several highest-energy vertices spawn a new job running the Parallel Step. Figure 6 shows that the event triggered scheduling policy performs better than the eager Δ scheduling even when using the optimal Δ value for the naive scheduler.

4 Conclusions

In this work we implemented a parallel Radiosity rendering algorithm on a loosely-coupled cluster of PCs. The initial naive algorithm did not perform well, and in fact the computation time increased with the number of machines. We then applied a handful of tunings and optimization techniques which eventually made the algorithm execute efficiently, achieving speedups close to optimal.

This project was carried on the MILLIPEDE system which provides an easy to use environment for parallel programming on top of distributed clusters of PCs. In fact, the MILLIPEDE concept was designed to promote this precise modus operandi (in the process of parallel programming): a first easy phase of porting an existing sequential code to the parallel domain, and then, in a follow-up phase, a step-by-step gradual optimization towards efficiency and improved speedups.

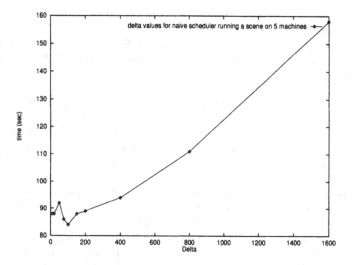

Fig. 5. The scheduler sleeps for Δ milliseconds between successive iterations. The optimal Δ for the scene benchmark turns out to be approximately 100 milliseconds. Execution time using five machines is very high when Δ values are much larger than the optimal one.

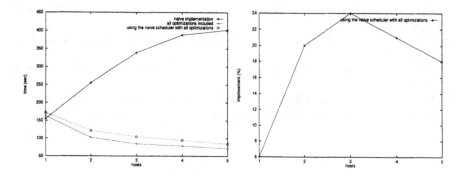

Fig. 6. The naive scheduler performs 15-20 percent worse than the event triggered scheduler. Results in the graph were taken with the optimal Δ values for the naive scheduler.

References

1. A. Keller. Instant Radiosity. In *Computer Graphics Proceedings*, pages 49–55, August 1997.
2. A. Bar-Lev, A. Itzkovich, A. Raviv, and A. Schuster. Vertex-to-Vertex Parallel Radiosity on a Cluster of PCs. In *Workshop on Algorithm Engineering*, Venice, September 1997.
3. K. Bouatouch and T. Priol. Data Management Scheme for Parallel Radiosity. *Computer Aided Design*, 26(12):876–882, 1994.

4. M. F. Cohen and J. R. Wallace. *Radiosity and Realistic Image Synthesis*. AP Professional, 1993.

5. R. Friedman, M. Goldin, A. Itzkovitz, and A. Schuster. Millipede: Easy Parallel Programming in Available Distributed Environments. *Software: Practice & Experience*, 27(8):929–965, August 1997. Preliminary version appeared in Proc. Euro–Par, Lyon, August 1996, pp. 84–87.

6. T.A. Funkhouser. Coarse-Grained Parallelism for Hierarchical Radiosity using Group Iterative Methods. In *ACM SIGGRAPH*, pages 343–352, 1996.

7. A. S. Glassner. *An Introduction to Ray-Tracing*. Academic Press, 1989.

8. A. Itzkovitz, A. Schuster, and L. Shalev. Millipede: Supporting Multiple Programming Paradigms on Top of a Single Virtual Parallel Machine. In *Proc. HIPS Workshop*, Geneve, April 1997.

9. A. Itzkovitz, A. Schuster, and L. Shalev. Thread Migration and its Applications in Distributed Shared Memory Systems. *To appear in the Journal of Systems and Software*, 1998. (Also: Technion TR LPCR-#9603).

10. L. Renambot and B. Arnaldi and T. Priol and X. Pueyo. Towards Efficient Parallel Radiosity for DSM-based Parallel Computer Using Virtual Interfaces. In *Proc. Symp. on Parallel Rendering*, pages 79–86, Phoenix, October 1997.

11. P. Lalonde. An Adaptive Discretization method for Progressive Radiosity. In *Proceedings Graphics Interface*, pages 78–86, May 1993.

12. N. L. Max and M. J. Allison. Linear Radiosity Approximation Using Vertex-To-Vertex Form Factors. *Graphics Gems*, 3:318–323, 1992.

13. A. Schuster and L. Shalev. Using Remote Access Histories for Thread Scheduling in Distributed Shared Memory Systems. Technical Report #9701, Technion/LPCR, Jan 1997.

14. E. M. Sparrow and R. D. Cess. *Radiation Heat Transfer*. Hemisphere Publishing Corp., 1978.

15. F. Tampieri. Accurate Form-Factor Computation. *Graphics Gems*, 3:329–333, 1992.

16. Y. Yu, O.H. Ibarra, and T. Yang. Parallel Progressive Radiosity with Adaptive Meshing. *Journal of Parallel and Distributed Computing*, 42:30–41, April 1997.

Load Balancing in Parallel Molecular Dynamics*

L. V. Kalé, Milind Bhandarkar and Robert Brunner

Dept. of Computer Science, and
Theoretical Biophysics Group, Beckman Institute,
University of Illinois,
Urbana Illinois 61801
{kale,milind,brunner}@ks.uiuc.edu

Abstract. Implementing a parallel molecular dynamics as a parallel application presents some unique load balancing challenges. Non-uniform distribution of atoms in space, along with the need to avoid symmetric redundant computations, produces a highly irregular computational load. Scalability and efficiency considerations produce further irregularity. Also, as the simulation evolves, the movement of atoms causes changes in the load distributions. This paper describes the use of an object-based, measurement-based load balancing strategy for a parallel molecular dynamics application, and its impact on performance.

1 Introduction

Computational molecular dynamics is aimed at studying the properties of biomolecular systems, and their dynamic interactions. As human understanding of biomolecules progresses, such computational simulations become increasingly important. In addition to their use in understanding basic biological processes, such simulations are used in rational drug design. As researchers have begun studying larger molecular systems, consisting of tens of thousands of atoms, (in contrast to much smaller systems of hundreds to a few thousands of atoms a few years ago), the computational complexity of such simulations has dramatically increased.

Although typical simulations may run for weeks, they consist of a large number of relatively small-grained steps. Each simulations step typically simulates the behavior of the molecular system for a few femtoseconds, so millions of such steps are required to generate several nanoseconds of simulation data required for understanding the underlying phenomena. As the computation involved in each timestep is relatively small, effective parallelization is correspondingly more difficult.

Although the bonds between atoms, and the forces due to them, have the greatest influence on the evolving structure of the molecular system, these forces do not constitute the largest computational component. The non-bonded forces,

* This work was supported in part by National Institute of Health (NIH PHS 5 P41 RR05969-04 and NIH HL 16059) and National Science Foundation (NSF/GCAG BIR 93-18159 and NSF BIR 94-23827EQ).

the van der Waals and electrostatic (Coulomb) forces between charged atoms consume a significant fraction of the computation time. As the non-bonded forces decrease as the square of interaction distance, a common approach taken in biomolecular simulations is to restrict the calculation of electrostatic forces within a certain radius around each atom (*cutoff* radius). Cutoff simulation efficiency is important even when full-range electrostatic forces are used, since it is common to perform several integration steps using only cutoff forces between each full-range integration step. This paper therefore focuses on cutoff simulations performance.

In this paper, we describe the load balancing problems that arise in parallelization of such applications in context of a production quantity molecular dynamics program, NAMD 2 [5], which is being developed in a collaboratory research effort. The performance of the load balancing strategies employed by this program are evaluated for a real biomolecular system. The load balancing strategy employed is based on the use of migratable objects, which are supported in Charm++ [7], a C++ based parallel programming system. It relies on actual measurement of time spent by each object, instead of predictions of their computational load, to achieve an efficient load distribution.

2 NAMD: A Molecular Dynamics Program

NAMD [8] is the production-quality molecular dynamics program we are developing in the Theoretical Biophysics group at the University of Illinois. From inception, it has been designed to be a scalable parallel program. The latest version, NAMD 2 implements a new load balancing scheme to achieve our performance goals.

NAMD simulates the motions of large molecules by computing the forces acting on each atom of the molecule by other atoms, and integrating the equations of motions repeatedly over time. The forces acting on atoms include bond forces, which are spring-like forces approximating the chemical bonds between specific atoms, and non-bonded forces, simulating the electrostatic and van der Waals forces between all pairs of atoms. NAMD, like most other molecular dynamics program, uses a cutoff radius for computing non-bonded forces (although it also allows the user to perform full-range electrostatics simulations computation using DPMTA library[9]).

Two decomposition schemes are usually employed to parallelize molecular dynamics simulations: force decomposition and spatial decompositon. In the more commonly used scheme, force decomposition, a list of atom pairs is distributed evenly among the processors, and each processor computes forces for its assigned pairs. The advantage of this scheme is that perfect load balance is obtained trivially by giving each processor the same number of forces to evaluate. In practice, this method is not very scalable because of the large amount of communication that results from the distribution of atom-pairs without regard for locality. NAMD uses a spatial decomposition method. The simulation space is divided into cubical regions called *patches*. The dimensions of the patches are chosen to be slightly larger than the cutoff radius, so that each patch only needs the coordinates from the 26 neighboring patches to compute the non-bonded forces. The disadvantage of spatial decomposition is that the load represented by a patch varies considerably because of the variable density of atoms in space, so a smarter load balancing strategy is necessary.

The original version of NAMD balanced the load by distributing patches among processors and giving each patch responsibility for obtaining forces exerted upon its constituent atoms. (In practice, Newton's third law insures that the force exerted by atom i due to atom j is the same as that exerted on j by i. Such pair interactions are computed by one of the owning patches, and the resulting force sent to the other patch, halving the amount of computation at the expense of extra communication.) We discovered that this method did not scale efficiently to larger numbers of processors, since out of a few hundred patches, only a few dozen containing the densest part of a biomolecular system accounted for the majority of the computation. The latest version of NAMD adds another abstraction, *compute* objects (see figure 1). Each compute object is responsible for computing the forces between one pair of neighboring patches. These compute objects may be assigned to any processor, regardless of where the associated patches are assigned. Since there is a much larger number of compute objects than patches, the program achieves much finer control over load balancing than previously possible. In fact, we found that the program could reach a good load balance just by moving compute objects, so we also refer to them as *migratable* objects. Migratable objects represent work that is not bound to a specific processor.

Most communication in NAMD is handled by a third type of object, called *proxy patches*. A proxy patch is an object responsible for caching patch data on remote processors for compute objects. Whenever data from a patch X is needed on a remote node, a proxy PX for patch X is created on that node. All compute objects on that remote node then access the data of patch X through proxy patch PX, so that the patch data is sent to a particular processor only once per simulation step. The main implication of this for load balancing is that once a proxy patch is required on a particular processor, placing additional compute objects on that processor does not result in increased communication cost.

3 Static Load Balancing

NAMD uses a predictive computational load model for initial load balancing. The computational load has two components. One component, proportional to the number of atoms, accounts for communication costs, integration, and bond force computation. The second component resulting from the non-bonded force computation, is based on the number of atom pairs in neighboring patches. Profiling indicates that this component can consume as much as eighty percent of typical simulations, so good overall load balance depends on good distribution of this work.

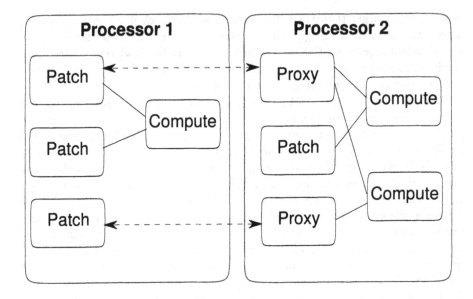

Fig. 1. This is a diagram showing object interactions in NAMD 2. Compute objects use data held by either patches or proxies. Each proxy receives and buffers position data from its owner patch on another processor, and returns forces calculated by compute objects to the owner patch. In order to move a Compute object to another processor, it is only necessary to create proxies on the new processor for all patches from which the compute receives position data.

The first stage of load balancing uses a recursive-bisection algorithm to distribute the patches among the processors so that the number of atoms on each processor is approximately equal. The recursive bisection algorithm ensures that adjacent patches are usually assigned to the same processor.

The second stage is the distribution of compute objects that carry out the non-bonded force computations. First, the compute objects responsible for self-interactions (interactions between a pair of atoms where both atoms are owned by the same patch) are assigned to the processor where the patch has been assigned, and the load for the processor incremented by the N_{atoms}^2. Then the compute objects for each pair of neighboring patches are considered. If the patches reside on the same processor, the compute object is assigned to that processor; otherwise the compute object is assigned to the least-loaded of the two processors. Then the load-balancer increments the load for that processor by $weight \times N_{atoms_1} \times N_{atoms_2}$. The weights take into account the geometric relationship between the two patches. There are three weights corresponding to whether the patches touch at a corner, edge or face, since patches which have an entire face in common contribute more pairs which actually must be computed than patches which share only a corner.

This method yields better load balance than earlier implementations, which only balanced the number of atoms. However, it does not take advantage of the freedom to place compute objects on processors not associated with either patch. It is possible to build a more sophisticated static load balancing algorithm that uses that degree of freedom, and even tries to account for some communication related processor-load. However, the geometric distributions of atoms within a patch impacts the load considerably, making it harder to predict by static methods. So, we decided to supplement this scheme with a dynamic load balancer.

4 Dynamic Load Balancing

Dynamic load balancing has been implemented in NAMD using a distributed object, the *load balance coordinator*. This object has one branch on each processor which is responsible for gathering load data and implementing the decisions of the load balancing strategy.

4.1 The Load Balancing Mechanism

NAMD was originally designed to allow periodic load rebalancing to account for imbalance as the atom distribution changes over the course of the simulation. We soon observed that users typically break multi-week simulations into a number of shorter runs, and that rebalancing at the start of each of these runs is sufficient to handle changes in atom distribution. However, difficulties with getting a good static load balance suggested that the periodic rebalancing could be used to implement a measurement-based load balancer.

If the initial patch assignment makes reasonable allowances for the load represented by each patch, compute object migration alone provides a good final load balance. During the simulation, each migratable object informs the load balance coordinator when it begins and ends execution, and the coordinator accumulates the total time consumed by each migratable object. Furthermore, the Converse runtime system [6] provides callbacks from its central message-driven scheduler, which allows the coordinator to compute idle time for each processor during the same period. All other computation, including the bond-force computations and integration, is considered background load. The time consumed by the background load is computed by subtracting the idle time and the migratable object times from the total time.

After simulating several time steps, the load balance coordinator takes this data and passes it to the selected load balancing strategy object (see section 4.2). The strategy object returns new processor assignments for each migratable object. The coordinator analyzes this list and determines where new proxy patches are required. The coordinator creates these new proxy patches, moves the selected migratable objects, and then resumes the simulation.

The first rebalancing results in many migratable object reassignments. The large number of reassignments usually results in changes to the background load, due to (difficult to model) changes in communication patterns. Therefore, after a few more steps of timing, a second load balancing step is performed, using an algorithm designed to minimize changes in assignments (and therefore in communication load). The second balancing pass produces a small number of additional changes, which do not change the background load significantly, but result in an improved final load distribution.

4.2 The Load Balancing Strategy

The load balancing strategy object receives the following pieces of information from the load balance coordinator (all times/loads are measured for several recent timesteps):

- The background (non-migratable) load on each processor.
- The idle time on each processor.
- The list of migratable objects on each processor, along with the computation load each contributes.
- For each migratable object, a list of patches it depends on.
- For each patch, its home processor, as well as the list of all processors on which a proxy must exist for non-migratable work.

Based on this information, the strategy must create a new mapping of migratable objects to processors, so as to minimize execution time while not increasing the communication overhead significantly. We implemented several load balancing strategies to experiment with this scenario. Two of the most successful ones are briefly described below.

It is worth noting that a simple greedy strategy is adequate for this problem if balancing computation were the sole criterion. In a standard greedy strategy, all the migratable objects are sorted in order of decreasing load. The processors are organized in a *heap* (i.e. a prioritized queue), so that the least loaded processor is at the top of the heap. Then, in each pass, the heaviest unassigned migratable object is assigned to the least loaded processor and the heap is reordered to account for the affected processor's load.

However, such a greedy strategy totally ignores communication costs. Each patch is involved in the electrostatic force computations with 26 neighboring patches in space. In the worst-case, the greedy strategy may end up requiring each patch to send 26 messages. In contrast, even static assignments, using reasonable heuristics, can lead to at most six or seven messages per patch. In general, since more than one patch resides on each processor, message-combining and multicast mechanisms can further reduce the number of messages per patch if locality is considered. Since the communication costs (including not just the cost of sending and receiving messages, but also the cost of managing various data structures related to proxies) constitute a significant fraction of the overall execution time in each timestep, it is essential that the load balancing algorithm also considers these costs.

One of the strategies we implemented modifies the greedy algorithm to take communication into account. Processors are still organized as a heap, and the migratable objects sorted in order of decreasing load. At each step, the "heaviest" unassigned migratable object is considered. The algorithm iterates through all the processors, and selects three candidate processors: (1) the least loaded processor where both patches (or proxies) the migratable object requires already exist (so no new communication is added), (2) the least loaded processor on which one of the two patches exists (so one additional proxy is incurred), and (3) the least loaded processor overall. The algorithm assigns the migratable object the best of these three candidates, considering both the increase in communication and the load imbalance. This step is biased somewhat towards minimizing communication, so the load balance obtained may not be best possible.

After all the migratable objects are tentatively assigned, the algorithm uses a refinement procedure to reduce the remaining load imbalance. During this step, all the overloaded processors (whose computed load exceeds the average by a certain amount) are arranged in a heap, as are all the under-loaded processors. The algorithm repeatedly picks a migratable object from the highest loaded processor, and assigns it to a suitable under-loaded processor.[2]

The result of implementing these new object assignments, in addition to anticipated load changes, creates new communication load shifts. Also, a part of the background work depends on the number (and size) of proxy patches on each processor, and increases or decrease as a result of the new assignment. As a result, the new load balance is not as good as the load balance strategy object expected. Rather than devising more complex heuristics to account for this load shift, we chose a simpler option. As described in section 4.1, a second load balancing pass is performed to correct for communication-induced load imbalance. This pass uses the refinement procedure only. Since the load is already fairly well balanced, only a few migratable objects are moved, improving load balance further without significant change in communication load. We find two passes sufficient to produce good load balance.

5 Performance

Performance measurements of various runs of NAMD were done using Projections, a performance tracing and visualization tool developed by our research group. Projections consists of an API for tracing system as well as user events and a graphical tool to present this information to the user in addition to modules to analyze the performance. To suit our needs for this particular task, Projections was extended with capabilities such as generating histograms, displaying user-defined markers, and tracing thread-events.

Our performance results are from an actual molecular system (ER-GRE, an estrogen receptor system) being studied by the Theoretical Biophysics Group. The system is composed of protein and DNA fragments in a sphere of water. Approximately 37,000 atoms are simulated using an 8.5 Å cutoff. The simulations are run on 16 processors of CRAY T3E.

---

[2] The techniques used to select the appropriate pair of candidates is somewhat involved and have been omitted.

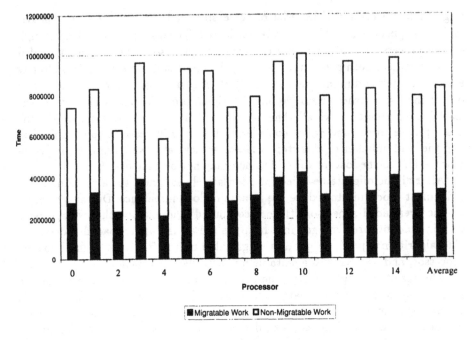

Fig. 2. Load with static load balancing

Figures 2 and 3 show a typical improvement obtained with our load balancer. Figure 2 shows measured execution times for the first eight steps of the simulation (where load balancing is based on a predictive static load balancing strategy described in section 3), and figure 3 shows times for the eight steps after load balancing. Since total execution time is determined by the time required by the slowest processor, reducing the maximum processor time by ten percent decreases execution time by the same amount. We have also observed that for larger numbers of processors, the static load balancer produces even more load variation. The measurement-based load balance does not produce such variation, producing greater performance improvement. Comparisons of the average times (the rightmost bar on each plot) shows that the new load distribution does not increase computational overhead to obtain better balance.

Figure 4 shows the speedup compared to one processor for the same simulation, using measurement-based load balancing. Since the number of computational objects is fixed, and total communication increases with increasing number of processors, load balancing grows more complex. Our load balancer exhibits good speedup even for larger numbers of processors.

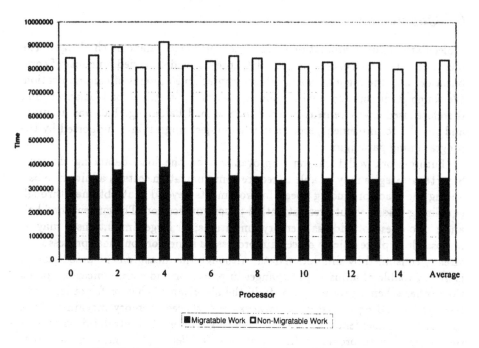

Fig. 3. Load after measurement-based load balancing

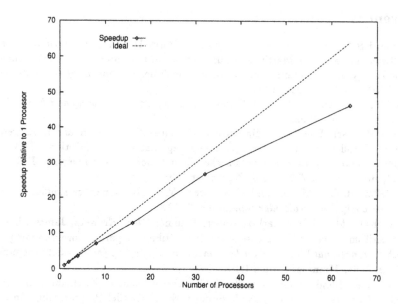

Fig. 4. Speedup for simulation steps after load balancing.

6 Conclusion

Molecular dynamics simulation presents complex load balancing challenges. The pieces of work which execute in parallel represent widely varied amounts of computation, and the algorithm exhibits an appreciable amount of communication. We found it difficult to devise good heuristics for a static load balancer, and therefore used measurement-based dynamic load balancing that was accurate and simpler to implement, especially when object migration is supported by the programming paradigm.

Future work will focus on improvements on the load balancing strategy. Smarter strategies may be able to reduce communication costs in addition to balancing the load, producing speed improvements beyond that obtained through equalizing load without great attention to communication. The current load balancer also does not address memory limitations; for large simulations it may be impossible to place objects on certain processors if memory on those processors is already consumed by simulation data. Although the measurement-based scheme is more capable of adjusting to changes in processor and communication speeds than other schemes, we will also study the algorithms behavior for other architectures, including workstation networks and shared memory machines. Also, the current implementation of load-balancing strategy is centralized on processor 0. We plan to provide a framework for parallel implementation for future load-balancing strategies.

References

1. Bernard R. Brooks, Robert E. Bruccoleri, Barry D. Olafson, David J. States, S. Swaminathan, and Martin Karplus. CHARMM: A program for macromolecular energy, minimization, and dynamics calculations. *Journal of Computational Chemistry*, 4(2):187–217, 1983.
2. Axel T. Brünger. *X-PLOR, A System for X-ray Crystallography and NMR*. Yale University Press, 1992.
3. Terry W. Clark, Reinhard v. Hanxleden, J. Andrew McCammon, and L. Ridgeway Scott. Parallelizing molecular dynamics using spatial decomposition. Technical report, Center for Research on Parallel Comutation, Rice University, P.O. Box 1892, Houston, TX 77251-1892, November 1993.
4. H. Heller, H. GrubMuller, and K. Schulten. Molecular dynamics simulation on a parallel computer. *Molecular Simulation*, 5, 1990.
5. L. V. Kalé, Milind Bhandarkar, Robert Brunner, Neal Krawetz, James Phillips, and Aritomo Shinozaki. A case study in multilingual parallel programming. In *10th International Workshop on Languages and Compilers for Parallel Computing*, Minneapolis, Minnesota, June 1997.
6. L. V. Kalé, Milind Bhandarkar, Narain Jagathesan, Sanjeev Krishnan, and Joshua Yelon. Converse: An Interoperable Framework for Parallel Programming. In *Proceedings of the 10th International Parallel Processing Symposium*, pages 212–217, Honolulu, Hawaii, April 1996.
7. L.V. Kale and S. Krishnan. Charm++: A portable concurrent object oriented system based on C++. In *Proceedings of the Conference on Object Oriented Programming Systems, Languages and Applications*, September 1993.

8. Mark Nelson, William Humphrey, Attila Gursoy, Andrew Dalke, Laxmikant Kalé, Robert D. Skeel, and Klaus Schulten. NAMD— A parallel, object-oriented molecular dynamics program. *Intl. J. Supercomput. Applics. High Performance Computing*, 10(4):251–268, Winter 1996.

9. W. Rankin and J. Board. A portable distributed implementation of the parallel multipole tree algorithm. *IEEE Symposium on High Performance Distributed Computing*, 1995. [Duke University Technical Report 95-002].

10. W. F. van Gunsteren and H. J. C. Berendsen. *GROMOS Manual*. BIOMOS b. v., Lab. of Phys. Chem., Univ. of Groningen, 1987.

11. P. K. Weiner and P. A. Kollman. AMBER: Assisted model building with energy refinement. a general program for modeling molecules and their interactions. *Journal of Computational Chemistry*, 2:287, 1981.

COMPASSION: A Parallel I/O Runtime System Including Chunking and Compression for Irregular Applications *

Jesús Carretero[1], Jaechun No [**], Alok Choudhary [***], and Pang Chen [†]

jcarrete@fi.upm.es
Arquitectura y Tecnología de Sistemas Informáticos,
Universidad Politécnica de Madrid, Spain.

Abstract. In this paper we present an experimental evaluation of COMPASSION, a runtime system for irregular applications based on collective I/O techniques. It provides a "Collective I/O" model, enhanced with "Pipelined" operations and compression. All processors participate in the I/O simultaneously, alone or grouped, making scheduling of I/O requests simpler and providing support for contention management. In-memory compression mechanisms reduce the total execution time by diminishing the amount of I/O requested and the I/O contention. Our experiments, executed on an Intel Paragon and on the ASCI/Red teraflops machine, demonstrate that COMPASSION can obtain significantly high-performance for I/O above what has been possible so far.
Keywords: irregular applications, runtime systems, parallel I/O, compression.

1 Introduction

Parallel computers are being used increasingly to solve large computationally intensive as well as data-intensive applications. A large number of these applications are *irregular* applications, where accesses to data are performed through one or more level of indirections. Sparse matrix computations, particle codes, and many CFD applications where geometries and meshes are described via indirections, exhibit these characteristics. Most of these applications must manage large data sets residing on disk, thus having tremendous I/O requirements [3]. To circumvent the I/O problems, most applications currently simply output each processor-local mesh to a local file, and rely on some sequential post-processor to recombine the data files into a single data file corresponding to the global

---
* This work was supported in part by Sandia National Labs award AV-6193 under the ASCI program, and in part by NSF Young Investigator Award CCR-9357840 and NSF CCR-9509143. Jesús Carretero is a postdoctoral fellow at NWU supported by the NATO Science Fellowships Programme.
** Dept. of Electrical Engineering and Computer Science, Syracuse University, USA
*** Electrical and Computer Engineering, Northwestern University, USA
† Sandia National Laboratories, USA

mesh. As the number of processors and the grid points scales, sequential re-combination of the MBytes/GBytes of data into the anticipated terabytes of data simply becomes infeasible.

In a previous paper [5] we presented the design and implementation of COM-PASSION, a high-performance runtime system for irregular applications on large-scale systems. The objective of COMPASSION is to provide a high-level interface to manage I/O data in the order imposed by the global array, hiding indirection management, data location, and internal optimizations. It always maintains the global data in some canonical order, so that the explosive number of different file formats can be avoided and data can be shared among different platforms and processor configurations without reorganizing the data. In this paper we present an experimental evaluation of COMPASSION on two large-scale parallel sys-tems: Intel PARAGON at California Institute of Technology and ASCI/Red Teraflops machine at Sandia National Lab. The rest of the paper is organized as follows. Section 2 presents an overview of COMPASSION. Section 3 discusses the metrics and experiments used to evaluate COMPASSION. Section 4 shows the experimental performance results of COMPASSION. Finally, we conclude in sec-tion 5, proving that COMPASSION can obtain significantly high-performance for I/O above what has been possible so far, and significatively reducing the storage space needed for the applications' data and the contention at the I/O system.

2 COMPASSION Overview

COMPASSION includes two alternative schemes for collective I/O: *"Collective I/O"* and *"Pipelined Collective I/O"*. The main difference among them is that in the pipelined collective scheme I/O requests are staggered to avoid contention. Both schemes includes three basic steps. 1) Schedule construction, 2) I/O op-erations, and 3) data redistribution among processors using the two-phase I/O strategies [2]. The basic idea behind two-phase collective I/O is to reorder at runtime the access patterns seen by the I/O system such that a large number of small and disjoint I/O requests can be transformed into a small number of large contiguous requests. Collective I/O requests are enhanced by using chunking and compression [6] to decrease the total execution time and the amount of storage needed by the application.

The schedule describes the communication and I/O patterns required for each node participating in the I/O operation. The following two steps are involved in schedule information construction:

1. Based on the chunk size, each processor is assigned a data domain for which it is responsible for reading or writing. For a chunk, each processor computes its part of data to be read or written while balancing I/O workload. Next, with each index value in its local memory, processor first decides from which chunk the appropriate data must be accessed and then determines which processor is responsible for reading the data from or writing the data to the chunk.

2. Index values in the local memory are rearranged into the *reordered-indirection array* based on the order of destination processors to receive them. Therefore, we can communicate consecutive elements between processors (communication coalescing).

Note that once it is built, the schedule information can be used repeatedly in the irregular problems whose access pattern does not change during computation, and thereby amortizing its cost.

In the *Parallel Collective* model of COMPASSION, a processor involved in the computation is also responsible for writing(reading) data to(from) files. In case of writing, each processor collects data belonging to its domain from other processors and then writes it to the file. By performing I/O this way, the workload can be evenly balanced across processors. Figure 1 (a) shows the steps in a collective write operation optimized with compression. First, the indirection array is ordered following the scheduling information. Second, the data array pointed by the indirection array is also ordered. Third, data is distributed among the participating processors using the canonical order. Fourth, data is compressed on each processor. Fifth, each compressed chunk is written after creating an associated chunk index element. Six, the global chunk index is gathered and written to disk. Using compression in COMPASSION means that the length of the I/O buffers cannot be computed in advance, because it depends on the input data and the compression algorithm used. Thus, to write a global file in a canonical order, a non-blocking receive operation is set on each processor to get the new offset to write each compressed chunk. The chunk is compressed is parallel in an in-memory buffer. Once the new offset is received, the compressed buffer is written to the file and a new entry is set on the index vector of this processor storing the features of the chunk. This entire process repeats if the chunk size is less than the amount of the data to be read/written. For example, if the amount of data to be written is 1Mbyte per processor, and if the buffer space available to each processor is only 0.5Mbyte, then two iterations of the write operation will be performed. When all the processes have finished the data I/O, they asynchronously send the local indexes to the first process, that generates a global index and writes it to the index file in a single write operation. In case of a collective read operation the chunk index is read first, and then the reverse procedure is applied: chunk reading, uncompression, and data distribution. In this model, all processors issue I/O requests to the I/O system simultaneously. As a result, contention at the I/O system may occur if there are large number of compute nodes compared to the number of I/O nodes. This is particularly possible in a parallel machine where the number of I/O nodes and disks are unbalanced with respect to the number of compute nodes. The pipelined collective I/O model solves this problem.

In the *Pipelined Collective* model, the processors are divided into multiple processor groups so that I/O contention may be managed dynamically overlapping communication and I/O from different groups: only processors in a group issue I/O requests simultaneously to reduce I/O node contention. Thus, if there are G groups, then there will be G interleaved communication and I/O steps, where in step g, $0 \leq g < G$, group g is responsible for the I/O operation. Figure 1 (b) illustrates the steps involved in the pipelined collective read operation when compression is used. The schedule for the read operation is constructed as before, except that there is a schedule for each group. In Figure 1 (b), it is clearly shown that first the chunk index file is read. Second, the first group reads data file and executes uncompression, while the second group issues a non-blocking I/O request. Third, while the read is being performed, communication for the first group takes place. This process continues until all groups have performed their read operation.

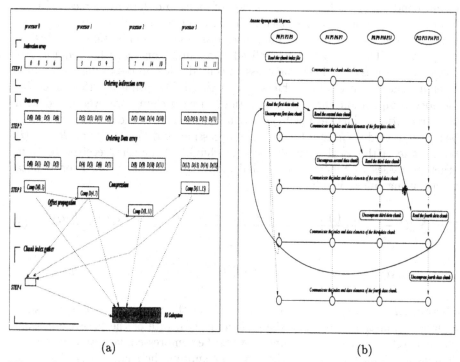

(a)   (b)

Fig. 1. Collective writing with compression (a). Pipelined reading with compression (b).

3 Performance Metrics and Experiments

In this section we describe the performance metrics and experiments used to obtain the selected performance results shown in further sections. The most important metric is the the overall cost to perform I/O, which is given by

$$T_{overall} = T_{schedule} + T_{io} \tag{1}$$

where $T_{schedule}$ is the cost to construct schedule information and T_{io} is the cost to perform I/O operations. $T_{schedule}$ is one-time processing cost; unless the data access pattern is changed during computation the schedule information is repeatedly used in the application. The overall bandwidth can be computed as:

$$B = \frac{D}{T_{overall}} \tag{2}$$

where D is the total size of data in bytes. $T_{schedule}$ can be neglected because it is much more smaller than the I/O time. T_{io} is defined below for each model of the runtime system.

Let P, and C be the number of processors and number of data chunks. Let $S_{indx}(i)$, $S_{data}(i)$ be the size of index and data to be sent to others in data chunk i and $R_{indx}(i)$, $R_{data}(i)$ be the size of index and data to be received from others in data chunk i. In the collective I/O model, each processor exchanges the index and data values in each chunk with other processors by taking exclusive-or operation. This all-to-all personalized communication algorithm requires P-1 steps for P processors. When compression is used, each processor must wait until receiving the file offset of the chunk to be written. To optimize performance, in-memory compression and index operations of each chunk are overlapped with offset reception. Thus, let Let $S_{off}(i)$ and $R_{off}(i)$ be the size of the file offset to be sent/received to/from others after compression, and $S_{lindex}(i)$ and $R_{lindex}(i)$ be the size of the local index of the data chunk i to be sent/received to/from other processes to generate the global index file. The time taken for C data chunks using collective I/O is given by

$$T_{io} = \sum_{i=1}^{C} [(P-1)(\frac{S_{indx}(i) + S_{data}(i) + R_{indx}(i) + R_{data}(i)}{r} + t_\alpha) + \frac{S_{off}(i) + R_{off}(i)}{r} + \tag{3}$$

$$t_{comp}(i) + t_{lindex}(i) + t_{wr}(c_i) + t_\beta] + \sum_{i=1}^{C} [\frac{R_{lindex}(i)}{r} + t_\gamma] + t_{wrindex}$$

where t_α, t_β, and t_γ are the times to perform the remaining operations such as a copy function. Also, r is the communication rate of the message-passing parallel system, $t_{wr}(c_i)$ is the time to write the compressed data to its own data domain in chunk i, $t_{comp}(i)$ is the time to compress the chunk, $t_{lindex}(i)$ is the time to set the values of the chunk in the local index, and $t_{wrindex}$ is the time required to write the global index to the index file. While we do not consider the I/O activity of other applications, $R_{off}(1)$ and $S_{off}(C)$ are zero. $t_{wrindex}$ is also zero for all processors but the one generating the global index file. It should be noted that, in this example, each communication step performs the complete exchange operation for index and data values. A similar metric can be defined for the Pipelined Collective I/O model changing process indexes by process group indexes.

Let B_i and CR_i be the bandwidth and the compression ratio used in experiment i. The joint bandwidth increasing and disk space saving can be measured with the following metric:

$$F_i = \frac{B_i * CR_i}{M}, where \ \forall i, \ M = min(B_i * CR_i) \tag{4}$$

It makes benefits relative to the worst case (no compression), to have a clear knowledge of the joint benefits of the compression techniques.

3.1 Performance Experiments

The parameters considered for performance evaluation include total size of data (D), number of processors (P), number of processor groups (PG) for the pipelined collective I/O implementation, stripe unit (SU), representing the amount of logically contiguous data read/written from each I/O node, chunk size (C), and compression ratio (CR). Their main goals are to know whether the runtime design is appropriated for irregular applications, to know whether the compression optimizations enhance the performance results of COMPASSION by reducing the I/O time, and to optimize the implementation factors influencing the performance of the basic COMPASSION design.

Four major performance experiments were executed to test the library. First, application overall bandwidth measurement varying the parameters described above. Metrics 2 and 3 were used. Note that when data is written, it is globally sorted before actually being stored, and when it is read, it is reordered and redistributed according to the distribution specified by the indirection arrays. That is, the bandwidth measure takes into account communication, computation overhead and the actual I/O operation. Second, evaluation of the compression ratio influence on performance, applying metrics 2 and 3. A variation of experiment 1 including also different compression ratios. Compression ratio 1:1 is equivalent to experiment 1. Third, evaluation of the influence of the grouping factor in the pipeline model. The application bandwidth was measured using 128 processors, a 1024 MB data size, a stripe unit of 512 KB, and varying the number of groups from 1 to 8. Four, measuring the combined performance of bandwidth and disk space using metric 4. The values for this experiment were obtained by executing experiments 1 and 5 and comparing relative results.

In the next two sections, the performance results of the runtime library are shown. The experiments were executed on **TREX**, an Intel PARAGON machine located at Caltech, and on the Intel **ASCI/Red** machine, located at Sandia National Labs. The Intel Paragon TREX is a 550 node Paragon XP/S with 448 GP (with 32 MB memory), 86 MP (with 64 MB memory) nodes, and 16 or 64 I/O nodes with 64 MB memory and a 4 GB Seagate disk. The filesystem used is Intel PFS striped across 16 I/O nodes with a stripe unit of 64 KBytes. Intel's TFLOPS (ASCI/Red) is a 4500+ compute nodes and 19 I/O nodes machine [4]. Each I/O node also has two 200 MHz Pentium Pro and 256 MB, having access to two 32 GB RAIDs. The partition used had 900 compute nodes and a PFS striped across 9 I/O nodes with a stripe unit of 512KB.

4 Performance Evaluation of COMPASSION

The results shown in this section were obtained executing the former experiments on TREX and the ASCI/Red machines. The overall application I/O bandwidth, the breakdown of the COMPASSION execution time, and the influence of several factor such as chunk size and stripe unit. We also evaluate the combined benefits in bandwidth and storage space savings generated by the compression techniques. Be aware, that we are not going to evaluate any compression algorithms. Several were evaluated previously [1], so the one with better results, *lzrw3*, is used in all the experiments.

4.1 Overall I/O Bandwidth

Figure 2 shows the overall bandwidth of the collective and pipelined models of COMPASSION obtained by running experiments 1 and 2 on **TREX** for both write (a) and read (b) operations, using different compression ratios and data sizes. The bandwidth of the compressed version is always higher than the original one (ratio 1:1). For a data set of 1 GB with a compression ratio of 8:1, the overall bandwidth has been enhanced by almost 200% in both models. As expected, there are few differences between both models, although the pipeline model provides better results for low compression ratios in write operations, being always slower for read operations. The reason is that, by having the index in advance, all the processes can read buffers in parallel with less contention than in the non compressed version. The reduction of the file size reduces both the data to be read and the disk head movement on each seek operation. The higher the compression ratio, the lower the I/O time and the I/O contention, and the lower the benefits derived from staggering I/O. Moreover, as a side effect, the communication time is greatly reduced when compression is used, because the processors have to wait less time for the I/O data.

We measured the application B/W for Collective Operations on the **ASCI/Red** executing the experiments with two different data types: double precision floats (Dt2) and a structure including double precision floats and integers (Dt1). Both data types were used to show the influence of a complex data type, where memory alignment is important, on the compression algorithm. Figure 3 shows the overall bandwidth of the collective and pipelined models of COMPASSION, for both write (a) and read (b) operations. The bandwidth of the compressed version is always higher than the original one (ratio 1:1). For a data set of 1 GB with a compression ratio of 6:1, the overall bandwidth has been enhanced by almost 25 % in both models. This increase is lower than on TREX because the I/O and communication systems of the ASCI/Red machine are faster than the TREX ones. Moreover, compressing the data on the ASCI/Red generated two new effects which have a negative influence of performance:

- *small writes* on the RAIDs (only 3 disks of 6 occupied), and
- compression ratio reduction (6:1 ASCI/Red, 8:1 Paragon) due to block alignment: 256KB are *wasted* per chunk against the 32 KB on the Paragon.

Both effects could be enhanced by tuning the parallel file system. The collective write operation performance is again better than the pipelined model. The reason is that the I/O and communication systems are not saturated at all, and

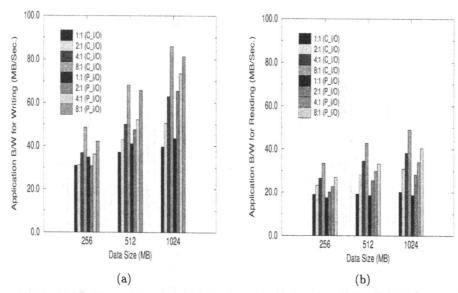

Fig. 2. Application B/W for Collective and Pipelined I/O Operations on TREX for write (a) and read (b) operations.

staggering I/O with the pipeline model is not beneficial. As in TREX, tests with more processors have to be executed to clearly state the pipeline influence on the write operations. However, the overall bandwidth of the read operations of the library in the ASCI/Red machine is better than the TREX one by more than 200 % in any case. Again, using compression is always beneficial (up to 25% increase of performance), but the data type used is less important. Compression influence is less beneficial here than in TREX because the communication time is very low: hundreds of msec on ASCI/Red, more than 20 sec on TREX for 256 processors. Thus, the reduction in communication time observed on TREX can be neglected here and its effect on the reduction of the total execution time is smaller than on TREX. It can be concluded that the performance increase due to compression is mostly achieved in the I/O part of COMPASSION when the network is very fast, while using compression on system with slower networks also improves communication and I/O time.

The data type size influences the I/O run-time performance when compression is used. Better performance is achieved using complex data types, as Dt1, because the two-phase step executed to reorder and to redistribute data involves less communication when the data type size is large (actually $\frac{chunk\_size}{datatype\_size*processors\_no}$) . As the number of processors is increased, the influence of the data type size is smaller, and the results are similar in both read and write operations. This effect can be seen in figure 3 with 1024KB data size, which means 256 processors and 4 MB chunk size.

Figures 4 and 5 show the performance results obtained by running experiment 3 on both machines. On **TREX**, with 64 I/O nodes, having 2 groups (group size:

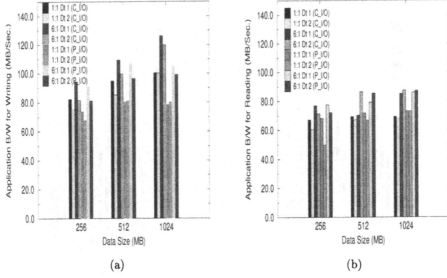

Fig. 3. Application B/W for Collective and Pipelined I/O Operations on the ASCI/Red for write (a) and read (b) operations.

64 processors) is optimum for write operations, while having 8 groups is optimum for read operations. The reason for this disparity is that read operations have more contention on the I/O nodes, thus staggering I/O reduces the contention. However, for writing, large groups are not beneficial because the I/O contention is smaller. On the **ASCI/Red** machine, the network and the CPUs are faster than on TREX, so I/O and communication is smaller, and thus the pipeline does not provide any benefits unless the number of processors raise up to 512.

4.2 Breakdown of Execution Time

Even when it is already very clear that compression increases the application performance, we evaluated the breakdown of execution time on the Collective and Pipelined model to know in more detail where were the benefits of compression, and where the overheads. To study the behavior of the applications, experiments 1 and 2 was executed several times using a 1 GB data set size, 256 processors, 16 I/O nodes in TREX, and 64 KB as stripe unit. Figure 6 shows the breakdown of execution time achieved on **TREX** for both the Write and Read I/O Operations of the Collective model. As may be seen, the effect of compression on execution is always beneficial, showing a direct relation between compression ratio and bandwidth increasing. The more sparse the data set, the lower the compression time, and the higher the compression ratio and the bandwidth. The effect of compression in the execution time of write operation is smaller because there is less actual disk activity involved than for read operations. Non-blocking I/O operations and I/O node caches play a major role

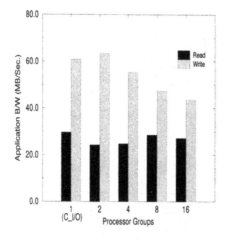 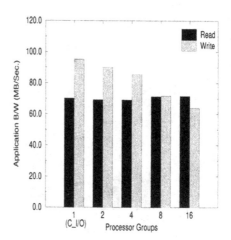

Fig. 4. Application B/W for Collective and Pipelined Collective I/O on TREX as a function of processor groups.

Fig. 5. Application B/W for Collective and Pipelined Collective I/O Operations on ASCI/Red Teraflops as a function of processor groups.

in write operations execution time, but, all in all, the I/O time is reduced by almost 50% with a 8:1 compression ratio. However, the reduction of execution time in read operations is important, raising up by almost 200% with a 8:1 compression ratio. Where are these benefits? Almost 50% of them belong to the communication time, because reducing the I/O time also reduces the waiting time of the processes before data distribution can be accomplished. The results for the Pipelined model on TREX were very similar. On the **ASCI/Red**, that has a very high communication bandwidth, the improvement achieved in the communication time was not relevant for the application performance because the communication time expended in COMPASSION is very low (milliseconds.) However, the I/O time was reduced with the same slope than on TREX. The time overheads due to compression are mainly two: compression algorithm and index management. The first depends on the compute nodes' CPU. Experimental values of the *lzrw3* algorithm, with a 4 MB buffer, are 2 sec on TREX and less than 200 msec on the ASCI/Red. The second is very small with the solution designed. The average time measured, including local index management, global index management, and index file writing, is lower than 200 msec in any system. So, the contribution of this time to the total execution time is not relevant, even when the number of processors is increased to 256.

4.3 Joint Increase of Bandwidth and Disk_SpaceSaving

Figure 7 shows the combined effect of disk space saving and overall I/O bandwidth of the application for the Collective Model (C_IO) and the Pipelined Col-

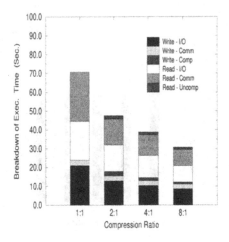 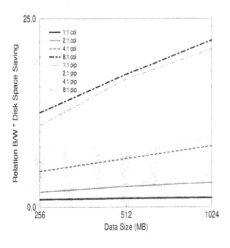

Fig. 6. Breakdown of execution time for collective operations.

Fig. 7. Relative increase of *Bandwidth*Disk_SpaceSaving* factor for write operations in Collective and Pipelined Models

lective Model (P_IO) in a 16 I/O node partition. All the values shown, obtained executing experiment 5 on **TREX**, are relatives to the case $(C\_I/O, 1 : 1)$, which is the smallest. They were calculated with the equation $\frac{BW(i)*CR(i)}{BW(C\_I/O,1:1)}$. As may be seen, both models are very similar in the original solution, and for compression with small data sets. Even when the Pipelined Model is better for smaller compression ratios, the Collective Model is better for higher ones. This results is related to the time needed to compress a buffer and the time needed to execute I/O. The bigger the compression time, the better the overlapping of the Pipeline Model. The values of this metric obtained on the **ASCI/Red** are smaller, because the execution time of the application is smaller than on TREX (10 to 22 in the best case.) Anyway, the amount of global resources consumed to execution the irregular application is substantially reduced in both machines.

5 Summary and Conclusions

In this paper we presented the experimental evaluation of COMPASSION, a runtime system for large-scale irregular problems. The main idea is to reorder data on the fly so that it is sorted before it is stored and reorganized before it is provided to processors for read operations. This eliminates expensive data post processing operations by eliminating dependences on the number of processors that created it. We briefly presented two COMPASSION models: collective I/O and pipelined collective I/O. The second one includes I/O staggering and overlapping of the communication and I/O operations. Both models have been enhanced using in-memory compression to reduce the amount of I/O data.

We presented performance results on both Intel Paragon **TREX** at Caltech and Sandia National Lab's **ASCI/Red** machine. Experimental results show that good performance can be obtained on a large number of processors. Performance results measured on the Intel Paragon show that we were able to duplicate the application level bandwidth for a 8:1 compression ratio using the *lzrw3* compression algorithm. Also, we were able to reduce the communication time of the application by almost 50%. The management of the meta-data (index file) associated to chunking and compression causes a minimum overhead. On the **TREX** machine, we were able to obtain application level bandwidth of up to 90MB/sec for write operations and almost 50 MB/sec for read operations. On the **ASCI/red** machine, we were able to obtain application level bandwidth of up to 130MB/sec for write operations and almost 100 MB/sec for read operations. In both systems, COMPASSION showed significantly high-performance for I/O above what has been possible so far for irregular applications, also reducing the storage space needed for the applications' data and the contention at the I/O system. There are many performance improvements possible, as enhancing prefetching using the chunked meta-data, using different file models, or using alternative file layouts, that we intend to develop in the future. Moreover, evaluations of COMPASSION are being performed on more MIMD machines, like the SBM/SP2.

References

1. J. Carretero, J. No, S. Park, A. Choudhary, and P. Chen. Compassion: a parallel i/o runtime system including chunking and compression for irregular applications. In *Proceedings of the International Conference on High-Performance Computing and Networking 1998*, Amsterdam, Holland, April 1998.
2. Alok Choudhary, Rajesh Bordawekar, Michael Harry, Rakesh Krishnaiyer, Ravi Ponnusamy, Tarvinder Singh, and Rajeev Thakur. PASSION: parallel and scalable software for input-output. Technical Report SCCS-636, ECE Dept., NPAC and CASE Center, Syracuse University, September 1994.
3. Juan Miguel del Rosario and Alok Choudhary. High performance I/O for parallel computers: Problems and prospects. *IEEE Computer*, 27(3):59–68, March 1994.
4. T. Mattson and G. Henry. The asci option red supercomputer. In *Intel Supercomputer Users Group. Thirteenth Annual Conference*, Albuquerque, USA, June 1997.
5. J. No, S. Park, J. Carretero, A. Choudhary, and P. Chen. Design and implementation of a parallel i/o runtime system for irregular applications. In *Proceedings of the 12th International Parallel Processing Symposium*, Orlando, USA, March 1998.
6. K. E. Seamons and M. Winslett. A data management approach for handling large compressed arrays in high performance computing. In *Proceedings of the Fifth Symposium on the Frontiers of Massively Parallel Computation*, pages 119–128, February 1995.

Transformations of Cauchy Matrices, Trummer's Problem and a Cauchy-Like Linear Solver[*]

V. Y. Pan[1], M. Abu Tabanjeh[2], Z. Q. Chen[2],
S. Providence[2], A. Sadikou[2]

[1] Department of Math. & Computer Sci., Lehman College,
CUNY, Bronx, NY 10468, Internet: vpan@lcvax.lehman.cuny.edu
[2] Ph.D. Programs in Math. and Computer Sci.
Graduate Center, CUNY, New York, NY 10036

1   Abstract and Introduction

Computations with dense structured matrices are ubiquitous in sciences, communication and engineering. Exploitation of structure simplifies the computations dramatically but sometimes leads to irregularity, e.g. to numerical stability problems. We will follow the line of [P90], [PLST93], [PZHY97] by exploiting the transformations among the most celebrated classes of structured matrices (that is, among the matrices of Toeplitz, Hankel, Cauchy and Vandermonde types), as a general tool for avoiding irregularity problems and improving computations with the structured matrices of the above classes. In this paper we apply the transformation techniques to Trummer's problem of multiplication of a Cauchy matrix by a vector and to multipoint polynomial evaluation, which should demonstrate the power of this general approach. In both cases (of Trummer's problem and polynomial evaluation) the resulting algorithms use only a few arithmetic operations, allow their effective parallelization and are stable numerically. The new transformations are described in sections 2–6. Subsequent sections reduce Cauchy-like matrix computations to recursive solution of Trummer's problem.

2   Structured Matrices: Correlations to Polynomial Computations

Definition 1. *We define Toeplitz, Hankel, Vandermonde and Cauchy matrices $T = (t_{i,j})$, $H = (h_{i,j})$, $V(x) = (v_{i,j})$, $C(s,t) = (c_{i,j})$, respectively, where $t_{i+1,j+1} = t_{i,j}$, $h_{i+1,j-1} = h_{i,j}$, $v_{i,j} = x_i^j$, $c_{i,j} = \frac{1}{s_i - t_j}$, $s = (s_i)$, $t = (t_i)$, $x = (x_i)$, $s_i \neq t_j$, for all i and j for which the above values are defined.*

The entries of the matrices are related to each other via some operators of displacement and/or scaling (e.g. $t_{i,j}$ and $h_{i,j}$ are invariant in their displacement along the diagonal or antidiagonal directions, respectively). All entries of such

[*] Supported by NSF Grant CCR 9625344 and PSC CUNY Awards 667340 and 668365.

an $n \times n$ structured matrix can be expressed via a few parameters (from n to $2n$, versus n^2 for a general $n \times n$ matrix). Such matrices can be multiplied by vectors fast as this task reduces to some basic operations with polynomials. For instance, polynomial product

$$\left(\sum_i u_i x^i\right)\left(\sum_j v_j x^j\right) = \sum_k w_k x^k,$$

Toeplitz-by-vector product $Tv = w$ and Hankel-by-vector product $Hv = w$ can be made equivalent by matching properly the matrices T, H and the vector $u = (u_i)$ (see [BP94], pp. 132-133). Likewise, the product

$$V(x)p = v \tag{2.1}$$

represents the vector $v = (v_i)$ of the values of the polynomial $p(x) = \sum_j p_j x^j$ on a node set $\{x_i\}$. Consequently, known fast algorithms for the relevant operations with polynomials also apply to the associated matrix operations and vice versa. In particular (cf. [BP94]), $O(n \log n)$ ops suffice for polynomial and Toeplitz(Hankel)-by-vector multiplication, and $O(n \log^2 n)$ ops suffice for multipoint polynomial evaluation (p.e.) as well as for the equivalent operations with the matrix $V(x)$ and the vectors p and v of (2.1), assuming the input size $O(n)$ in all cases. Here and hereafter, "ops" stand for "arithmetic operations", "p.e." for "multipoint polynomial evaluation".

In important special case of (2.1), $x = w = (w_n^i)_{i=0}^{n-1}$ is the vector of the n-th roots of 1, $w_n = \exp(2\pi\sqrt{-1}/n)$, $w_n^n = 1$. In this case, we write $F = V(w)/\sqrt{n}$, and p.e. turns into discrete Fourier transform, takes $O(n \log n)$ ops (due to FFT) and allows numerically stable implementation, in sharp contrast with the case of general $V(x)$. The numerical stability requirement is practically crucial and motivated the design of fast *approximation algorithms* for the p.e. [PLST93], [PZHY97], which rely on expressing $V(x)$ of (2.1) via Cauchy matrices, e.g. as follows:

$$V(x) = \frac{1}{\sqrt{n}}\text{diag}\,(1 - x_i^n)_{i=0}^{n-1}\, C(x, w)\text{diag}\,(w_i)_{i=0}^{n-1}\, F. \tag{2.2}$$

Here we use x and w defined above; here and hereafter diag $(h_i)_{i=0}^{n-1} = D(h)$ for $h = (h_i)_{i=0}^{n-1}$ denotes the $n \times n$ diagonal matrix with diagonal entries $h_0, \ldots, h_{n-1}$. ((2.2) is also the main result of [R97]; [R97] ignores [PZHY97] and briefly cites [PLST93] as a paper "to appear".) (2.2) reduces the operations of (2.1) with $V(x)$ to ones with $C(x, w)$, which brings us to Trummer's problem (that is, the problem of multiplication of an $n \times n$ Cauchy matrix by a vector), our next topic. Its solution by Multipole Algorithm leads to a p.e. algorithm based on (2.2), which is both fast in terms of ops used and (according to experimental tests of [PZHY97]) numerically stable even on the inputs where the known $O(n \log^2 n)$ algorithms fail numerically, due to roundoff errors.

Reducing p.e. to Trummer's problem for $C(x, w)$, we have an extra power of varying vector x, that is, we may linearly map x into

$$y = ax + be, \quad e = (1)_{i=0}^{n-1}, \tag{2.3}$$

where we may choose any scalars $a \neq 0$ and b.

3 Trummer's Problem: Fast Unstable Solution

The solution of Trummer's problem is required in many areas of scientific and engineering computing (see bibliography in [BP94], p.260; [PACLS,a]). The straightforward algorithm solves Trummer's problem in $O(n^2)$ ops. Let us next show $O(n \log^2 n)$ algorithms.

Definition 2. $H(t) = (h_{i,j})_{i,j=0}^{n-1}$, $h_{i,j} = t_{i+j}$ for $i + j \leq n - 1$, $h_{i,j} = 0$ for $i + j \geq n$. W^{-1}, W^T and W^{-T} denote the inverse, the transpose, and the transpose of the inverse of a matrix (or vector) W, respectively.

Definition 3. $p_t(x) = \prod_{j=0}^{n-1}(x - t_j)$, $p'_t(x) = \sum_{i=0}^{n-1} \prod_{j=0(j\neq i)}^{n-1}(x - t_j)$.

Definition 4. $D(s,t) = \text{diag}\,(p_t(s_i))_{i=0}^{n-1} = \text{diag}(\prod_{j=0}^{n-1}(s_i - t_j))_{i=0}^{n-1}$. $D'(t) = \text{diag}(p'_t(t_i))_{i=0}^{n-1} = \text{diag}(\prod_{j=0(j\neq i)}^{n-1}(t_i - t_j))_{i=0}^{n-1}$.

Theorem 1. [FHR93] Let $s_i \neq t_j$, $i,j = 0, 1, \ldots, n - 1$. Then

$$C(s,t) = D(s,t)^{-1}V(s)H(t)V(t)^T, \qquad (3.1)$$

$$C(s,t) = D(s,t)^{-1}V(s)V(t)^{-1}D'(t). \qquad (3.2)$$

Theorem 1 reduces Trummer's problem essentially to the evaluation of the coefficients of $p_t(x)$ and then the values of $p_t(s_i)$ and $p'_t(t_j)$ followed by multiplication of structured matrices $V(s)$, $V^T(t)$, $H(t)$ and $V(t)^{-1}$ by vectors. Both of (3.1) and (3.2) lead to $O(n \log^2 n)$ algorithms [Ger87]. As numerically unstable, however, they are non-practical for larger n.

4 Fast and Numerically Stable Approximate Solution. Its Limitations

Presently the algorithm of choice for practical solution of Trummer's problem is the celebrated Multipole Algorithm, which belongs to the class of hierarchical methods (for bibliography see [PACLS,a], [BP94], pp. 261–262). The algorithm approximates the solution in $O(n)$ ops in terms of n, and works efficiently for a large class of inputs but also has some "difficult" inputs. The basis for the algorithm is the following expansions:

$$\frac{1}{s_i - t_j} = \frac{1}{s_i}\sum_{k=0}^{\infty}(\frac{t_j}{s_i})^k = -\frac{1}{t_j}\sum_{k=0}^{\infty}(\frac{s_i}{t_j})^k.$$

The prefix of the former (latter) power series up to the term $(t_j/s_i)^\kappa$ for a fixed moderately large κ already approximates $\frac{1}{s_i - t_j}$ well if, say, $|t_j/s_i| < 1/2$

(resp. $|s_i/t_j| < 1/2$). Substitute such a prefix into the expressions $C(s,t)v = \left(\sum_{j=0}^{n-1} \frac{v_j}{s_i-t_j}\right)_{i=0}^{n-1}$ and obtain their approximations, say, by

$$-\sum_{j=0}^{n-1} \left(\frac{v_j}{t_j}\right) \sum_{k=0}^{\kappa} \left(\frac{s_i}{t_j}\right)^k = \sum_{k=0}^{\kappa} A_k s_i^k,$$

where $A_k = -\sum_{j=0}^{n-1} v_j/t_j^{k+1}$. For $n \times n$ matrix $C(s,t)$, the computation of such approximations for all i require $O(n\kappa)$ ops and is stable numerically.

The approximation errors are small already for moderate κ if one of the dual ratios $|t_j/s_i|$ and $|s_i/t_j|$ is small. "Difficult" inputs have these dual ratios close to 1 for some i, j. In such irregular cases, some tedious hierarchical techniques give partial remedy. We, however, will treat the irregularity by general regularization techniques of transformation of the input (Cauchy) matrix.

5 New Transformations of a Cauchy Matrix and Trummer's Problem

To improve convergence of Multipole Algorithm, we will reduce Trummer's problem for $C(s,t)$ to ones for $C(s,q)$ and/or $C(q,t)$ for a vector q of our choice. We will rely on the next theorem (cf. also Remark 6.2).

Theorem 2. [PACLS,a]. For a triple of n-dimensional vectors $q = (q_i)_{i=0}^{n-1}$, $s = (s_j)_{j=0}^{n-1}$, $t = (t_k)_{k=0}^{n-1}$, where $q_i \neq s_j$, $s_j \neq t_k$, $t_k \neq q_i$ for $i,j,k = 0,\ldots,n-1$, we have the following matrix equations:

$$C(s,t) = D(s,t)^{-1}V(s)V(q)^{-1}D(q,t)C(q,t), \tag{5.1}$$

$$C(s,t) = D(s,t)^{-1}D(s,q)C(s,q)D'(q)^{-1}D(q,t)C(q,t), \tag{5.2}$$

$$C(s,t) = C(s,q)D(q,s)V(q)^{-T}V(t)^{T}D(t,s)^{-1}, \tag{5.3}$$

$$C(s,t) = -C(s,q)D(q,s)D'(q)^{-1}C(q,t)D(t,q)D(t,s)^{-1}. \tag{5.4}$$

6 Some Algorithmic Aspects

The expressions (5.2) and (5.4) for $C(s,t)$ are Vandermonde-free and Hankel-free, but they enable us to transform the basis vectors s and t for $C(s,t)$ into the two pairs of basis vectors s, q and q, t for any choice of the vector $q = (q_j)$, $q_j \neq s_i$, $q_j \neq t_k$, $i,j,k = 0,\ldots,n-1$. The associated Trummer's problem is reduced to

a) the evaluation of the diagonal matrices $D'(q)^{-1}$, $D(f,g)$ and/or $D(f,g)^{-1}$, for (f,g) denoting (s,t), (q,t), (s,q), (q,s), (t,q) and/or (t,s),

b) recursive multiplication of these matrices and the matrices $C(q,t)$ and $C(s,q)$ by vectors.

Let us specify parts a) and b) in the two next paragraphs.

To compute the matrices $D'(g)$, $D(f,g)$ and $D(f,g)^{-1}$ for given (f,g), we first compute the coefficients of the polynomial $p_g(x) = \prod_{j=0}^{n-1}(x - g_j)$ and then $p_g(f_i)$ and $p'_g(g_i)$, $i = 0,\ldots,n-1$. We compute the coefficients by the fan-in method, that is, we pairwise multiply at first the linear factors $x - g_j$ and then, recursively, the computed products (cf. [BP94], p. 25). The computation is numerically stable and uses $O(n \log^2 n)$ ops. Multipoint polynomial evaluation in $O(n \log^2 n)$ ops ([BP94], p. 26) is not stable numerically, but the fast and numerically stable approximation techniques of [R88], [P95], [PLST93], [PZHY97] can be used instead. We may much simplify the evaluation of the matrices $D(f,g)$, $D(f,g)^{-1}$ and $D'(q)$, where $f = q$ or $g = q$ if we may choose any vector $q = (q_i)_{i=0}^{n-1}$. For instance, let us fill this vector with the scaled n-th roots of 1, so that

$$q_i = aw_n^i, \ i = 0,1,\ldots,n-1, \tag{6.1}$$

for a scalar a and $w_n = \exp(2\pi\sqrt{-1}/n)$. Then $p_q(x) = \prod_{i=0}^{n-1}(x - aw_n^i) = x^n - a^n$, $p'_q(x) = nx^{n-1}$, and the matrices $D(f,q)$ and $D(q)$ can be immediately evaluated in $O(n \log n)$ flops. Furthermore, any polynomial $p(x)$ of degree n can be evaluated at the scaled n-th roots of 1 in $O(n \log n)$ ops, by means of FFT.

The multiplication of $C(q,t)$ or $C(s,q)$ by a vector is Trummer's problem, its solution can be simplified under appropriate choice of the vector q. In particular, even if we restrict q by (6.1), the scaling parameter a still controls fast convergence of the power series of the Multipole Algorithm. The above study can be extended to the expressions (5.1) and (5.3) for $C(s,t)$. Each of them involves two Vandermonde matrices, but one of these matrices in each expression is defined by a vector q of our choice, and this enables us to yield simplification. In particular, for two given vectors $u = (u_i)_{i=0}^{n-1}$ and $q = (q_i)_{i=0}^{n-1}$, the vector $v = V(q)^{-1}u$ is the coefficient vector of the polynomial $v(x)$ that takes on the values u_k at the points q_k, $k = 0,\ldots,n-1$. For q_k being a scaled n-th roots of 1, as in (6.1), the computation of v takes $O(n \log n)$ ops due to the inverse FFT. Similar comments apply to the multiplication of the matrix $V(q)^{-T}$ by a vector. Effective parallelization is immediate at all steps of the computation.

Remark 1. Trummer's problem frequently arises for Cauchy degenerate matrices $C(s) = (c_{i,j})$, $c_{i,i} = 0$, $c_{i,j} = \frac{1}{s_i - s_j}$ for all pairs of distinct i and j. We have $C(s) = \frac{1}{h}\sum_{g=0}^{h-1} C(s, s + \epsilon w_h^g e) + O(\epsilon^h)$ as $\epsilon \to 0$, where $e = (1)_{j=0}^{n-1}$, $s = (s_i)$, ϵ is a scalar parameter. Indeed, $\sum_{g=0}^{h-1} \frac{1}{s_i - s_j - \epsilon w_h^g} = \frac{1}{s_i - s_j}\sum_{l=0}^{\infty}\sum_{g=0}^{h-1}(\frac{\epsilon w_h^g}{s_i - s_j})^l = \frac{h}{s_i - s_j}(1 + O(\epsilon^h))$ because $\sum_{l=0}^{n-1} w_n^{gl} = 0$ for $g = 1,\ldots,n-1$.

Remark 2. A distinct transformation of Trummer's problem may rely on substitution of (2.2) into (3.1) or (3.2), where we may use map (2.3).

7 Definitions and Basic Facts for a Divide-and-Conquer Cauchy-Like Linear Solver

The transformations among the listed classes of structured matrices as a general means of improving the known algorithms for computations with such matrices were first proposed in [P90]. Along this line, the known practical Toeplitz and Toeplitz-like linear solvers were improved substantially in [GKO95] by reduction to Cauchy-like solvers, which enhanced the importance of the latter ones. A known explicit formula for the inverse of a Cauchy matrix (cf. e.g. [BP94], p.131) produces a good Cauchy solver but this is not enough in the application to Toeplitz linear solvers.

In the remainder of this paper, we will follow [PZ96] and [OP98] to show a distinct Cauchy-like linear solver, which extends the well known divide-and-conquer MBA algorithm, proposed in [M74], [M80], [BA80] as a Toeplitz-like linear solver. Every recursive divide-and-conquer step of the algorithm reduces to Trummer's problem, which relates this algorithm to the previous sections. We will start with definitions and basic facts [GO94a], [H94], [PZ96], [OP98].

Definition 5. *For a pair of n-dimensional vectors $q = (q_i)$ and $t = (t_j)$, $q_i \neq t_j$, $i, j = 0, \ldots, n-1$, an $n \times n$ matrix A is a Cauchy-like matrix if*

$$F_{[D(q), D(t)]}(A) = D(q)A - AD(t) = GH^T, \qquad (7.1)$$

G, H are $n \times r$ matrices, and $r = O(1)$. The pair of matrices (G, H^T) of (7.1) is a $[D(q), D(t)]$-generator (or a scaling generator) of a length (at most) r for A (s.g.$_r$(A)). The minimum r allowing the representation (7.1) is equal to rank $F_{[D(q), D(t)]}(A)$ and is called the $[D(q), D(t)]$-rank (or the scaling rank) of A.

Lemma 1. *Let A, q, t, $G = [g_1, \ldots, g_r] = (u_i^T)_{i=0}^{n-1}$, $H = [h_1, \ldots, h_r] = (v_j^T)_{j=0}^{n-1}$ be as in Definition 5, such that (7.1) holds. Then*

$$A = \sum_{m=1}^{r} diag(g_m)C(q, t)diag(h_m) = \left(\frac{u_i^T v_j}{q_i - t_j}\right)_{i,j=0}^{n-1}, \qquad (7.2)$$

where $C(q, t)$ is a Cauchy matrix.

It follows from (7.2) that (7.1) is satisfied if A is of the form $\left(\frac{u_i^T v_j}{q_i - t_j}\right)_{i,j=0}^{n-1}$, where u_i and v_j are r-dimensional vectors for $i, j = 0, 1, \ldots, n-1$. A Cauchy matrix $C(q, t)$ and a Loewner matrix $\left(\frac{r_i - s_j}{q_i - t_j}\right)_{i,j=0}^{n-1}$ are two important special cases of Cauchy-like matrices; they have $[D(q), D(t)]$-ranks 1 and 2, respectively.

Lemma 2. *Let $T(n)$ ops suffice to multiply an $n \times n$ Cauchy matrix by a vector v. Let A be an $n \times n$ Cauchy-like matrix given with an s.g.$_r$(A). Then the product Av can be computed in $(3n + T(n))r$ ops.*

Lemma 3. *Let A_1, A_2 denote a pair of $n \times n$ matrices, where $F_{[D(q_i),D(q_{i+1})]}(A_i) = G_i H_i^T$, G_i, H_i are $n \times r_i$ matrices, $i = 1,2$, and the vectors q_1 and q_3 share no components. Then $F_{[D(q_1),D(q_3)]}(A_1 A_2) = GH^T$, $G = [G_1, A_1 G_2]$, $H = [A_2^T H_1, H_2]$, G, H are $n \times r$ matrices, $r = r_1 + r_2$, and $O(r_1 r_2 T(n))$ ops suffice to compute G and H.*

Lemma 4. *[H94]. Let A denote an $n \times n$ nonsingular matrix with*

$$F_{[D(q),D(t)]}(A) = GH^T, \quad \text{where} \quad G = [g_1, \ldots, g_r] \text{ and } H = [h_1, \ldots, h_r].$$

Then $F_{[D(t),D(q)]}(A^{-1}) = -UV^T$, where the $n \times r$ matrices U, V satisfy $AU = G$, $V^T A = H^T$.

Corollary 1. *Under assumptions of Lemma 4, rank $F_{[D(t),D(q)]}(A^{-1}) \leq r$.*

Lemma 5. *Let an $n \times n$ matrix A satisfy (7.1) and let $B_{I,J}$ be its $k \times d$ submatrix formed by its rows $i_1, \ldots, i_k$ and columns $j_1, \ldots, j_d$. Then $B_{I,J}$ has a $[D(q_I), D(t_J)]$-generator of a length at most r, where $I = [i_1, \ldots, i_k]$, $J = [j_1, \ldots, j_d]$, $D(q_I) = diag(q_{i_1}, \ldots, q_{i_k})$, $D(t_J) = diag(t_{j_1}, \ldots, t_{j_d})$.*

8 Recursive Factorization of a Strongly Nonsingular Matrix

Definition 6. *A matrix W is strongly nonsingular if all its leading principal submatrices are nonsingular.*

I and O denote the identity and null matrices of appropriate sizes,

$$A = \begin{pmatrix} I & O \\ EB^{-1} & I \end{pmatrix} \begin{pmatrix} B & O \\ O & S \end{pmatrix} \begin{pmatrix} I & B^{-1}C \\ O & I \end{pmatrix}, \quad (8.1)$$

$$A^{-1} = \begin{pmatrix} I & -B^{-1}C \\ O & I \end{pmatrix} \begin{pmatrix} B^{-1} & O \\ O & S^{-1} \end{pmatrix} \begin{pmatrix} I & O \\ -EB^{-1} & I \end{pmatrix}, \quad (8.2)$$

where A is an $n \times n$ strongly nonsingular matrix,

$$A = \begin{pmatrix} B & C \\ E & J \end{pmatrix}, \quad S = J - EB^{-1}C, \quad (8.3)$$

B is a $k \times k$ matrix, and $S = S(B, A)$ is called the *Schur complement* of B in A. (8.1) represents block Gauss-Jordan elimination applied to the 2×2 block matrix A of (8.3). If the matrix A is strongly nonsingular, then the matrix S of (8.3) can be obtained in $n - k$ steps of Gaussian elimination.

Lemma 6. *(cf. [BP94], Exercise 4, p. 212). If A is strongly nonsingular, so are B and S.*

Lemma 7. *(cf. [BP94], Proposition 2.2.3). Let A be an $n \times n$ strongly nonsingular matrix and let S be defined by (8.3). Let A_1 be a leading principal submatrix of S and let S_1 denote the Schur complement of A_1 in S. Then S^{-1} and S_1^{-1} form the respective southeastern blocks of A^{-1}.*

Lemma 8. *If (8.1) holds, then det $A = (\det B)\det S$.*

Due to Lemma 6, we may extend factorization (8.1) of A to B and S and then recursively continue this factorization process until we arrive at 1×1 matrices ([St69], [M74], [M80], [BA80]). We invert these matrices and then reverse the decomposition process by using only matrix multiplications and subtractions until we arrive at A^{-1}. As a by-product, we compute all the matrices defined in the recursive descending process. The entire computation will be called the *CRF (or complete recursive factorization)* of A.

We will always assume *balanced* CRFs, where B of (8.1) is the $\lfloor \frac{n}{2} \rfloor \times \lfloor \frac{n}{2} \rfloor$ submatrix of A, and similar balancing property holds in all the subsequent recursive steps. The balanced CRF has depth at most $d = \lceil \log_2 n \rceil$.

Let us summarize our description in the form of a recursive algorithm.

Algorithm 1 *Recursive triangular factorization and inversion.*
Input: *a strongly nonsingular $n \times n$ matrix A.*
Output: *balanced CRF of A, including the matrix A^{-1}.*
Computations:

1. Apply Algorithm 1 to the matrix B (replacing A as its input) to compute the balanced CRF of B (including B^{-1}).

2. Compute the Schur complement $S = J - EB^{-1}C$.

3. Apply Algorithm 1 to the matrix S (replacing A as its input) to compute the balanced CRF of S (including S^{-1}).

4. Compute A^{-1} from (8.2).

Clearly, given A^{-1} and a vector $\boldsymbol{b}$, we may immediately compute the vector $\boldsymbol{x} = A^{-1}\boldsymbol{b}$. If we also seek det A, then it suffices to add the request for computing det B, det S, and det A at stages 1, 3, and 4, respectively.

9 Recursive Factorization of a Strongly Nonsingular Cauchy-Like Matrix

Hereafter, we will assume for simplicity that $n = 2^d$ is an integer power of 2. We write $\boldsymbol{q} = (q_i)_{i=0}^{n-1}$, $\boldsymbol{q}^{(1)} = (q_i)_{i=0}^{\frac{n}{2}-1}$, $\boldsymbol{q}^{(2)} = (q_i)_{i=\frac{n}{2}}^{n-1}$, $\boldsymbol{t} = (t_i)_{i=0}^{n-1}$, $\boldsymbol{t}^{(1)} = (t_i)_{i=0}^{\frac{n}{2}-1}$, $\boldsymbol{t}^{(2)} = (t_i)_{i=\frac{n}{2}}^{n-1}$. We will start with some auxiliary results.

Lemma 9. *[GO94], [OP98]. Let A be a Cauchy-like matrix of Lemma 1, partitioned into blocks according to (8.3). Let (G_0, H_0), (G_0, H_1), (G_1, H_0), (G_1, H_1) and (G_S, H_S) denote the induced five scaling generators of the blocks B, C, E, J of A and of the Schur complement S of (8.3), respectively. Then $G_S = G_1 - EB^{-1}G_0$, $H_S^T = H_1^T - H_0^T B^{-1}C$.*

Lemma 10. *Let A be an $n \times n$ strongly nonsingular Cauchy-like matrix with $F_{[D(q),D(t)]}(A) = GH^T$, for $n \times r$ matrices G, H. Let A, B, C, E, J, and S satisfy (8.3). Then*

$$rankF_{[D(t),\ D(q)]}(A^{-1}) \le r, \qquad rankF_{[D(t^{(1)}),\ D(q^{(1)})]}(B^{-1}) \le r, \qquad (9.1)$$

$$rankF_{[D(q^{(2)}),\ D(t^{(2)})]}(S) \le r, \qquad\qquad (9.2)$$

$$rankF_{[D(t^{(2)}),\ D(q^{(2)})]}(S^{-1}) \le r, \qquad\qquad (9.3)$$

$$rankF_{[D(q^{(1)}),\ D(t^{(2)})]}(C) \le r, \qquad rankF_{[D(q^{(2)}),\ D(t^{(1)})]}(E) \le r. \qquad (9.4)$$

Proof. Deduce (9.2) and (9.4) from Lemmas 5 and 9. Apply Corollary 1, obtain (9.1) and (9.3).

Fact 1 *(cf. Proposition A.6 of [P92b], [P93a]). Given an $s.g._{r^*}(A) = (G, H)$ and the scaling rank r of A, $r < r^* \le n$, one can compute an $s.g._r(A)$ by using $O((r^*)^2 n)$ ops.*

Now we are ready to present the computational complexity estimates.

Theorem 3. *Let A denote an $n \times n$ strongly nonsingular Cauchy-like matrix with its F-generator of a length r for the operator $F = F_{[D(q),D(t)]}$. Then the respective F-generators of all the matrices encountered in the balanced CRF of A (including an $s.g._r(A^{-1})$) and $\det A$ can be computed in $O(r^2 T(n) \log n) = O(nr^2 \log^3 n)$ ops (for $T(n)$ of Lemma 2) and can be stored by using $O(nr \log n)$ words of storage space.*

Proof. Let us apply the fast version of Algorithm 1 to the matrix A of Theorem 3, that is, instead of slower computations with matrices, let us perform faster computations with their short scaling generators. Let $\phi_r(n)$ ops be involved in computing the balanced CRF of A (including the computation of an $s.g._r(A^{-1})$). Furthermore, let $\sigma_r(n)$ ops be used for computing an $s.g._r(S)$ from given $s.g._r(B^{-1})$, $s.g._r(C)$, $s.g._r(-E)$, and $s.g._r(J)$ (cf. Lemma 9, and let $\mu_r(n)$ ops be required for computing an $s.g._r(A^{-1})$ from given $s.g._r(B^{-1})$, $s.g._r(C)$, $s.g._r(E)$, and $s.g._r(S^{-1})$ (cf. (8.2)). This is summarized below.

Input	$s.g._r(A)$	$s.g._r(B^{-1}), s.g._r(C),$ $s.g._r(-E), s.g._r(J)$	$s.g._r(B^{-1}), s.g._r(C),$ $s.g._r(E), s.g._r(S)$
Output	CRF of A	$s.g._r(S)$	$s.g._r(A^{-1})$
ops	$\phi_r(n)$	$\sigma_r(n)$	$\mu_r(n)$

Let $\phi_r(k)$, $\sigma_r(k)$, and $\mu_r(k)$ denote the similar estimates for a strongly nonsingular $k \times k$ input matrix W given with an $s.g._r(W)$. Then from Algorithm 1,

$$\phi_r(n) \le 2\phi_r\left(\frac{n}{2}\right) + \sigma_r(n) + \mu_r(n). \qquad (9.5)$$

Now we apply (8.2), Lemmas 3, 5, 9, and 10 and deduce that

$$\sigma_r(n) = O(r^2 T(n)), \quad \mu_r(n) = O(r^2 T(n)). \tag{9.6}$$

Substitute (9.6) into (9.5), recursively extend (9.5), and deduce that

$$\phi_r(n) = O(r^2 T(n)),$$

which gives us the arithmetic time bound of Theorem 3. The storage space bound follows similarly when we inspect Algorithm 1 applied to the matrix A and apply Lemmas 3, 5, 6, 9 and 10.

Remark 3. In [OP98], Lemma 9, was extended to the case of Hankel-like matrices H associated with operators $ZH - HZ^T$, Z being a shift matrix. This enabled practical improvement of the MBA Hankel/Toeplitz linear solver.

10 Ensuring Strong Nonsingularity

To extend our algorithm to any nonsingular matrix A, we will seek a strongly nonsingular matrix X, such that the matrix AX is strongly nonsingular. Then we may apply our machinery to the matrices X and AX or XA, compute $(AX)^{-1} = X^{-1}A^{-1}$ or $(XA)^{-1} = A^{-1}X^{-1}$, det $(AX)=$det (XA), det X, and then $A^{-1} = X(AX)^{-1} = (XA)^{-1}X$ and det $A = $ det $(AX)/$det X. If A is singular, the same algorithm will involve a division by 0 and thus will show us that det $A = 0$. In [PZ96] the algorithm is extended to computing rank A and solving consistent singular Cauchy-like linear systems.

Computing with reals we may set $X = A^T$. Indeed, the matrix $XA = A^T A$ is strongly nonsingular provided that A is nonsingular. Moreover, in this case the condition numbers of all diagonal blocks of the CRF do not exceed the condition number of A (cf. [BP94], Fact 2.1.4 and page 237). As a by-product, we immediately arrive at a least-squares (normal equations) solution $(A^T A)^{-1} A^T b$ to a Cauchy-like linear system $Ax = b$ for a Cauchy-like $m \times n$ rectangular matrix A having full rank n, $n \le m$.

For the operators $F_{[D(t),D(t)]}$ and $F_{[D(q),D(q)]}$ associated with the matrices $W = A^T A$ and $U = AA^T$, respectively, the assumption $q_i \ne t_j$ of Definition 5 is not extended, but we will operate with W represented as the product $C^{-1}(q,t)Y$, where $Y = C(q,t)W$ and $C^{-1}(q,t)$ are Cauchy-like matrices, and similarly for U.

References

[BA80] R.R. Bitmead, B.D.O. Anderson, Asymptotically Fast Solution of Toeplitz and Related Systems of Linear Equations, *Linear Algebra Appl.*, **34**, 103-116, 1980.

[BP94] D. Bini, V. Y. Pan, *Polynomial and Matrix Computations, Volume 1: Fundamental Algorithms*, Birkhäuser, Boston, 1994.

[FHR93] T. Fink, G. Heinig, K. Rost, An Inversion Formula and Fast Algorithms for Cauchy-Vandermonde Matrices, *Linear Algebra Appl.*, **183**, 179–191, 1993.

[Ger87]   A. Gerasoulis, A Fast Algorithm for the Multiplication of Generalized Hilbert Matrices with Vectors, *Math. Comp.*, **50, 181**, 179–188, 1987.

[GKO95] I. Gohberg, T. Kailath, V. Olshevsky, Fast Gausian Elimination with Partial Pivoting for Matrices with Displacement Structure, *Math. of Computation*, **64**, 1557-1576, 1995.

[GO94]   I. Gohberg, V. Olshevsky, Fast State Space Algorithms for Matrix Nehari and Nehari-Takagi Interpolation Problems, *Integral Equations and Operator Theory*, **20**, 1, 44-83, 1994.

[GO94a]   I. Gohberg, V. Olshevsky, Complexity of Multiplication with Vectors for Structured Matrices, *Linear Algebra Appl.*, **202**, 163-192, 1994.

[H94]   G. Heinig, Inversion of Generalized Cauchy Matrices and the Other Classes of Structured Matrices, *Linear Algebra for Signal Processing, IMA Volume in Math. and Its Applications*, **69**, 95-114, Springer, 1994.

[OP98]   V. Olshevsky, V. Y. Pan, A Superfast State-Space Algorithm for Tangential Nevanlinna-Pick Interpolation Problem, preprint, 1998.

[M74]   M. Morf, Fast Algorithms for Multivariable Systems, PhD Thesis, Standford University, Standford, CA, 1974.

[M80]   M. Morf, Doubling Algorithms for Toeplitz and Related Equations, *Proc. IEEE Internat. Conf. on ASSP*, 954-959,1980.

[P90]   V.Y. Pan, On Computations with Dense Structured Matrices, *Math. of Computation*, **55, 191**, 179-190, 1990.

[P92b]   V.Y. Pan, Parametrization of Newton's Iteration for Computations with Structured Matrices and Applications, *Computers and Mathematics (with Applications)*, **24**, 3, 61-75, 1992.

[P93a]   V.Y. Pan, Decreasing the Displacement Rank of a Matrix, *SIAM J. Matrix Anal. Appl.*, **14**, 1, 118-121, 1993.

[P95]   V. Y. Pan, An Algebraic Approach to Approximate Evaluation of a Polynomial on a Set of Real Points, *Advances in Computational Mathematics*, **3**, 41–58, 1995.

[PACLS,a] V. Y. Pan, M. Abu Tabanjeh, Z. Q. Chen, E. Landowne, A. Sadikou, New Transformations of Cauchy Matrices and Trummer's Problem, accepted by *Computers & Math. (with Applics.)*.

[PLST93] V.Y. Pan, E. Landowne, A. Sadikou, O. Tiga, A New Approach to Fast Polynomial Interpolation and Multipoint Evaluation, *Computers & Math. (with Applics.)*, **25**, 9, 25-30, 1993.

[PZ96]   V. Y. Pan, A. Zheng, Fast Algorithms for Cauchy-like Matrix Computations and an Extension to Singular Toeplitz-like Computations, Preprint, 1996.

[PZHY97] V. Y. Pan, A. Zheng, X. Huang, Y. Yu, Fast Multipoint Polynomial Evaluation and Interpolation via Computations with Structured Matrices, *Annals of Numerical Math.*, 4, 483-510, 1997.

[R88]   V. Rokhlin, A Fast Algorithm for the Discrete Laplace Transformation, *J. of Complexity*, 4, 12–32, 1988.

[R97]   J. H. Reif, *Proc. 29th ACM STOC*, 30–39, 1997.

[St69]   V. Strassen, Gaussian Elimination is Not Optimal, *Numer. Math.*, **13**, 354-356, 1969.

A Parallel GRASP for the Steiner Problem in Graphs

Simone L. Martins, Celso C. Ribeiro, and Mauricio C. Souza

Department of Computer Science, Catholic University of Rio de Janeiro, Rio de Janeiro, RJ 22453-900, Brazil. E-mail: {simone,celso,souza}@inf.puc-rio.br.

Abstract. A greedy randomized adaptive search procedure (GRASP) is a metaheuristic for combinatorial optimization. Given an undirected graph with weights associated with its nodes, the Steiner tree problem consists in finding a minimum weight subgraph spanning a given subset of (terminal) nodes of the original graph. In this paper, we describe a parallel GRASP for the Steiner problem in graphs. We review basic concepts of GRASP: construction and local search algorithms. The implementation of a sequential GRASP for the Steiner problem in graphs is described in detail. Feasible solutions are characterized by their non-terminal nodes. A randomized version of Kruskal's algorithm for the minimum spanning tree problem is used in the construction phase. Local search is based on insertions and eliminations of nodes to/from the current solution. Parallelization is done through the distribution of the GRASP iterations among the processors on a demand-driven basis, in order to improve load balancing. The parallel procedure was implemented using the Message Passing Interface library on an IBM SP2 machine. Computational experiments on benchmark problems are reported.

1 Introduction

Let $G = (V, E)$ be a connected undirected graph, where V is the set of nodes and E denotes the set of edges. Given a non-negative weight function $w : E \to \mathbb{R}_+$ associated with its edges and a subset $X \subseteq V$ of terminal nodes, the Steiner problem $SPG(V, E, w, X)$ consists in finding a minimum weighted connected subgraph of G spanning all terminal nodes in X. The solution of $SPG(V, E, w, X)$ is a Steiner minimal tree (SMT). The non-terminal nodes that end up in the SMT are called Steiner nodes.

The Steiner problem in graphs is a classic NP-complete (Karp [13]) combinatorial optimization problem, see e.g. Hwang, Richards and Winter [10], Maculan [16], and Winter [27] for surveys on formulations, special cases, exact algorithms, and heuristics. Applications can be found in many areas, such as telecommunication network design, VLSI design, and computational biology, among others. Due to its NP-hardness, several heuristics have been developed for its approximate solution, see e.g. Duin and Voss [6], Hwang, Richards and Winter [10], and Voss [25]. Among the most efficient approaches, we find implementations of metaheuristics such as genetic algorithms (Esbensen [7], and Kapsalis, Rayward-Smith and Smith [12]), tabu search (Gendreau, Larochelle and Sansó [9], Ribeiro and Souza [24], and Xu, Chiu and Glover [28]), and simulated annealing (Dowsland [4]).

A greedy randomized adaptive search procedure (GRASP) is a metaheuristic for combinatorial optimization. A GRASP [8] is an iterative process, where each iteration consists of two phases: construction and local search. The construction phase builds a feasible solution, whose neighborhood is explored by local search. The best solution over all iterations is returned as the result. In this paper we present a parallel GRASP for the Steiner problem in graphs. In the next section, we review the basic components of this approach and we present a sequential GRASP for the Steiner problem in graphs. Feasible solutions are characterized by their non-terminal nodes. A randomized version of Kruskal's algorithm for the minimum spanning tree problem is used in the construction phase. Local search is based on insertions and eliminations of Steiner nodes to/from the current solution. In Sec. 3 we describe the parallelization of this GRASP, performed through the distribution of its iterations among the processors on a demand-driven basis, in order to improve load balancing. The parallel procedure was implemented using the Message Passing Interface library on an IBM SP2 machine. Computational experiments on benchmark problems are reported. Concluding remarks are made in Sec. 4.

2 Sequential GRASP

Approximate solutions for the Steiner minimal tree problem can be obtained by either spanning tree or path based approaches. In this section, we apply the concepts of GRASP to the approximate solution of the Steiner problem in graphs, using a spanning tree based approach. We summarize below the basic concepts of GRASP, as presented in Resende and Ribeiro [23] and Prais and Ribeiro [22].

GRASP can be seen as a metaheuristic which captures good features of pure greedy algorithms (e.g. fast local search convergence and good quality solutions) and also of random construction procedures (e.g. diversification). Each iteration consists of the construction phase, the local search phase and, if necessary, the incumbent solution update. In the construction phase, a feasible solution is built, one element at a time. At each construction iteration, the next element to be added is determined by ordering all elements in a candidate list with respect to a greedy function that estimates the benefit of selecting each element. The probabilistic component of a GRASP is characterized by randomly choosing one of the best candidates in the list, but usually not the top one.

The solutions generated by a GRASP construction are not guaranteed to be locally optimal. Hence, it is almost always beneficial to apply local search to attempt to improve each constructed solution. A local search algorithm works in an iterative fashion by successively replacing the current solution by a better one from its neighborhood. It terminates when there are no better solutions in the neighborhood. Success for a local search algorithm depends on the suitable choice of a neighborhood structure, efficient neighborhood search techniques, and the starting solution. The GRASP construction phase plays an important role with respect to this last point, since it produces good starting solutions for local search. The customization of these generic principles into an approximate algorithm for the Steiner problem in graphs is described in the following.

2.1 Solution Characterization

Let $G = (V, E)$ be an undirected graph with node set V and edge set E. A graph $H = (V(H), E(H))$ is said to be a subgraph of G if $V(H) \subseteq V$ and $E(H)$ is any subset of the edges in E having both extremities in $V(H)$. This graph H is said to span a subset $U \subseteq V$ of the nodes in G if $U \subseteq V(H)$. For any subset of nodes $W \subseteq V$, the edge subset $E(W) = \{(i, j) \in E \mid i \in W, j \in W\}$ defines the induced subgraph $G(W) = (W, E(W))$ in G by the nodes in W.

We denote by $T_{V,E,w}(X)$ the Steiner minimal tree solving the Steiner problem $SPG(V, E, w, X)$ formulated in Sec. 1. Given the graph $G = (V, E)$ with non-negative weights w associated with its edges, the minimum spanning tree problem $MSTP(V, E, w)$ consists in finding a minimum weighted subtree of G spanning all nodes in V. This problem can be seen as particular case of the Steiner problem $SPG(V, E, w, X)$, in which we have $X = V$. Accordingly, we denote by $T_{V,E,w}(V)$ the minimum spanning tree solving $MSTP(V, E, w)$. We can associate a feasible solution of the Steiner problem $SPG(V, E, w, X)$ with each subset $S \subseteq V \setminus X$ of Steiner nodes, given by a minimum spanning tree solving problem $MSTP(S \cup X, E(S \cup X), w)$. Let S^* be the set of Steiner nodes in the optimal solution of $SPG(V, E, w, X)$. The optimal solution $T_{V,E,w}(X)$ is a minimum spanning tree of the graph induced in G by the node set $S^* \cup X$, i.e., the solution to the minimum spanning tree problem $MSTP(S^* \cup X, E(S^* \cup X), w)$.

In the following, solutions of the Steiner problem $SPG(V, E, w, X)$ will be characterized by the associated set of Steiner nodes and one of the corresponding minimum spanning trees. Accordingly, the search for the Steiner minimal tree $T_{V,E,w}(X)$ will be reduced to the search for the optimal set S^* of Steiner nodes.

2.2 Construction Phase

The construction phase of our GRASP is based on the distance network heuristic, suggested by Choukmane [3], Iwainsky, Canuto, Taraszow and Villa [11], Kou,

Markowsky and Berman [14], and Plesník [21], with time complexity $O(|X||V|^2)$. Lately, Mehlhorn [18] proposed a modification of the original version, leading to a procedure using simple data structures and presenting an improved time complexity. For every terminal node $i \in X$, let $N(i)$ be the subset of non-terminal nodes of V that are closer to i than to any other terminal node. The first step of this procedure consists in the computation of a graph $G' = (X, E')$, where $E' = \{(i,j), i,j \in X \mid \exists(k,\ell) \in E, k \in N(i), \ell \in N(j)\}$. Then, we associate a weight $w'_{ij} = \min\{d(i,k)+w_{k\ell}+d(\ell,j) \mid (k,\ell) \in E, k \in N(i), \ell \in N(j)\}$ with each edge $(i,j) \in E'$, where $d(a,b)$ denotes the shortest path from a to b in the original graph $G = (V, E)$ in terms of the weights w. Next, Kruskal's algorithm [15] is used to solve the minimum spanning tree problem $MSTP(X, E', w')$. Finally, the edges in the minimum spanning tree $T_{X,E',w'}(X)$ so obtained are replaced by the edges in the corresponding shortest paths in the original graph G. Neighborhoods $N(i), \forall i = 1, \ldots, |X|$ can be computed in time $O(|E| + |V| \log |V|)$, which is the same complexity of the minimum spanning tree computation. Then, the overall complexity of the distance heuristic network with Mehlhorn improvements is only $O(|E| + |V| \log |V|)$, which is much better than the original bound.

As discussed in the previous section, the construction phase of GRASP relies on randomization to build different solutions at different iterations. Graph $G' = (X, E')$ is created once for all and does not change throughout all computations. In order to add randomization to Mehlhorn's version of the distance network heuristic, we make the following modification in Kruskal's algorithm. Instead of selecting the feasible edge with the smallest weight, we build a restricted candidate list (RCL) with all edges $(i,j) \in E'$ such that $w'_{ij} \leq w'_{min} + \alpha(w'_{max} - w'_{min})$, where $0 \leq \alpha \leq 1$ and w'_{min} and w'_{max} denote, respectively, the smallest and the largest weights among all edges still unselected to form the minimum spanning tree. Then, an edge is selected at random from the restricted candidate list. The operations associated with the construction phase of our GRASP are implemented in lines 2 and 4 of the pseudo-code of algorithm GRASP_for_SPG outlined in Fig. 1.

2.3 Local Search

Since the solution produced by the construction phase is not necessarily a local optimum, local search can be applied as an attempt to improve it. The first step towards the implementation of a local search procedure consists in identifying an appropriate neighborhood definition.

Let S be the set of Steiner nodes in the current Steiner tree. We have noticed in Sec. 2.1 that each subset S of Steiner nodes can be associated with a feasible solution of the Steiner problem $SPG(V, E, w, X)$, given by a minimum spanning tree $T_{S \cup X, E(S \cup X), w}(S \cup X)$ solving problem $MSTP(S \cup X, E(S \cup X), w)$. Moreover, let $W(T)$ denote the weight of a tree T. The neighbors of a solution characterized by its set S of Steiner nodes are defined by all sets of Steiner nodes which can be obtained either by adding to S a new non-terminal node, or by eliminating from S one of its Steiner nodes.

Given the current tree $T_{S \cup X, E(S \cup X), w}(S \cup X)$ and a non-terminal node $s \in \{V \setminus X\} \setminus S$, the computation of neighbor $T_{S \cup \{s\} \cup X, E(S \cup \{s\} \cup X), w}(S \cup \{s\} \cup X)$ obtained by the insertion of s into the current set S of Steiner nodes can be done in $O(|V|)$ average time, using the algorithm proposed by Minoux [19]. For each non-terminal node $t \in S$, neighbor $T_{S \setminus \{t\} \cup X, E(S \setminus \{t\} \cup X), w}(S \setminus \{t\} \cup X)$ obtained by the elimination of t from the current set S of Steiner nodes is computed by Kruskal's algorithm as the solution of the minimum spanning tree problem $MSTP(S \setminus \{t\} \cup X, E(S \setminus \{t\} \cup X), w)$.

In order to speedup the local search, since the computational time associated with the evaluation of all insertion moves is likely to be much smaller than that of the elimination moves, only the insertion moves are evaluated in a first pass. The evaluation of elimination moves is performed only if there are no improving insertion moves.

2.4 Algorithm Description

The pseudo-code with the complete description of procedure GRASP_for_SPG for the Steiner problem in graphs is given in Fig. 1. The procedure takes as input the original graph $G = (V, E)$, the set X of terminal nodes, the edge weights w, the restricted candidate list parameter α ($0 \leq \alpha \leq 1$), a seed for the pseudo random number generator, and the maximum number of GRASP iterations to be performed. The value of the best solution found is initialized in line 1. The preprocessing computations associated with Mehlhorn's version of the distance network heuristic are performed in line 2, as described in Sec. 2.2. The procedure is repeated max_iterations times. In each iteration, a greedy randomized solution T is constructed in line 4 using the randomized version of Kruskal's algorithm. Let S be the set of Steiner nodes in T.

Next, the local search attempts to produce a better solution. In line 5 we initialize the best set of Steiner nodes as those in the current solution, and the weight of the best neighbor as that of the current solution. The loop from line 6 to 11 searches for the best insertion move. In line 7 we compute the minimum spanning tree T^{+s} associated with problem $MSTP(S \cup \{s\} \cup X, E(S \cup \{s\} \cup X), w)$ defined by the insertion of node s into the current set of Steiner nodes. Let $W(T^{+s})$ be its weight. In line 8 we check whether the new solution T^{+s} improves the weight of the current best neighbor. The best set of Steiner nodes and the weight of the best neighbor are updated in line 9. Once all insertion moves have been evaluated, we check in line 12 whether an improving neighbor has been found. If this is the case, the set of Steiner nodes, the current Steiner tree and its weight are updated in line 13, and the local search resumes from this new current solution.

```
procedure GRASP_for_SPG(V, E, w, X, α, seed, max_iterations)
1    best_value ← ∞;
2    Compute graph G' = (X, E') and weights w'ᵢⱼ, ∀(i,j) ∈ E';
3    for k = 1,..., max_iterations do
                                               /* Construction phase */
4        Apply a randomized version of Kruskal's algorithm to obtain a
         spanning tree T of G' = (X, E') with S as its set of Steiner nodes;
                                               /* Insertion moves */
5        best_set ← S; best_weight ← W(T);
6        for all s ∈ (V \ X) \ S do
7            Compute the minimum spanning tree T⁺ˢ;
8            if W(T⁺ˢ) < best_weight then do
9                best_set ← S ∪ {s}; best_weight ← W(T⁺ˢ);
10           end then;
11       end for;
12       if best_weight < W(T) then do
13           S ← S ∪ {s}; T ← T⁺ˢ; W(T) ← W(T⁺ˢ); go to line 5;
14       end then;
                                               /* Elimination moves */
15       best_set ← S; best_weight ← W(T);
16       for all t ∈ S do
17           if G⁻ᵗ = ((S \ {t}) ∪ X, E((S \ {t}) ∪ X)) is connected then do
18               Compute the minimum spanning tree T⁻ᵗ;
19               if W(T⁻ᵗ) < best_weight then do
20                   best_set ← S \ {t}; best_weight ← W(T⁻ᵗ);
21               end then;
22           end then;
23       end for;
24       if best_weight < W(T) then do
25           S ← S \ {t}; T ← T⁻ᵗ; W(T) ← W(T⁻ᵗ); go to line 5;
26       end then;
                                               /* Best solution update */
27       if W(T) < best_value then do
28           S* ← S; T* ← T; best_value ← W(T);
29       end then;
30   end for;
31   return S*, T*;
end GRASP_for_SPG;
```

Fig. 1. Pseudo-code of the sequential GRASP procedure for the Steiner problem in graphs

If no improving insertion moves have been found, then the elimination moves are evaluated. In line 15 we reinitialize the best set of Steiner nodes as those in the current solution, and the weight of the best neighbor as that of the current solution. We check in line 17 whether the graph $G^{-t} = ((S\setminus\{t\})\cup X, E((S\setminus\{t\})\cup X))$ obtained by the elimination of node t is connected or not. If it is connected, we compute in line 18 the minimum spanning tree T^{-t} associated with problem $MSTP((S\setminus\{t\})\cup X, E((S\setminus\{t\})\cup X), w)$ defined by the elimination of node t from the current set of Steiner nodes. Again, let $W(T^{-t})$ be its weight. In line 19 we check whether the new solution T^{-t} improves the weight of the current best neighbor. Once again, the best set of Steiner nodes and the weight of the best neighbor are updated in line 20. Once all elimination moves have been evaluated, we check in line 24 whether an improving neighbor has been found. If this is the case, the set of Steiner nodes, the current Steiner tree and its weight are updated in line 25, and the local search resumes from this new current solution.

If the solution found at the end of the local search is better than the best solution found so far, we update in line 28 the best set of Steiner nodes, the current Steiner tree and its weight. The best set S^* of Steiner nodes and the best Steiner tree T^* are returned in line 31.

2.5 Acceleration

In order to accelerate the local search phase, we implemented a faster evaluation scheme for insertion and deletion moves. The basic idea consists in keeping candidate lists with promising moves of each type, which are periodically updated.

In the first GRASP iteration, we build a list containing the k_best improving insertion moves, which is kept in nondecreasing order of the associated move values. At each following iteration, let S be the set of Steiner nodes in the current solution. Instead of reevaluating all insertion moves, we just take the node s corresponding to the first move in this candidate list and we reevaluate the weight of a minimum spanning tree associated with the insertion of this node into the current set of Steiner nodes. If this move reveals itself to be an improving one, then the current solution is updated, the move is eliminated from the candidate list, and the local search resumes from the new set $S\cup\{s\}$ of Steiner nodes. Otherwise, if the move is not an improving one, it is eliminated from the candidate list and the next candidate move is evaluated. Once the candidate list becomes empty, a new full iteration is performed, all insertion moves are evaluated, and the candidate list is rebuilt. A similar procedure is implemented for deletion moves.

We present in Table 1 some computational results illustrating the efficiency of the above acceleration scheme on an IBM RISC/6000 370 processor with 256 Mbytes of RAM. For each of the 20 series C test problems from the OR-Library [2], we present the weight $W(T^*)$ of the best solution found and the computation time in seconds (sec's) obtained by (i) a straightforward version of algorithm GRASP_for_SPG with a fixed value for $\alpha = 0.5$, and (ii) the same algorithm using the above acceleration scheme for insertion and deletion moves with k_best = 40. These results show that this technique significantly reduced the computational times, even attaining a reduction of up to 81% in the case of problem C.05, with very small losses in terms of solution quality.

Table 1. Effect of the acceleration scheme on OR-Library problems of series C

	GRASP_for_SPG		Acceleration scheme	
Problem	$W(T^*)$	sec's	$W(T^*)$	sec's
C.01	85	149.8	85	76.3
C.02	144	154.1	144	88.4
C.03	754	3143.4	766	703.0
C.04	1080	5009.3	1089	947.1
C.05	1579	8058.6	1579	1569.7
C.06	55	108.1	55	80.4
C.07	103	200.3	102	91.7
C.08	509	3290.9	523	664.0
C.09	709	5412.6	737	1112.0
C.10	1094	11378.8	1100	7160.7
C.11	32	110.0	32	82.8
C.12	46	248.1	46	100.5
C.13	260	3912.0	280	1179.4
C.14	325	5611.5	336	2575.1
C.15	559	28825.7	561	20479.3
C.16	12	93.6	12	84.2
C.17	18	146.4	18	86.6
C.18	124	3052.5	134	1256.9
C.19	167	5126.0	186	2123.22
C.20	267	30056.6	268	28063.7

3 Parallelization and Computational Results

The most straightforward GRASP parallelization scheme consists in the distribution of the iterations among the available processors, using a master-slave scheme. Each slave processor performs a fixed number of GRASP iterations, equal to the total number of GRASP iterations divided by the number of processors. Once all processors have finished their computations, the best solution among those found by each processor is collected by the master processor. To reduce communication times, each slave processor should have its own copy of all problem data. Each slave processor is involved with exactly one communication step at the end of its computation, when it communicates to the master the best solution it has found and its weight.

However, since some slave processors may be slower than others (they can even be different, in the case of a heterogeneous system) and since the time spent on the local search may be quite different from one iteration to another, this kind of strategy leads very often to critical load unbalancing and small efficiency values. It has been recently shown in Alvim [1] that a very simple strategy, consisting in the distribution of the iterations among the slave processors on a demand-driven basis, can significantly improve load balancing and leads to smaller computation (elapsed) times. In fact, this strategy allowed reductions in the order of 20% in the case of the parallelization of a GRASP for a scheduling problem arising in the context of traffic assignment in TDMA systems [1, 22].

We use the same strategy in the parallelization of the GRASP_for_SPG sequential algorithm, whose pseudo-code was presented in Fig. 1. The max_iterations GRASP iterations to be performed are divided into fixed batches of batch_size iterations each. In the beginning, each slave processor receives a batch of iterations to perform. Whenever a slave processor finishes its current batch, it requests a new one to the master. If the total number of GRASP iterations has already been performed, then this slave finishes its computations. Otherwise, it performs a new batch of iterations.

The computational experiments have been performed on a set of 60 benchmark problems from series C, D and E of the OR-Library [2]. The graphs have been previously reduced using the "smaller special distance test" proposed by Duin and Volgenant [5]. The above parallelization scheme was implemented using the Message Passing Interface (MPI) library [20, 26] on an IBM SP2 machine with 16 RISC/6000 370 processors with 256 Mbytes of RAM each.

In the case of series C and D, the parallel GRASP was implemented with the following parameter settings: $\alpha = 0.1$, max_iterations $= 400$, k_best equal to the number of terminal nodes, and batch_size $= 25$. The computational results obtained using five processors (with four of them acting as slaves) are reported in Table 2. For each test problem, we present the number of nodes, terminal nodes and edges of the graph, the weight *Best_value* of the optimal Steiner tree, the weight *Par-GRASP* of the best solution found by the parallel GRASP algorithm, an indication on whether this solution is optimal or not, and the local index of the iteration on which the best solution was found (*#_iter*).

Since the problems in series E are larger, fewer iterations have been performed. We used max_iterations $= 100$ with batch_size $= 2$. The computational results obtained using eleven processors (with ten of them as slaves) are reported in Table 3.

These results illustrate the effectiveness of the proposed GRASP procedure for the Steiner problem in graphs. The parallel GRASP found the optimal solution for 18 out of 20 problems of series C. For the other two problems, the best

Table 2. Run statistics on OR-Library problems of series C and D

Problem	$|V|$	$|X|$	$|E|$	Best_value	Par-GRASP	optimal?	#_iter
C.01	500	5	625	85	85	yes	1
C.02	500	10	625	144	144	yes	1
C.03	500	83	625	754	754	yes	7
C.04	500	125	625	1079	1079	yes	1
C.05	500	250	625	1579	1579	yes	1
C.06	500	5	1000	55	55	yes	1
C.07	500	10	1000	102	102	yes	1
C.08	500	83	1000	509	509	yes	3
C.09	500	125	1000	707	707	yes	1
C.10	500	250	1000	1093	1093	yes	1
C.11	500	5	2500	32	32	yes	1
C.12	500	10	2500	46	46	yes	1
C.13	500	83	2500	258	258	yes	20
C.14	500	125	2500	323	323	yes	1
C.15	500	250	2500	556	556	yes	1
C.16	500	5	12500	11	11	yes	1
C.17	500	10	12500	18	18	yes	1
C.18	500	83	12500	113	114	no	25
C.19	500	125	12500	146	147	no	1
C.20	500	250	12500	267	267	yes	1
D.01	1000	5	1250	106	106	yes	1
D.02	1000	10	1250	220	220	yes	1
D.03	1000	167	1250	1565	1565	yes	3
D.04	1000	250	1250	1935	1935	yes	1
D.05	1000	500	1250	3250	3250	yes	1
D.06	1000	5	2000	67	68	no	1
D.07	1000	10	2000	103	103	yes	1
D.08	1000	167	2000	1072	1072	yes	24
D.09	1000	250	2000	1448	1449	no	1
D.10	1000	500	2000	2110	2110	yes	3
D.11	1000	5	5000	29	29	yes	1
D.12	1000	10	5000	42	42	yes	1
D.13	1000	167	5000	500	500	yes	8
D.14	1000	250	5000	667	667	yes	17
D.15	1000	500	5000	1116	1116	yes	6
D.16	1000	5	25000	13	13	yes	1
D.17	1000	10	25000	23	23	yes	1
D.18	1000	167	25000	223	227	no	6
D.19	1000	250	25000	310	314	no	7
D.20	1000	500	25000	537	537	yes	1

Table 3. Run statistics on OR-Library problems of series E

Problem	$|V|$	$|X|$	$|E|$	Best_value	Par-GRASP	optimal?	#_iter
E.01	2500	5	3125	111	111	yes	1
E.02	2500	10	3125	214	214	yes	1
E.03	2500	417	3125	4013	4013	yes	1
E.04	2500	625	3125	5101	5102	no	1
E.05	2500	1250	3125	8128	8128	yes	1
E.06	2500	5	5000	73	73	yes	1
E.07	2500	10	5000	145	145	yes	1
E.08	2500	417	5000	2640	2647	no	1
E.09	2500	625	5000	3604	3606	no	1
E.10	2500	1250	5000	5600	5600	yes	2
E.11	2500	5	12500	34	34	yes	1
E.12	2500	10	12500	67	67	yes	1
E.13	2500	417	12500	1280	1286	no	2
E.14	2500	625	12500	1732	1736	no	2
E.15	2500	1250	12500	2784	2787	no	2
E.16	2500	5	62500	15	15	yes	1
E.17	2500	10	62500	25	25	yes	1
E.18	2500	417	62500	564	575	no	2
E.19	2500	625	62500	758	764	no	2
E.20	2500	1250	62500	1342	1342	yes	1

solutions found weight only one unit more than the optimal values. For what concerns series D, the optimal solution was found for 16 out of 20 problems. Among the four problems not solved to optimality, for two of them the best solutions found weight only one unit more than the corresponding optimal values. The solutions found for the remaining two problems are approximately within 2% from optimality. For series E, the optimal solution was found for 12 out of 20 problems. Seven of the remaining problems are less than 1% from optimality and the last one is approximately within 2% from optimality. We also notice that, in most of the cases, the best solution was found in very few iterations: for 42 out of 60 test problems the best solution was found by at least one of the processors in its first iteration.

4 Concluding Remarks

In this paper we described a parallel greedy randomized adaptive search procedure for finding approximate solutions to the Steiner problem in graphs. Feasible solutions are characterized by their non-terminal nodes. A randomized version of Kruskal's algorithm for the minimum spanning tree problem is used in the construction phase. Local search is based on insertions and eliminations of Steiner nodes to/from the current solution. Parallelization is performed through the distribution of the iterations among the processors on a demand-driven basis, in order to improve load balancing. The parallel procedure was implemented using the Message Passing Interface library on an IBM SP2 machine.

Computational experiments on a set of benchmark problems from series C, D and E of the OR-Library illustrate the effectiveness of the proposed parallel GRASP procedure, which found the optimal solution for 46 out of the 60 test problems. The best solution found is only one unit from optimality for five among the remaining problems.

However, the observed execution times per iteration are quite high and should be improved in order to allow for the execution of more GRASP iterations. Since local search seems to be the bottleneck of the current implementation in terms of computational times, one promising strategy is the combination of the randomized Kruskal-like construction phase with a local search procedure based on key-paths, as recently proposed by Pardalos and Resende [17]. Another hybrid strategy also currently under investigation is the application of a variable neighborhood search procedure starting from the best solution found by each processor.

Acknowledgements. The authors acknowledge the *Laboratório Nacional de Computação Científica* (Rio de Janeiro, Brazil) for making available their computational facilities and the IBM SP system on which the computational experiments were performed.

References

1. A.C. ALVIM, *Evaluation of GRASP parallelization strategies* (in Portuguese), M.Sc. Dissertation, Department of Computer Science, Catholic University of Rio de Janeiro, 1998.
2. J.E. BEASLEY, "OR-Library: Distributing test problems by electronic mail", *Journal of the Operational Research Society* 41 (1990), 1069–1072.
3. E.-A. CHOUKMANE, "Une heuristique pour le problème de l'arbre de Steiner", *RAIRO Recherche Opérationnelle* 12 (1978), 207–212.
4. K.A. DOWSLAND, "Hill-climbing simulated annealing and the Steiner problem in graphs", *Engineering Optimization* 17 (1991), 91–107.
5. C.W. DUIN AND A. VOLGENANT, "Reduction tests for the Steiner problem in graphs", *Networks* 19 (1989), 549–567.
6. C.W. DUIN AND S. VOSS, "Efficient path and vertex exchange in Steiner tree algorithms", *Networks* 29 (1997), 89–105.
7. H. ESBENSEN, "Computing near-optimal solutions to the Steiner problem in a graph using a genetic algorithm", *Networks* 26 (1995), 173–185.
8. T.A. FEO AND M.G. RESENDE, "Greedy randomized adaptive search procedures", *Journal of Global Optimization* 6 (1995), 109–133.
9. M. GENDREAU, J.-F. LAROCHELLE AND B. SANSÓ, "A tabu search heuristic for the Steiner tree problem in graphs", GERAD, Rapport de recherche G-96-03, Montréal, 1996.

10. F.K. HWANG, D.S. RICHARDS AND P. WINTER, *The Steiner tree problem*, North-Holland, Amsterdam, 1992.

11. A. IWAINSKY, E. CANUTO, O. TARASZOW AND A. VILLA, "Network decomposition for the optimization of connection structures", *Networks* 16 (1986), 205–235.

12. A. KAPSALIS, V.J. RAYWARD-SMITH AND G.D. SMITH, "Solving the graphical Steiner tree problem using genetic algorithms", *Journal of the Operational Research Society* 44 (1993), 397–406.

13. R.M. KARP, "Reducibility among combinatorial problems", in *Complexity of Computer Computations* (E. Miller and J.W. Thatcher, eds.), 85–103, Plenum Press, New York, 1972.

14. L.T. KOU, G. MARKOWSKY AND L. BERMAN, "A fast algorithm for Steiner trees", *Acta Informatica* 15 (1981), 141–145.

15. J.B. KRUSKAL, "On the shortest spanning subtree of a graph and the traveling salesman problem", *Proceedings of the American Mathematical Society* 7 (1956), 48–50.

16. N. MACULAN, "The Steiner problem in graphs", in *Surveys in Combinatorial Optimization* (S. Martello, G. Laporte, M. Minoux, and C.C. Ribeiro, eds.), *Annals of Discrete Mathematics* 31 (1987), 185–212.

17. S.L. MARTINS, P. PARDALOS, M.G. RESENDE, AND C.C. RIBEIRO, "GRASP procedures for the Steiner problem in graphs", presented at the *DIMACS Workshop on Randomization Methods in Algorithm Design*, research report in preparation, 1997.

18. K. MEHLHORN, "A faster approximation for the Steiner problem in graphs", *Information Processing Letters* 27 (1988), 125–128.

19. M. MINOUX, "Efficient greedy heuristics for Steiner tree problems using reoptimization and supermodularity", *INFOR* 28 (1990), 221–233.

20. MESSAGE PASSING INTERFACE FORUM, "MPI: A new message-passing interface standard (version 1.1)", Technical report, University of Tennessee, Knoxville, 1995.

21. J. PLESNÍK, "A bound for the Steiner problem in graphs", *Math. Slovaca* 31 (1981), 155–163.

22. M. PRAIS AND C.C. RIBEIRO, "Reactive GRASP: An application to a matrix decomposition problem in TDMA traffic assignment", Research paper submitted for publication, Catholic University of Rio de Janeiro, Department of Computer Science, 1998.

23. M.G. RESENDE AND C.C. RIBEIRO, "A GRASP for Graph Planarization", *Networks* 29 (1997), 173–189.

24. C.C. RIBEIRO AND M.C. SOUZA, "An improved tabu search for the Steiner problem in graphs", Working paper, Catholic University of Rio de Janeiro, Department of Computer Science, 1997.

25. S. VOSS, "Steiner's problem in graphs: Heuristic methods", *Discrete Applied Mathematics* 40 (1992), 45–72.

26. D.W. WALKER, "The design of a standard message passing interface for distributed memory concurrent computers", *Parallel Computing* 20 (1994), 657–673.

27. P. WINTER, "Steiner problem in networks: A survey", *Networks* 17 (1987), 129–167.

28. J. XU, S.Y. CHIU AND F. GLOVER, "Tabu search heuristics for designing a Steiner tree based digital line network", Working paper, University of Colorado at Boulder, 1995.

A New Simple Parallel Tree Contraction Scheme and Its Application on Distance-Hereditary Graphs

Sun-Yuan Hsieh[1] and Chin-Wen Ho[2] and Tsan-Sheng Hsu[3] and Ming-Tat Ko[3] and Gen-Huey Chen[1]

[1] Department of Computer Science and Information Engineering, National Taiwan University, Taipei, Taiwan, ROC
[2] Department of Computer Science and Information Engineering, National Central University, Chung-Li, Taiwan, ROC
[3] Institute of Information Science, Academia Sinica, Taipei, Taiwan, ROC

Abstract. We present a new parallel tree contraction scheme which takes $O(\log n)$ contraction phases to reduce a tree to its root, and implement this scheme in $O(\log n \log \log n)$ time using $O(n/\log \log n)$ processors on an arbitrary CRCW PRAM. We then show a data structure to represent a connected distance-hereditary graph G in the form of a rooted tree. Applying our tree contraction scheme on the above data structure together with graph theoretical properties, we solve the problems of finding a minimum connected γ-dominating set and finding a minimum γ-dominating clique on G in $O(\log n \log \log n)$ time using $O((n + m)/\log \log n)$ processors on an arbitrary CRCW PRAM, where n and m are the number of vertices and edges in G, respectively.

1 Introduction

A graph is *distance-hereditary* [1, 16] if the distance stays the same between any of two vertices in every connected induced subgraph containing both (where the *distance* between two vertices is the length of a shortest path connecting them). Distance-hereditary graphs form a subclass of perfect graphs [9, 14, 16] that are graphs G in which the maximum clique size equals the chromatic number for every induced subgraph of G [2, 13]. Properties of distance-hereditary graphs are studied by researchers [1, 5, 9, 10, 14, 16] which resulted in sequential algorithms to solve quite a few interesting graph-theoretical problems on this special class of graphs. However, few results [7, 8, 17] are known in the parallel context.

Let G be a distance-hereditary graph in which an integer value $\gamma(v_i)$ is assigned to each vertex v_i. In this paper, we focus on various generalizations of the γ-domination problem, where a γ-domination in G is a minimum subset of vertices such that for every vertex $v \in G$ there is a vertex in the γ-domination with their distance within $\gamma(v)$. The concept of domination is used to model many location problems in operations research and game theory [4, 15]. We will study the *connected γ-domination problem*, i.e., the problem of finding a minimum cardinality γ-domination which induces a connected subgraph, and the *γ-dominating clique problem*, i.e., the problem of finding a minimum γ-domination which induces a clique. The sequential linear time algorithms to solve the connected γ-domination problem and the γ-dominating clique problem have been presented in [5, 10].

The approach we use to solve the above problems is first to show a data structure to represent a connected distance-hereditary graph in the form of a rooted tree. We then present a new parallel tree contraction scheme, which may be of independent interest, to prune such a tree in $O(\log n \log \log n)$ time using $O(n/\log \log n)$ processors on an arbitrary CRCW PRAM, where n is the number of nodes in the given tree. We show this scheme can be applied to solve the above problems in efficient parallel complexities. Our parallel algorithms run in $O(\log n \log \log n)$ time using $O((n + m)/\log \log n)$ processors on an arbitrary CRCW PRAM, where n and m are the number of vertices and edges in the input graph, respectively.

2 Preliminaries

This paper considers a finite, simple, undirected and connected graph $G = (V, E)$, where V and E are the vertex and edge sets of G, respectively. Let $n = |V|$ and $m = |E|$. For graph-theoretic terminologies and notations not mentioned here, we refer to [13].

Let v be a vertex of G. We denote the number of edges incident to v by $deg_G(v)$ and let $deg(G) = max\{deg_G(v)|\ v \in G\}$. We also denote the *neighborhood* of v, consisting of all vertices adjacent to v, by $N_G(v)$, and the *closed neighborhood* of v, the set $N_G(v) \cup \{v\}$, by $N_G[v]$.

Let S be a subset of V. We denote $N_G(S)$ the *open neighborhood of* S, that is the set of vertices in G, exclusive of S, which are adjacent to any vertex in S. We also denote $N_G[S] = N_G(S) \cup S$. The subscript G in the notations can be omitted when no ambiguity arises. The *subgraph induced by* S, denoted by $\langle S \rangle$, consists of the vertices of S where (x, y) is an edge of $\langle S \rangle$ if x and y are in S and $(x, y) \in E$.

For any two vertices u and v, let $dist(u, v)$ denote the distance between u and v in G. Given a vertex $u \in V$, the *hanging* of a connected graph $G = (V, E)$ rooted at u, denoted by h_u, is the collection of sets $L_0(u), L_1(u), \ldots, L_t(u)$ (or simply $L_0, L_1, \ldots, L_t$ when no ambiguity arises), where $t = max_{v \in V}\ dist(u, v)$ and $L_i(u) = \{v \in V | dist(u, v) = i\}$ for $0 \le i \le t$. For any vertex $v \in L_i$ and any vertex set $S \subseteq L_i, 1 \le i \le t$, let $N'(v) = N(v) \cap L_{i-1}$ and $N'(S) = N(S) \cap L_{i-1}$. Any two vertices $x, y \in L_i\ (1 \le i \le t-1)$ are said to be *tied* if x and y have a common neighbor in L_{i+1}.

Let $\gamma(v_1), \gamma(v_2), \ldots, \gamma(v_n)$ be non-negative integers associated with n vertices $v_1, v_2, \ldots, v_n$ of G. A subset $D \subseteq V$ is a *γ-dominating set* in G if for every $v \in V \backslash D$, there is a $u \in D$ with $dist(u, v) \le \gamma(v)$. D is a *connected γ-dominating set* of G if D γ-dominates G and the subgraph induced by D is connected. It is a *γ-dominating clique* in G if D is also a clique. The *connected γ-domination problem* (respectively, *γ-dominating clique problem*) is the problem of finding a minimum cardinality connected γ-dominating set (respectively, γ-dominating clique) of G.

3 A New Simple Tree Contraction

Our scheme bases upon two abstract parallel tree contraction operations, namely RAKE and SHRINK. The scheme works in phases: during each phase, one RAKE and then one SHRINK operation are performed consecutively.

Let $T = (V, E)$ be a rooted tree with n vertices and $\{v_1, v_2, \ldots, v_k\} \subseteq V$, where $k \geq 2$. We say that $C = [v_1, v_2, \ldots, v_k]$ is a *chain* of length $k - 1$ if v_1 is not the root, the degree of v_1 is 2, v_{i+1} is the only child of v_i, $1 \leq i < k$, and v_k is a leaf. A chain is said to be *maximal* if it is not possible to add any vertex to form a longer chain. Further, we say that a maximal chain $C = [v_1, v_2, \ldots, v_k]$ is *reduced* if the vertices $v_2, v_3, \ldots, v_k$ are removed from T. The following two operations are defined in T.

1. SHRINK: An operation reduces all the maximal chains of T.
2. RAKE: An operation removes all the leaves from T.

We define a *contraction phase* of the current tree by first applying a RAKE operation and then applying a SHRINK operation. Our tree contraction scheme, called $R\&S$, applies a sequence of contraction phases to the original tree until it being reduced to its root. It is not difficult to show the following result.

Lemma 1. *After $O(\log n)$ contraction phases, T is reduced to a single vertex which is its root.*

In what follows, we present a method to implement our tree contraction scheme $R\&S$. We assume the tree is ordered so that, for each vertex v in the original tree, the children of v are ordered $v_1, v_2, \ldots, v_{l(v)}$, where $l(v)$ is the number of children of v, and each child knows its index. That is, let $index[v_i] = i$ be the index of v_i in this ordering of children. For each vertex v we set aside $l(v)$ locations $label[v, i]$, $i = 1, \ldots, l(v)$ in the shared memory such that $label[v, j]$ is marked by v_j if v_j is deleted from the current tree using tree contraction. Initially each $label[v, i]$ is empty or *unmarked*. We use $arg(v)$ to denote the current number of unmarked labels for v. Thus, initially, $arg(v) = l(v)$.

For a node v in a rooted tree T with n nodes, let $child_T(v)$ denote the children of v and $par_T(v)$ denote the parent of v in T. Throughput this section, we also use $child(v)$ and $par(v)$ to denote the children and the parent of v in the current tree when no ambiguity occurs. Our method works in $O(\log n)$ phases corresponding to $O(\log n)$ contraction phases. Assume that the children of the root r is given by $(u_0, u_1, \ldots, u_{l(r)-1})$. To begin, we compute the Euler tour for T which is represented by an array $ET = [e_1 = (r, u_0), e_2, \ldots, e_{2(n-1)} = (u_{l(r)-1}, r)]$ (i.e., $ET[i]$ records the ith edge in the tour constructed). This can be achieved in $O(\log n)$ time using $O(n/\log n)$ processors on an EREW PRAM by applying the list ranking technique [18].

For each vertex $v \neq r$ in T, assume $ET[i_v] = (par_T(v), v)$ for some $1 \leq i_v \leq 2(n-1)$. We construct an array with $2(n-1)$ entries, denoted by $\mathcal{D}_{ET}[1, ..., 2(n-1)]$, corresponding to $2(n-1)$ entries of array ET. This array will be updated in each phase as follows. We set $\mathcal{D}_{ET}[i_v] = 1$, or 2, or 3, if the degree of v in the current tree is 1 or 2, or at least 3, respectively. We also set $\mathcal{D}_{ET}[j] = 0$, where $j \neq i_v$ for each $v \in V(T)$, or $j = i_v$ for each $v \in V(T)$ being not in the current tree.

In a phase, we first show how to implement RAKE. For each vertex w with $arg(w) = 0$, it is a leaf of the current tree. So we delete all the nodes w with $arg(w) = 0$ (corresponding to a RAKE operation) and modify $\mathcal{D}_{ET}$ as follows. We first set $\mathcal{D}_{ET}[i_w] = 0$. Assume $f = par_T(w)$. We use the following method to compute $\mathcal{D}_{ET}[i_f]$ in $O(1)$ time using totally $O(n)$ processors. Assume each node is assigned with one processor. Under the arbitrary CRCW PRAM model, we start by setting aside a memory location $argindex[f] = null$. Each processor assigned to an unmarked child writes its index into the memory location $argindex[f]$. Assume that one of these children succeeds in writing its index. If $argindex[f] = null$, we set $\mathcal{D}_{ET}[i_f] = 1$ because all the children of f are deleted by executing a RAKE operation. To test whether $\mathcal{D}_{ET}[i_f] = 2$, each processor assigned to an unmarked child reads $argindex[f]$ and if the value is not the same as its own index, it rewrites its index to $argindex[f]$. If the value of $argindex[f]$ does not change, then $\mathcal{D}_{ET}[i_f] = 2$. Otherwise, $\mathcal{D}_{ET}[i_f] = 3$. The above modification can be done in $O(1)$ time using $O(n)$ processors.

We then show how to implement SHRINK. Let H denote the current tree and $N_{d=\omega} = \{v \in V(H)| \deg_H(v) = \omega\}$. To find a maximal chain in the current tree, we first compute the set $Q = \{v \in N_{d=2}| \deg_H(par_H(v)) \geq 3$, or $par_H(v)$ is the root of $H\}$ in $O(1)$ time using $O(n)$ processors. We call each vertex $v \in Q$ *chain-leader*. For each chain-leader v, we define the v-interval $I_v = [a, b]$ for v with $ET[a] = (par_T(v), v)$ and $ET[b] = (v, par_T(v))$. In particular, a (respectively, b) is said to be the *left endpoint* (respectively, *right endpoint*) of I_v. For any two $I_x = [a, b]$ and $I_y = [a', b']$, we say I_x *covers* I_y if $a < a'$ and $b' < b$, and we say both intervals are *disjoint* if $a < b < a' < b'$. According to the property of trees, the following result can be shown.

Lemma 2. *Let* $\mathcal{I} = \{I_{v_1}, I_{v_2}, \ldots, I_{v_q}\}$ *be the set of* v_j-*intervals for all the chain-leaders* v_j *in a phase. For any two intervals* I_{v_x} *and* I_{v_y} *in* $\mathcal{I}$, *either one covers the other or both are disjoint.*

Given two intervals I_x and I_y, we define $I_x \preceq_{\mathcal{I}} I_y$ if I_y covers I_x. Note that $\preceq_{\mathcal{I}}$ is a partial order. According to the definition of a maximal chain, we have the following lemma.

Lemma 3. *Let* $\mathcal{I} = \{I_{v_1}, I_{v_2}, \ldots, I_{v_q}\}$ *denote the set of* v_j-*intervals for all the chain-leaders* v_j *in a contraction phase, and let* $\mathcal{I}^*$ *denote the set of minimal elements in* $(\preceq_{\mathcal{I}}, \mathcal{I})$. *For each* $I_w = [a, b] \in \mathcal{I}^*$, *the subtree rooted at* w *in the current tree forms a maximal chain if and only if* $max\{\mathcal{D}_{ET}[i]| a \leq i \leq b\} \leq 2$.

Based on the parallel scheme to solve the *all nearest smaller values (ANSV) problem* [3], we can find $\mathcal{I}^*$ in $\mathcal{I}$ in $O(\log \log n)$ time using $O(n/\log \log n)$ processors on a common CRCW PRAM. Thus all maximal chains can be identified with the above complexities.

After the above computation, a contraction phase can be completed. At this time, we need to update $\mathcal{D}_{ET}$ with the new degree of each vertex after executing a SHRINK operation. For each $I_w = [a, b] \in \mathcal{I}^*$, recall that $ET[a] = (par_T(w), w)$. After reducing the maximal chain to the chain-leader w, we set $\mathcal{D}_{ET}[a] = 1$ because the degree of w in the current tree is 1, and set each $\mathcal{D}_{ET}[j] = 0$, where $a < j < b$. This takes $O(1)$ time using $O(n)$ processors. After completing a contraction phase and updating $\mathcal{D}_{ET}$, we then go on executing the next phase. The above discussion leads to the following result.

Theorem 4. *Algorithm R&S can be implemented correctly in $O(\log n \log \log n)$ time using $O(n/\log \log n)$ processors on an arbitrary CRCW PRAM, where n is the number of vertices of the input tree.*

4 Connected γ-Dominations on Distance-Hereditary Graphs

In this section, a sequential algorithm is presented to find a minimum connected γ-dominating set on a distance-hereditary graph. Throughout this paper, G is used to denote a connected distance-hereditary graph whenever no ambiguity occurs. We first give some previous known properties of distance-hereditary graphs.

Proposition 5. *[1, 9, 14] Suppose $h_u = (L_0, L_1, \ldots, L_t)$ is a hanging of G. If $x, y \in L_i$ $(1 \leq i \leq t)$ are in the same connected component of $\langle L_i \rangle$ or tied, then $N'(x) = N'(y)$.*

Proposition 6. *[1, 14] Suppose $h_u = (L_0, L_1, \ldots, L_t)$ is a hanging of G. For any two vertices $x, y \in L_i$ with $i \geq 1$, $N'(x)$ and $N'(y)$ are either disjoint, or one of them is contained in the other.*

Given a graph $G = (V, E)$, a vertex subset S is *homogeneous* in a graph $G = (V, E)$ if and only if every vertex in $V \setminus S$ is adjacent to either all or none of the vertices of S. The set S is further said to be *proper homogeneous* if $2 \leq |S| \leq n - 2$. Note that every vertex $v \in V \setminus S$ has equal distance from the vertices of S. We call a family of subsets *arboreal* if every two subsets of the family are either disjoint or comparable (by set inclusion). For a hanging $h_u = (L_0, L_1, \ldots, L_t)$, Hammer and Maffray [14] defined an equivalence relation $\equiv_i$ between vertices of L_i by $x \equiv_i y$ means x and y are in the same connected component of L_i or x and y are tied. Let $\equiv_a$ be defined on $V(G)$ by $x \equiv_a y$ means $x \equiv_i y$ for some i.

Proposition 7. *[14] Let h_u be the hanging of G rooted at u and let $R_1, R_2, \ldots, R_r$ be the equivalence classes with respect to h_u. The family $\{N'(R_k) | N'(R_k) \subseteq R_i\}$, for $1 \leq i \leq r$, is an arboreal family of homogeneous subsets of $\langle R_i \rangle$.*

4.1 The Equivalence-Hanging Tree

Let h_u be the hanging of a distance-hereditary graph rooted at u. Let R be an equivalence class of G with respect to h_u and $\Gamma_R = \{S \subset R|$ there is an equivalence class R' with $N'(R') = S\}$. We call Γ_R the upper neighborhood system in R and call each $S(= N'(R'))$ in Γ_R the upper neighborhood of R'. By Proposition 7, Γ_R is an arboreal family of homogeneous subsets of $\langle R \rangle$. We define a partial order $\preceq$ between two different sets S_p and S_q in Γ_R by $S_p \preceq S_q \Leftrightarrow S_p \subset S_q$. Let $R_1, R_2, \ldots, R_k$ be the equivalence classes of G with respect to h_u. We define a graph $T_{h_u} = (V(T_{h_u}), E(T_{h_u}))$, as follows. For each S in $\Gamma_{R_1} \cup \Gamma_{R_2} \cup \cdots \cup \Gamma_{R_k} \cup \{R_i| i = 1, 2, \ldots, k\}$, we creat a node for T_{h_u} to represent S. There are totally $|\Gamma_{R_1} \cup \Gamma_{R_2} \cup \cdots \cup \Gamma_{R_k} \cup \{R_i| i = 1, 2, \ldots, k\}|$ created nodes. For each node $\omega \in V(T_{h_u})$, let S_ω denote the vertex set represented by ω. There is an edge $(\alpha, \beta) \in E(T_{h_u})$ if (a) $S_\alpha \preceq S_\beta$ and no other S_ω satisfies $S_\alpha \preceq S_\omega \preceq S_\beta$ or (b) $S_\beta = N'(S_\alpha)$. We call type (a) edge the abnormal edge and call type (b) edge the normal edge. The following lemma can be shown based on the characterization of T_{h_u}.

Lemma 8. The graph T_{h_u} is a tree and $|V(T_{h_u})| = O(n)$.

We call T_{h_u} the equivalence-hanging tree of the given distance-hereditary graph G with respect to h_u. For the rest of this paper, we assume T_{h_u} is a tree rooted at the node representing u.

Consider the process of removing nodes of T_{h_u} one at a time from the leaves to the root. Let i be an integer that is less than $|V(T_{h_u})|$. We define $T_{h_u}^i$ to be the tree resulted after removing i nodes from T_{h_u} under the above process, and define G^i to be the subgraph of G induced by $\cup_{\nu \in V(T_{h_u}^i)} S_\nu$. For a tree T, let $leaf(T)$ denote the leaves of T. The following lemmas relate distance-hereditary graphs with their equivalence-hanging trees.

Lemma 9. Graph G^i, is a distance-hereditary graph.

Lemma 10. If $\nu \in leaf(T_{h_u}^i)$, then S_ν is a homogeneous set of G^i.

Lemma 11. Suppose ν is a node of $T_{h_u}^i$. Let $A = \{\alpha_1, \alpha_2, \ldots, \alpha_k\}$ and $B = \{\beta_1, \beta_2, \ldots, \beta_l\}$ be the sets of children of ν in $T_{h_u}^i$ such that (α_i, ν) is an abnormal edge and (β_i, ν) is a normal edge. If $A = \emptyset$ or each $\alpha_i \in A$ is a leaf in $T_{h_u}^i$, then S_ν is a homogeneous set of G^i.

Given a distance-hereditary graph G, T_{h_u} can be constructed in $O(\log n \log \log n)$ time using $O((n+m)/\log \log n)$ processors on an arbitrary CRCW PRAM based on the techniques described in [6, 17].

4.2 A Sequential Algorithm

A linear time sequential algorithm is first described in [5]. Here we present another sequential algorithm which will be used latter to be parallelized. Let G' be an induced subgraph of G with a new γ value γ' assigned to each of its vertices. Let $D(G)$ and $D(G')$ denote a minimum connected γ-dominating set of G and a minimum connected γ'-dominating set G', respectively. Proposition 12 shows that by properly choosing G' and setting γ' values, we can reduce the problem of computing $D(G)$ to the problem of computing $D(G')$.

For a vertex x in a homogeneous set Q of graph G, let $tag(x) = 1$ if there is a vertex $y \in V(G) \setminus Q$ with $dist(x, y) > \gamma(y)$, and let $tag(x) = 0$ for otherwise.

Proposition 12. *[5] Let Q be a vertex subset of the given graph $G = (V, E)$.*
(a) Assume that $Q \subset V$ is a proper homogeneous set of G. Let x be a vertex of Q with $\gamma(x) = min\{\gamma(y)|\ y \in Q\}$. Also let $G' = \langle (V \setminus Q) \cup \{x\}\rangle$ and $\gamma'(v) = \gamma(v)$ for all $v \in G'$. (i) If $\gamma(x) \geq 2$, or $\gamma(x) = 1$ and $tag(x) = 1$, then $D(G) = D(G')$. (ii) If $\gamma(x) = 0$ and $\gamma(w) = 0$ for some vertex $w \in V \setminus Q$, then $D(G) = D(G') \cup \{y \in Q|\ \gamma(y) = 0\}$.
(b) Assume that Q contains only one vertex x such that $N(x)$ forms a homogeneous set in G. Let y be a vertex of $N(x)$ with $\gamma(y) = min\{\gamma(z)|\ z \in N(x)\}$. Also let $G' = \langle V \setminus \{x\}\rangle$ and $\gamma'(v) = \gamma(v)$ for all $v(\neq y) \in G$. (i) If $\gamma(x) \geq 2$, or $\gamma(x) = 1$ and $tag(x) = 1$, then $D(G) = D(G')$ with $\gamma'(y) = min\{\gamma(y), \gamma(x) - 1\}$. (ii) If $\gamma(x) = 0$ and $\gamma(w) = 0$ for some vertex $w \in V \setminus N[x]$, then $D(G) = D(G') \cup \{x\}$ with $\gamma'(y) = 0$.

Proposition 13. *[5] Suppose x is a vertex of a homogeneous set Q with $\gamma(x) = min\{\gamma(y)|y \in Q\} = 1$ and $tag(x) = 0$. If there is a vertex $w \in Q$ satisfying $N[w] \supseteq \{v \in Q|\ \gamma(v) = 1\}$, then $D(G) = \{w\}$. Otherwise, $D(G) = \{x, z\}$, where $z \in N(Q)$.*

Given a hanging $h_u = (L_0, L_1, \ldots, L_t)$ of a distance-hereditary graph G, the sequential algorithm in [5] processes G from L_t to L_0 based on propositions 12 and 13 as follows. When L_i is processed, the algorithm finds connected components of $\langle L_i \rangle$. By Proposition 5, each connected component is a homogeneous set of the graph $\langle V \setminus \cup_{j=i+1}^t L_j \rangle$. The connected components are ordered increasingly according to the cardinalities of their upper neighborhoods. Then the algorithm removes components from G one at a time starting from the one with the smallest order. According to the above propositions, the new γ-value is adjusted for the resulting graph after each removal, and the information of a minimum connected γ-dominating set of G is gathered on the γ-values.

To efficiently obtain a parallel solution, we apply a new order to reduce a distance-hereditary graph G represented by its equivalence-hanging tree T_{h_u}. Our algorithm (described latter) removes nodes of T_{h_u} one at a time from the leaves to the root in an arbitrary order. Recall that $T_{h_u}^i$ and G^i denote the current tree and current graph under the above process. Before proceeding to describe the detail algorithm, we first give some preliminaries. For a vertex v in a tree T, let $T(v)$ denote the subtree of T rooted at v. We define the *critical value* of α, denoted by $cri(\alpha)$, to be the smallest γ-value of S_α when $\alpha \in leaf(T_{h_u}^i)$ and G is reduced to G^i based on Proposition 12. Under the same assumption as above, let $\{v \in S_\alpha|\ \gamma(v) = cri(\alpha)\}$ be the *critical vertex set* of α.

We next describe the additional data structures and information associated with the given equivalence-hanging tree which are manipulated by our algorithm. For each node $\alpha \in V(T_{h_u}^0)$, we initially set $\gamma(\alpha) = min\{\gamma(z)|\ z \in S_\alpha \setminus \cup_\nu S_\nu$, where $\nu \in child_{T_{h_u}^0}(\alpha)$ and (ν, α) is abnormal$\}$. The γ-value associated with α will be updated in removing each node in $child_{T_{h_u}^0}(\alpha)$. It leads to $\gamma(\alpha) = cri(\alpha)$ when $\alpha \in leaf(T_{h_u}^i)$. During the execution of our algorithm, we also maintain a set for each node ν in current tree $T_{h_u}^i$, denoted by $\Lambda_\nu = \{v_1, v_2, \ldots, v_k\}$, where $\gamma(v_1) = \gamma(v_2) = \cdots = \gamma(v_k) = min\{\gamma(x)|\ x \in S_\nu \setminus \cup_\mu S_\mu$, where $\mu \in child_{T_{h_u}^i}(\nu)$ and (μ, ν) is abnormal $\}$. We also define $\delta(\Lambda_\nu) = \gamma(v_i)$ for arbitrary v_i in Λ_ν.

We now present a high level description of our sequential algorithm, called SCD, to find a minimum connected γ-domination $D(G)$. If there is a vertex $u \in V(G)$ whose γ-value is 0, we compute the hanging h_u. Otherwise, we compute a hanging rooted at an arbitrary vertex, u. Next, an equivalence-hanging tree T_{h_u} is constructed. Let α be an arbitrary leaf in $T_{h_u}^i$ which is not the root. Also let $par(\alpha) = \beta$ and $\gamma(\beta) = q$. Note that S_α is a homogeneous set of graph G^i by Lemma 10. We process α in the following two cases.

Case 1 $\gamma(v) > 0$ for all v in the current graph G^i.

Case 1.1 $\gamma(\alpha) = 0$. Let ω be a node in $child_{T_{h_u}^0}(\alpha)$ with $\gamma(\omega) = 1$ and (ω, α) is a normal edge. /*ω is not in the current tree.*/ Pick an arbitrary vertex x in Λ_ω. Let $tag(x) = 1$ if there is a vertex $v \in \cup_{\nu \in V(T_{h_u}^0)\setminus V(T_{h_u}^0(\omega))}S_\nu$ with $dist(x, v) > \gamma(v)$. Otherwise, let $tag(x) = 0$.

Case 1.1.1 $tag(x) = 1$. Pick a vertex $y \in \Lambda_\alpha$. Determine the hanging h_y of the subgraph of G induced by $(\cup_{\nu \in V(T_{h_u}^0)\setminus V(T_{h_u}^0(\alpha))}S_\nu) \cup S_\alpha$. Construct the equivalence-hanging tree T_{h_y}. Replace $T_{h_u}^i$ with T_{h_y}, and goto the next iteration.

Case 1.1.2 $tag(x) = 0$. /* x γ-dominates all vertex of $V(G) \setminus S_\omega$ */. If there is a vertex $w \in S_\omega$ satisfying $N[w] \supseteq \Lambda_\omega$, output $D(G) = \{w\}$ and terminate the execution. Otherwise, output $D(G) = \{x, z\}$, where z is an arbitrary vertex in $N'(S_\omega)$. Terminate the execution. /* based on Proposition 13. */

Case 1.2 $\gamma(\alpha) \geq 1$.

Case 1.2.1 (α, β) is normal. Let $min\{\gamma(\alpha) - 1, q\}$ be the new γ-value of β. Remove α and S_α from $T_{h_u}^i$ and $V(G^i)$, respectively. If α is the only child of β in $T_{h_u}^i$ and $\gamma(\beta) < \delta(\Lambda_\beta)$, then pick an arbitrary vertex g in Λ_β to be the new set of β, and let $\gamma(g) = \gamma(\beta)$.

Case 1.2.2 (α, β) is abnormal. Let $min\{\gamma(\alpha), q\}$ be the new γ-value of β. If $\delta(\Lambda_\alpha) = \delta(\Lambda_\beta)$, let $\Lambda_\alpha \cup \Lambda_\beta$ be the new set of β. If $\delta(\Lambda_\alpha) < \delta(\Lambda_\beta)$, let Λ_α be the new set of β. Remove α from $T_{h_u}^i$. If α is the only child of β in $T_{h_u}^i$ and $\gamma(\beta) < \delta(\Lambda_\beta)$, then pick an arbitrary vertex g in Λ_β to be the new set of β, and let $\gamma(g) = \gamma(\beta)$.

Case 2 $\gamma(v) = 0$ for some v in the current graph G^i /* note that $\gamma(u) = 0$, where u is the root of the current hanging */. If (α, β) is normal, let $max\{0, min\{\gamma(\alpha) - 1, q\}\}$ be the new γ-value of β. Otherwise, let $max\{0, min\{\gamma(\alpha), q\}\}$ be the new γ-value of β. Using the same method as **Case 1.2** to maintain the set of β and remove nodes from the current tree and current graph.

Algorithm SCD works by repeatedly executing the above two cases until either T_{h_u} is reduced to its root or $D(G)$ is found. Assume T_{h_u} is reduced to its root. If there exists no vertex whose γ-value is set to 0 during the execution, then $D(G)$ is the root of the given hanging. Otherwise, $D(G) = \{w|\ \gamma(w) = 0\}$.

5   Finding a Connected γ-Dominating Set in Parallel

Let $\nu \in leaf(T_{h_u}^i)$ and $child_{T_{h_u}^o}(\nu) = \{\mu_1, \mu_2, \ldots, \mu_k\}$. In executing Algorithm SCD, the situation that $\gamma(\nu)$ equals to $cri(\nu)$ occurs after deleting $child_{T_{h_u}^o}(\nu)$. In deleting each μ_i, we provide $\gamma(\mu_i) = cri(\mu_i)$ as an *input value* to update $\gamma(\nu)$ depending on the type of (μ_i, ν). When $child_{T_{h_u}^o}(\nu)$ are all deleted, we say ν is in a *complete state*. Moreover, we say ν is in an *almost complete state without* μ if $cri(\nu)$ can be computed by giving $cri(\mu)$.

5.1   Algorithms for RAKE and SHRINK

We first briefly describe how our parallel algorithm works as follows. If there is a vertex $u \in V(G)$ with $\gamma(u) = 0$, we construct T_{h_u}. Otherwise, we construct an equivalence-hanging tree T_{h_u}, where u is an arbitrary vertex. The initial values and data structure maintained for T_{h_u} are the same as Algorithm SCD. We then design algorithms executed with RAKE and SHRINK to adjust γ-values of the current tree and graph such that $D(G)$ can be generated consequently using our tree contraction scheme. The algorithms for RAKE are described as follows. Suppose $W = \{\alpha_1, \alpha_2, \ldots, \alpha_k\}$ is a maximal set of leaves in $T_{h_u}^i$ which have the common parent, denoted by $par(W) = par(\alpha_i) = \beta$. We will refer W by a *maximal common-parent leaf set* for convenience. We also assume (α_i, β) is abnormal (respectively, normal) for $1 \leq i \leq j$ (respectively, $j + 1 \leq i \leq k$) and $\gamma(\beta) = q$ before executing R1. Let $r = min\{\gamma(\alpha_1), \gamma(\alpha_2), \ldots, \gamma(\alpha_j)\}$ and $s = min\{\gamma(\alpha_{j+1}), \gamma(\alpha_{j+2}), \ldots, \gamma(\alpha_k)\}$. Below are two algorithms applied with RAKE on W.

Algorithm R1 /* *works on* W *when* $\gamma(v) > 0$ *for all* v *in the current graph* G^i */

Case 1 $min\{r, s\} = 0$. Find a node $\alpha \in \{\alpha_1, \alpha_2, \ldots, \alpha_k\}$ such that $\gamma(\alpha) = 0$. Let ω be a node in $child_{T_{h_u}^o}(\alpha)$ satisfying $\gamma(\omega) = cri(\omega) = 1$ and (ω, α) is normal. Pick an arbitrary vertex x in Λ_ω. Determine $tag(x)$ as Case 1.1 of Algorithm SCD. If $tag(x) = 1$, execute Case 1.1.1 of Algorithm SCD. Otherwise, execute Case 1.1.2 of Algorithm SCD.

Case 2 $min\{r, s\} \geq 1$. Let $min\{q, r, s - 1\}$ be the new γ-value associated with β. Assume $min\{\delta(\Lambda_\beta), r\} = l$. Let $\mathcal{P} = \{\Lambda_y|\ y \in \{\beta, \alpha_1, \alpha_2, \ldots, \alpha_j\}$ and $\delta(\Lambda_y) = l\}$, and $\Lambda = \cup_{X \in \mathcal{P}} X$.

Case 2.2.1 $W = child_{T^i_{h_u}}(\beta)$ and $\gamma(\beta) < l$. /* In this case, $cri(\beta) = \gamma(\beta)$.*/ Select an arbitrary vertex x from Λ to be the new set of β, and let $\gamma(x) = \gamma(\beta)$.

Case 2.2.2 $W \subset child_{T^i_{h_u}}(\beta)$ or $\gamma(\beta) \geq l$. Let Λ be the new set of β.

Algorithm R2 /* works on W when $\gamma(u) = 0$ where u is the root of the given hanging.*/

Determine l value and create the set Λ as Case 2 of R1 does. If $s = 0$, let 0 be the γ-value of β. If $s > 0$, let $min\{q, r, s-1\}$ be the new γ-value of β. Maintain the new set of β using the method of Case 2 of R1.

In the following, we describe the algorithms executed for SHRINK. Suppose T_{h_u} is the equivalence-hanging tree with respect to $h_u = (L_0, L_1, \ldots, L_t)$. Let $C = [\alpha_1, \alpha_2, \ldots, \alpha_k]$ be a maximal chain of $T^i_{h_u}$, where α_k is a leaf. Note that α_k is in a complete state and each α_i, $i < k$, is in an almost-complete state without α_{i+1}. We define $level(\alpha_i) = q$ if $L_q \supseteq S_{\alpha_i}$. A node α_i in C is said to be a *jumped 0-node* if C contains a node α_j, $j \geq i$, such that $\gamma(\alpha_j) - (level(\alpha_j) - level(\alpha_i)) = 0$. Note that C may contain more than one jumped 0-nodes. We further say α_i is the *lowest jumped 0-node* if C contains no other jumped 0-node α_j with $j > i$. The following lemma can be shown by induction.

Lemma 14. *Suppose* $C = [\alpha_1, \alpha_2, \ldots, \alpha_k]$ *is a chain. Let* $d_i = level(\alpha_i) - level(\alpha_1)$ *and* $l = min\{\gamma(\alpha_1) - d_1, \gamma(\alpha_2) - d_2, \gamma(\alpha_3) - d_3, \ldots, \gamma(\alpha_k) - d_k\}$. *Then,* $cri(\alpha_1) = l$ *if one of the following two conditions is satisfied. (1)* $\gamma(u) \neq 0$ *and* $\gamma(\alpha_i) - d_i \geq 1$, *(2)* $\gamma(u) = 0$ *and* $\gamma(\alpha_i) - d_i \geq 0$.

The above lemma can be generalized as follows.

Lemma 15. *Suppose* $C = [\alpha_1, \alpha_2, \ldots, \alpha_k]$ *is a chain. Given an integer* j *such that* $1 \leq j \leq k$, *let* $d_i = level(\alpha_i) - level(\alpha_j)$ *for all* $j + 1 \leq i \leq k$ *and* $l = min\{\gamma(\alpha_j) - d_j, \gamma(\alpha_{j+1}) - d_{j+1}, \ldots, \gamma(\alpha_k) - d_k\}$. *Then,* $cri(\alpha_j) = l$ *if one of the following two conditions is satisfied. (1)* $\gamma(u) \neq 0$ *and* $\gamma(\alpha_i) - d_i \geq 1$, *(2)* $\gamma(u) = 0$ *and* $\gamma(\alpha_i) - d_i \geq 0$.

Lemma 16. *Suppose* $C = [\alpha_1, \alpha_2, \ldots, \alpha_k]$ *is a chain and* α_t *is the lowest jumped 0-node of* C. *Then,* α_t *is the largest-level node whose critical value is 0.*

According to Lemma 15 and the computation of Algorithm SCD, we have the following lemma.

Lemma 17. *Suppose* $C = [\alpha_1, \alpha_2, \ldots, \alpha_k]$ *is a chain of* $T^i_{h_u}$ *with* $\gamma(u) = 0$ *and* α_t *being its lowest jumped 0-node. Then,* $cri(\alpha_1) = cri(\alpha_2) = \cdots = cri(\alpha_t) = 0$.

Below are two algorithms applied with SHRINK on $C = [\alpha_1, \alpha_2, \ldots, \alpha_k]$.

Algorithm S1 /* works on C when $\gamma(v) > 0$ for all v in the current graph G^i*/

Case 1 C contains a jumped 0-node. Find the lowest jumped 0-node α_t and compute its critical vertex set Λ_{α_t}. If $t \neq k$, (α_{t+1}, α_t) is normal, and $\gamma(\alpha_{t+1}) = 1$, then compute the critical vertex set of α_{t+1}.

Let ω be an arbitrary node in $child_{T_{h_u}^0}(\alpha_t)$ satisfying $\gamma(\omega) = cri(\omega) = 1$ and (ω, α_t) is normal. Pick an arbitrary vertex x in Λ_ω. Determine $tag(x)$ as Case 1.1 of Algorithm SCD does. If $tag(x) = 1$, let $\gamma(y) = 0$ for a vertex $y \in \Lambda_{\alpha_t}$ and execute Case 1.1.1 of Algorithm SCD to determine a hanging h_y. Otherwise, execute Case 1.1.2 of Algorithm SCD.

Case 2 C contains no jumped 0-node. Let $d_i = level(\alpha_i) - level(\alpha_1)$. Let
$$\gamma(\alpha_1) = cri(\alpha_1) = min\{\gamma(\alpha_1) - d_1, \gamma(\alpha_2) - d_2, \gamma(\alpha_3) - d_3, \ldots, \gamma(\alpha_k) - d_k\}$$
according to Lemma 14. Compute the critical vertex set Λ_{α_1} and let $\gamma(x) = cri(\alpha_1)$ for each $x \in \Lambda_{\alpha_1}$ if $\gamma(x) \neq cri(\alpha_1)$.

Algorithm S2 /* *works on C when $\gamma(u) = 0$, where u is the root of the given hanging.*/

Case 1 C contains a jumped 0-node. Find the lowest one α_t. Let $\gamma(\alpha_1) = \gamma(\alpha_2) = \cdots = \gamma(\alpha_t) = 0$ according to Lemma 17. Compute the critical vertex set Λ_{α_1} and let $\gamma(x) = 0$ for each $x \in \Lambda_{\alpha_1}$ if $\gamma(x) \neq 0$. Let $Z = \{\alpha_i |$ S_{α_i} is an equivalence class, $1 \leq i \leq t\}$. For each $\alpha \in Z$, compute the critical vertex sets Λ_α and let $\gamma(x) = 0$ for each $x \in \Lambda_\alpha$ if $\gamma(x) \neq 0$.

Case 2 C contains no jumped 0-node. The computation is the same as Case 2 of S1.

According to Lemma 15 and the technique described in [12], we can show the following lemma which implements Case 2 of S1, Case 2 of S2 and Case 1 of S1.

Lemma 18. *Suppose $C = [\alpha_1, \alpha_2, \ldots, \alpha_k]$ is a chain of $T_{h_u}^i$ (not necessary maximal). If $\gamma(\alpha_i) > 0$, $2 \leq i \leq k$, the critical vertex set of α_1 can be found in $O(\log \log k)$ time using $O(k / \log \log k)$ on an arbitrary CRCW PRAM when C is reduced.*

Besides, the implementation of Case 1 of S2 can be done in $O(\log \log k)$ time using $O(k / \log \log k)$ processors on a common CRCW PRAM. Using the techniques described in [18], we have the following result.

Lemma 19. *Given a maximal chain $C = [\alpha_1, \alpha_2, \ldots, \alpha_k]$ after preprocessing, the lowest jumped 0-node can be computed in $O(\log \log k)$ time using $O(k / \log \log k)$ processors on a common CRCW PRAM.*

From the above discusion, we can efficiently parallelize our sequential algorithm and obtain the following theorem.

Theorem 20. *The connected γ-domination problem on distance-hereditary graphs can be solved in $O(\log n \log \log n)$ time using $O((n + m) / \log \log n)$ processors on an arbitrary CRCW PRAM.*

In [10, 11], a method was presented to solve the γ-dominating clique problem. Since the method is based on the reduction scheme similar to the one shown in Proposition 12, the γ-dominating clique problem on distance-hereditary graphs can also be solved in $O(\log n \log \log n)$ time using $O((n + m) / \log \log n)$ processors on an arbitrary CRCW PRAM.

References

1. H. J. Bandelt and H. M. Mulder. Distance-hereditary graphs. *Journal of Combinatorial Theory Series B*, 41(1):182-208, Augest 1989.
2. C. Berge. *Graphs and hypergraphs*. North-Holland, Amsterdam, 1973.
3. O. Berkman, B. Schieber, and U. Vishkin. Optimal doubly logarithmic parallel algorithms based on finding all nearest smaller values. *Journal of Algorithms.*, vol. 14, pp. 344-370, 1993.
4. A.Brandstadt. Special graph classes-a survey. *Technical Report SM-DU-199*, University of Duisburg, 1993.
5. A. Branstadt and F. F. Dragan. A linear time algorithm for connected γ-domination and Steiner tree on distance- hereditary graphs. Technical Report, Gerhard-Mercator- Universität-Gesamthochschule Duisburg SM-DU-261, 1994.
6. R. Cole. Parallel merge sort. *SIAM Journal on Computing*, 17(4):770-785, August 1988.
7. E. Dahlhaus, "Optimal (parallel) algorithms for the all-to-all vertices distance problem for certain graph classes," Lecture notes in computer science 657, pp. 60-69, 1993.
8. E. Dahlhaus. Efficient parallel recognition algorithms of cographs and distance-hereditary graphs. *Discrete Applied Mathematics*, 57(1):29-44, February 1995.
9. A. D'atri and M. Moscarini. Distance-hereditary graphs, steiner trees, and connected domination. *SIAM Journal on Computing*, 17(3):521-538, June, 1988.
10. F. F. Dragan. Dominating cliques in distance-hereditary graphs. Technical Report SM-DU-248, University of Duisburg, 1994.
11. F. F. Dragan and A. Brandstadt. γ-dominating cliques in Helly graphs and chordal graphs. Technical Report SM-DU-228, University of Duisburg, 1993. Proceedings of the 11th STACS, Caen, France, Springer, LNCS 775, pp. 735-746, 1994.
12. Joseph Gil and Larry Rudolph. Counting and packing in parallel. *Proceedings of the 1986 International Conference on Parallel Processing*, vol. 3, pp.1000-1002.
13. M. C. Golumbic. *Algorithmic graph theory and perfect graphs*, Academic press, New York, 1980.
14. P. L. Hammer and F. Maffray. Complete separable graphs. *Discrete Applied Mathematics*, 27(1):85-99, May 1990.
15. S. C. Hedetniemi and R. Laskar, (eds.) Topics on domination, *Annals of Discrete Mathematics*, 48, North-Holland, 1991.
16. E. Howorka. A characterization of distance-hereditary graphs. *Quarterly Journal of Mathematics (Oxford)*, 28(2):417-420. 1977.
17. S.-y. Hsieh, C. W. Ho, T.-s. Hsu, M. T. Ko, and G. H. Chen. Efficient parallel algorithms on distance-hereditary graphs. *Parallel Processing Letters*, to appear. A preliminary version of this paper is in *Proceedings of the International Conference on Parallel Processing*, pp. 20–23, 1997.
18. J. Ja'Ja'. *An Introduction to Parallel Algorithms*. Addison Wesley, 1992.

Cooperative Multi-thread Parallel Tabu Search with an Application to Circuit Partitioning

Renata M. Aiex, Simone de L. Martins, Celso C. Ribeiro, and
Noemi de la R. Rodriguez

Department of Computer Science, Catholic University of Rio de Janeiro, Rio de
Janeiro, RJ 22453-900, Brazil. E-mail: {rma,simone,celso,noemi}@inf.puc-rio.br.

Abstract. In this work, we propose a cooperative multi-thread parallel
tabu search heuristic for the circuit partitioning problem. This procedure is based on the cooperation of multiple search threads. Each thread
implements a different variant of a sequential tabu search algorithm,
using a different combination of initial solution algorithm and move attribute definition. These threads communicate by exchanging elite solutions. PVM and Linda are used in the implementation of the parallel
tabu search procedure. Numerical results reported for a set of ISCAS
benchmark circuits illustrate the effectiveness of the parallel tabu search
procedure. Comparative results illustrating the efficiency of the implementations in PVM and Linda are also assessed.

1   Introduction

The logical test of integrated VLSI circuits is one of the main phases of their
design and fabrication. Testing a circuit amounts to submitting it to different
input patterns and checking whether the observed outputs are exactly those
expected according to the design of the circuit, in order to evaluate if the logical
gates are behaving as expected and to ensure that faults do not occur. Several
approaches exist for the logical test: (i) exhaustive test, (ii) fault simulation, and
(iii) pseudo-exhaustive test. The pseudo-exhaustive approach for the logical test
of integrated circuits was introduced in the literature in the eighties [3, 8, 33].
It consists in partitioning the original circuit to be tested into non-overlapping
subcircuits with a small, bounded number of inputs, which are then exhaustively
tested in parallel. Although it does not cover all possible logical faults, this
approach does not depend on a fault simulation model and ensures a 100% fault
coverage for single stuck-at faults (lines always fixed at the same logical level).

Patashnik [34] has shown that the problem of optimally decomposing a combinational circuits into testable subcircuits is NP-complete. Circuit decomposition implies in cutting some lines (by selector circuits) and, consequently, in
the creation of new inputs and outputs, the so-called *pseudo-inputs* and *pseudo-
outputs*. Suitable algorithms are needed for partitioning the original circuit, in
order to get as few subcircuits as possible (to ensure a high fault coverage and to
minimize the number of testers required) and not too many cuts (to minimize the

cost of the additional hardware which has to be inserted at each point where the original circuit is cut). Roberts and Lala [37] proposed the first general heuristic for this problem. Improved algorithms have been proposed by Davis-Moradkhan and Roucairol [19], with better results in terms of the number of subcircuits in the partition. More recently, a tabu search algorithm was proposed by Andreatta and Ribeiro [2].

In this work, we propose a cooperative multi-thread parallel tabu search heuristic for the circuit partitioning problem. This procedure is based on the cooperation of multiple threads, each of which implementing a different variant of the basic sequential procedure. The search threads use different algorithms for the construction of the initial solution and different move attributes. This paper is organized as follows. The circuit decomposition problem is formulated in Sec.2, where the currently existing algorithms for this problem are reviewed. In Sec.3, we recall the basic elements of the sequential tabu search heuristic for the circuit partitioning problem. Issues such as the definition of solutions, moves and their attributes, tabu lists, initial solutions, and stopping criteria are discussed in details. The cooperative parallel tabu search based on the sequential heuristic is described in Sec.4. PVM and Linda are used in the implementation of the parallel tabu search procedure. These two parallel programming tools for distributed memory environments are introduced in Sec.5. We also give an overview of the implementations, emphasizing the differences resulting from the use of message-passing versus shared-memory paradigms. We present in Sec.6 the computational results obtained through the application of the parallel tabu search algorithm to a set of ISCAS benchmark circuits. The solutions obtained by this algorithm are compared with the best results found in the literature. Comparative results illustrating the efficiency of the implementations in PVM and Linda are also discussed. Concluding remarks are made in the last section.

2 Circuit Partitioning for Pseudo Exhaustive Logical Test

Given that the total duration of the pseudo-exhaustive test should not exceed a certain time T, let L be a parameter equal to the maximum number of inputs such that 2^L test patterns may be generated and applied to the largest subcircuit, and the outputs compared with the correct ones, in total time less or equal than T. The circuit partitioning problem consists in finding a decomposition of the circuit to be tested into non-overlapping circuits with no more than L inputs and at least one logical gate each. Different objectives may be associated with this decomposition, among them (i) the minimization of the number of cuts (minimization of the cost of the additional hardware inserted into the circuit to be tested), and (ii) the minimization of the number of subcircuits (maximization of the fault coverage rate and, moreover, minimization of hardware costs with external testers).

Let $G = (X, A)$ be the directed acyclic graph associated with a combinational circuit C, where X denotes the set of components (inputs, logical gates, and outputs) and A the set of lines used for signal propagation. Given a subset of nodes $V \subset X$, its *input-neighborhood* $\omega^-(V)$ is defined as the set of nodes which are not in V and have at least one successor in V. The set of nodes X is formed by three non-empty disjoint subsets E, P, and S, where E is the set of inputs, P is the set of logical gates, and S is the set of outputs of the combinational circuit C. Partitioning the combinational circuit C into testable subcircuits amounts to finding a partition of X into a non-fixed number of K subsets X_k, $k = 1, \ldots, K$, such that the induced subgraphs $G_k = (X_k, A_k)$ satisfy the following conditions:

- $X = \cup_{k=1}^{k=K} X_k$ and $X_k \cap X_\ell = \emptyset$, $\forall k \neq \ell$, $(k, \ell) \in \{1, \ldots, K\}^2$;
- $n_k + c_k \leq L$, $\forall k = 1, \ldots, K$, where $n_k = |X_k \cap E|$ is the number of inputs in G_k and $c_k = |\omega^-(X_k)|$ is the number of gates in the input-neighborhood of X_k which originated pseudo-inputs;
- $X_k \cap P \neq \emptyset$, $\forall k = 1, \ldots, K$; and
- $G_k = (X_k, A_k)$ is either a connected graph or formed by disjoint subgraphs satisfying the condition above, $\forall k = 1, \ldots, K$.

The second condition above ensures the testability of each subcircuit involved in the partition. Each time an arc (i.e., a line of the original circuit) is cut, both a pseudo-input and a pseudo-output are created. Let $G^+ = (X^+, A^+)$ be the augmented graph obtained by the partitioning algorithm, with $X^+ = X \cup E' \cup S'$, where E' is the set of pseudo-inputs and S' is the set of pseudo-outputs. Let $E^+ = E \cup E'$ and $S^+ = S \cup S'$ be, respectively, the set of inputs and outputs of G^+. The graph G^+ consists of K disjoint subgraphs $G_k^+ = (X_k^+, A_k^+)$, $k = 1, \cdots, K$, where the subsets X_k^+ satisfy the conditions above. The testability condition may be represented by the inequality $|X_k^+ \cap E^+| \leq L$, $k = 1, \ldots, K$.

As an example, consider the graph in Fig. 1. Nodes 1 to 6 are the inputs. Figure 2 illustrates a solution of the partitioning problem for the parameter $L = 4$, with $K = 6$ subcircuits: $X_1 = \{1, 8, 9, 13, 16\}$, $n_1 = 1$ and $c_1 = 3$; $X_2 = \{15, 22, 25\}$, $n_2 = 0$ and $c_2 = 2$; $X_3 = \{23, 26\}$, $n_3 = 0$ and $c_3 = 4$; $X_4 = \{2, 10\}$, $n_4 = 1$ and $c_4 = 1$; $X_5 = \{24, 27\}$, $n_5 = 0$ and $c_5 = 2$; and $X_6 = \{3, 4, 5, 6, 7, 11, 12, 14, 17, 18, 19, 20, 21, 28\}$, $n_6 = 4$ and $c_6 = 0$. The pseudo-inputs and pseudo-outputs are denoted by pi and po, respectively, and are indexed from 1 to 12.

The first heuristic for the circuit partitioning problem was proposed by Bhatt, Chung and Rosenberg [6]. Roberts and Lala [37] proposed a general heuristic based on the relaxation of the testability condition. This algorithm frequently obtains solutions which violate too much the testability condition. Moreover, this violation gets larger and many small subcircuits with a few inputs are created when the in-degree of the logical gates increases. Davis-Moradkhan and Roucairol [18–20] proposed two constructive heuristics asp and cep for this problem, which perform better than that of Roberts and Lala. Andreatta and Ribeiro [2] have more recently proposed the tabu search heuristic described in the next section, using the algorithms proposed by Davis-Moradkhan and Roucairol for the generation of initial solutions.

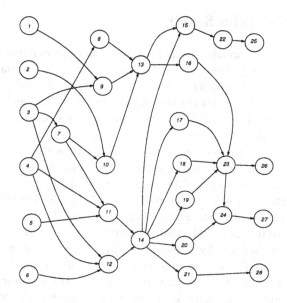

Fig. 1. Combinational circuit to be partitioned represented by a graph

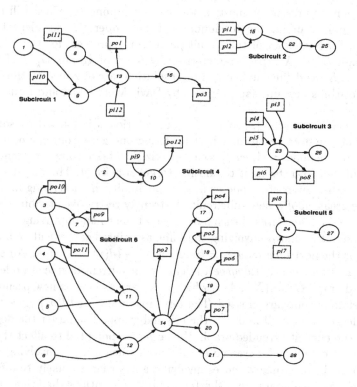

Fig. 2. A partition into $K = 6$ subcircuits

3 Sequential Tabu Search

Tabu search [26–29] is an adaptive procedure for solving combinatorial optimization problems, which guides a hill-descending heuristic to continue exploration without becoming confounded by an absence of improving moves. Briefly, it may be described as follows. Tabu search starts in the same way as local search, by choosing an initial solution and iteratively moving from one solution to another, until some termination condition is attained. At each iteration a move is applied to the current solution, leading to another solution within its neighborhood. Contrary to the basic local search scheme, moves towards solutions that deteriorate the cost function are permitted, when no improving moves are found. One of the main features of tabu search is the use of memory structures which are used to guide the search through the solution space. As part of this strategy, moves towards solutions already visited should be deemphasized or interdicted. These restrictions are usually implemented by means of a short-term memory function which prohibits reversing moves along some iterations, i.e., moves which would reestablish values characterizing previously visited solutions. The number of iterations along which a move is prohibited is called its *tabu tenure*.

The reader is referred to Andreatta and Ribeiro [2] for a complete description of the sequential tabu search algorithm for the circuit partitioning problem. In this section we present its most fundamental components, which will be necessary for understanding the development of the cooperative parallel tabu search strategy. Each *solution* of the circuit partitioning problem for the original circuit graph $G = (X, A)$ is characterized by a set of graphs $G_k^+ = (X_k^+, A_k^+)$, $k = 1, \ldots, K$, as defined in Sec. 2. Initial solutions are obtained by the constructive algorithms cep and asp proposed by Davis-Moradkhan and Roucairol [18–20].

The *neighborhood* $N(s)$ of the current solution s is formed by solutions $\bar{s}$ which may be obtained from s by transfering one gate from one of its subcircuits to another one. Each *move* is characterized by taking one gate from a source subcircuit and transfering it to another target subcircuit. The target subcircuit may be either an existing one or a new subcircuit, characterizing in the latter case the creation of a new subcircuit. A strongly restrictive attribute is used to determine the *tabu status* of each move: every time a non-improving move is performed, all other moves involving the same associated gate p will be made tabu active for the next $tabu\_tenure_1 = (\text{maximum}_{v \in X}\{d(v)\} - current\_degree(v)) \cdot \gamma_1$ iterations, where $current\_degree(v)$ denotes the current degree of node v in the extended graph $G^+ = (X^+, A^+)$, which may vary as long as new pseudo-inputs and pseudo-outputs are created. The term $\text{maximum}_{v \in X}\{d(v)\}$ gives the maximum degree among all nodes in the original graph. The larger the degree of a gate in the current extended graph, the larger its potential to affect the search, since many cuts may be created or destroyed when a move involving this gate is performed. Accordingly, moves involving gates that are likely to affect more the search are made tabu for a shorter number of iterations than those involving gates with a few adjacent nodes. The parameter γ_1 must be tuned and should assume larger values for larger graphs.

Two stopping criteria are used: (i) the overall number of iterations, and (ii) the number of iterations without improvement in the current best solution. These parameters have been empirically set as $max\_iterations = \lceil |P|^{(1.2)} \rceil$ and $max\_moves = 3 \cdot |P|$.

The description of the tabu search algorithm **TS-CPP** for the circuit partitioning problem is given in Fig. 3. Procedure `pack_together` works as a post-optimization step after the tabu search heuristic, packing together small subcircuits appearing in the best feasible solution which do not violate the testability condition. A bin-packing heuristic is used, coupled with a mechanism to evaluate the possible reduction in the number of cuts whenever two subcircuits are packed together.

Algorithm TS-CPP
begin
  Initialize the short term memory function
  Generate the initial solution s_0
  $s, s^* \leftarrow s_0$; $non\_improving\_moves, current\_iteration \leftarrow 0$; $best\_unfeasible \leftarrow \infty$
  Determine the set of candidate moves in the neighborhood of the current solution
  while ($non\_improving\_moves < max\_moves$ **and** $current\_iteration < max\_iterations$) **do**
  begin
    $best\_move\_value \leftarrow \infty$
    for each ($candidate\_move$) **do**
    begin
      if ($candidate\_move$ is admissible **or** satisfies the aspiration criterion) **then**
      begin
        Obtain the neighbor solution $\bar{s}$ by applying $candidate\_move$ to the current solution s
        $move\_value \leftarrow c(\bar{s}) - c(s)$
        if ($move\_value < best\_move\_value$) **then**
        begin
          $best\_move\_value \leftarrow move\_value$
          $s' \leftarrow \bar{s}$
        end_if
      end_if
    end_for
    if ($best\_move\_value \geq 0$) **then** update the short term memory function
    if ($c(s') < c(s^*)$) **then**
    begin
      $non\_improving\_moves \leftarrow 0$
      $s^* \leftarrow s'$
    end_if
    else $non\_improving\_moves \leftarrow non\_improving\_moves + 1$
    $s \leftarrow s'$
    Update the set of candidate moves in the neighborhood
    $current\_iteration \leftarrow current\_iteration + 1$
  end_while
  $s^* \leftarrow$ `pack_together`(s^*)
end_TS-CPP

Fig. 3. Tabu search algorithm for the circuit partitioning problem.

In the next section we show how different possibilities for the construction of initial solutions and tabu lists can be combined into an asynchronous parallel tabu search heuristic for the circuit partitioning problem.

4 Cooperative Multi-thread Parallel Tabu Search

We have reviewed in the previous section the main components of the sequential tabu search heuristic. We now focus our attention into different constructs which can be used for two of these components within the tabu search heuristic:

(i) *Initial solutions* can be constructed by different algorithms:
- algorithm cep;
- algorithm asp; and
- algorithm cir with subcircuits formed by just one gate or one output each.

(ii) Tabu lists can be defined by different *move atributes*:
- the gate which was moved;
- the target subcircuit to where a gate was moved; and
- the source subcircuit from where a gate was moved

Experimental results reported by Aiex [1] have shown that none of the constructs within each group dominate the others, in terms of the quality of the solutions found. Accordingly, different search strategies are entailed by taking different combinations of the pair (construction algorithm, move attribute), again without dominance of any particular combination. We make use of such different combination possibilities to construct a cooperative multi-thread parallel tabu search heuristic for the circuit partitioning problem.

Crainic and Toulouse [13] have recently reviewed metaheuristic parallel search methods. Although several applications appear in the literature, cooperative multi-thread tabu search strategies have not been broadly investigated yet. Such strategies may be classified as Single Point Different Strategies, Multiple Point Single Strategy, or Multiple Point Different Strategies, according to the taxonomy proposed by Crainic et al. [15]. These authors also developed and compared various synchronous and asynchronous multi-thread tabu search strategies for the multicommodity location-allocation problem with balancing requirements [14–16]. The quadratic assignment problem and the mapping problem were studied by De Falco et al. [21, 22]. Their multi-thread strategies were experimented on transputer networks (with 16, 32, and 64 processors), a MAsPAr MPP-1 SIMD computer, and a Convex Meta Series MIMD machine. Approaches to the development of cooperative multi-thread parallel tabu search methods based on adaptive memory were also developed for real-time routing and vehicle dispatching problems [25] and for the vehicle routing problem with time windows [4, 41]. Crainic and Gendreau [12] proposed a synchronous cooperative multi-thread parallel tabu search for the fixed cost, capacited, multicommodity network design problem. Computational results reported in the literature for the above applications show that better solutions were obtained when cooperation was included, outperforming independent thread strategies. Thus, cooperative multi-thread tabu search strategies seem to offer interesting perspectives for metaheuristics. However, several issues remain to be addressed. Toulouse et al. [17] attempt to identify the key questions to be considered in the design of cooperative multi-thread heuristics applied to methods such as tabu search. In particular, they show that questions related to inter-agent communications are a central element of the algorithmic design of these methods.

The parallel procedure proposed for the circuit partitioning problem in this work is based on the cooperation of multiple search threads, each of which implementing a different variant of the sequential procedure defined by a pair (construction algorithm, move attribute). These threads communicate by exchanging elite solutions, in order to improve the quality of the solutions that each of them would be able to find by itself working on a stand alone basis. We present below the main components of this parallel tabu search heuristic; the reader is referred to Aiex [1] for a more detailed description.

4.1 Search Threads

In the parallel algorithm presented in this work, several search processes collaborate exchanging elite solutions which are used as new initial solutions. Each search thread (a *searcher*) consists basically of the sequential tabu search **TS-CPP** algorithm using a different combination of the pair (construction algorithm, move attribute), with communication points where solutions are exchanged through a pool of elite solutions. Algorithms **asp** and **cep** for the construction of initial solutions are described in detail in [18–20]. Algorithm **cir** was proposed by Aiex [1] and constructs subcircuits formed by exactly one gate or one output each. The different possibilities for move attributes lead to the restriction of different moves that can be performed after the move defined by the transference of gate p from subcircuit j to subcircuit i:

- the gate which was moved: all moves involving the same gate p will be made tabu for the next $tabu\_tenure_1$ iterations, whose value is obtained as described in Sec. 3.
- the target subcircuit to where a gate was moved: all moves associated with removing a gate from the same subcircuit i will be made tabu for the next $tabu\_tenure_2 = (|E| + |P| + |S| - (|E^+ \cap X_i^+| + |S \cap X_i^+| + |P \cap X_i^+|)) \cdot \gamma_2$ iterations.
- the source subcircuit from where a gate was moved: all moves associated with inserting a gate into the same subcircuit j will be made tabu for the next $tabu\_tenure_3 = (|E^+ \cap X_j^+| + |S \cap X_j^+| + |P \cap X_j^+|) \cdot \gamma_3$ iterations.

As for constant γ_1 in Sec. 3, γ_2 and γ_3 are empirically defined constants controlling the restrictiveness of each search.

All communication is done through the pool of elite solutions. A master process (the *master*) handles pool management. Elite solutions are sent by the searchers to the pool, and the searchers request elite solutions from the pool whenever they are not able anymore to improve their own solutions after a certain number of iterations. Elite solutions are defined in this work as feasible local minima which improve the current best solution found by a searcher. Whenever a searcher finds an elite solution, it sends this solution to the pool. This characterization of elite solutions has been shown in practice to allow a good trade-off between solution quality and processing time, as far as less restrictive definitions imply in much more communication between the searchers and the master.

One of the stopping criteria used by the basic sequential tabu search **TS-CPP** algorithm described in Sec. 3 was the maximum number of iterations *max_moves* without improvement in the current best solution. A similar criterion is used in the parallel algorithm to define communication points where a searcher requests a solution from the pool. Whenever a searcher is not able to improve its current best solution after the last *max_moves* $= 3 \cdot |P|$ iterations, it requests a solution from the pool and resumes the search using this solution as a new starting point. In case the pool does not have a solution to return to the searcher, the latter resumes the search from its current solution.

4.2 Pool Management

A master processor handles the pool of elite solutions. A list of forbidden searchers is associated with each solution in the pool. A searcher is forbidden for a given solution in the pool either because it is one of those who sent this solution to the pool, or because it has already received this solution from the pool. All communication is done through the master processor, in order to reduce communication and updating costs when compared to a strategy based on direct communication between the searchers.

Whenever a searcher finds a new elite solution, it first sends the cost of this solution to the master. The master then checks whether this solution should be inserted in the pool or not. To this effect, its cost is compared with that of the best solution currently in the pool. If its cost is lower than $(1 + \alpha)$ times the cost of the best elite solution in the pool, then the master notifies the searcher that it can send the full solution itself. If it does not belong to the pool, it is accepted and placed in the pool if the latter is not yet full, or replaces the currently worst solution in the pool otherwise (even if its cost is higher than the currently worst). The goal of using this threshold is twofold: first, it avoids the insertion in the pool of solutions much worse than the currently best elite solution; second, and more important, it reduces the rate of solutions effectively sent to the pool, avoiding that the master be overcharged with respect to its maximum capacity of receiving and handling new solutions. In fact, extensive computational experiments reported in [1] revealed that hardware limitations are such that, if this threshold is not used (which amounts to taking $\alpha = \infty$), there is very soon congestion and queue formation in the master, deteriorating the performance of the parallel procedure. We stress that an accepted solution replaces the currently worst in the pool even if its cost is higher than that of the latter, in order to allow that the pool be frequently refreshed.

Whenever a searcher requests a solution to the master, the latter checks for the existence of elite solutions in the pool which do not have this searcher in its list of forbidden searchers, and randomly selects and returns one of them. The master also inserts the identification of the searcher which made the request in the list of forbidden searchers associated with this solution.

4.3 Stopping Criteria

We define a *search cycle* as all computations performed by a searcher since it received the last solution from the pool (or since the beginning of execution, in the case of the first cycle). Whenever a searcher has performed a given number of search cycles (which we took as equal to six in this work), we consider it *ready-to-stop*. It remains active, but will be interrupted as soon as all other searchers are also ready-to-stop or when one of the stopping criteria is attained.

Two stopping criteria are implemented. By the first one, a searcher is immediatelly interrupted if the master is not able to return an admissible solution to it for two consecutive requests. The second criterion corresponds to the execution of a maximum number of iterations $total\_iterations = 3 \cdot max\_iterations$ encompassing all search cycles, where $max\_iterations = \lceil |P|^{(1.2)} \rceil$ is the maximum number of iterations used as one of the stopping criteria implemented in the sequential tabu search algorithm presented in Sec. 3. Whenever a searcher attains one of these two stopping criteria, it communicates its interruption to the master.

The parallel tabu search computations finish when all searchers are either inactive or ready-to-stop. At this point, the master is the only active processor. It then applies the post-optimization procedure **pack_together** to all solutions currently in the pool, and the best resulting solution is retained.

5 Parallel Implementations

Two different implementations were developed for the parallel tabu search described in Sec. 4. Linda [10] and PVM [40] were chosen as development tools for these implementations for their wide acceptance in the scientific community [11, 30, 32, 35, 39] and also for reflecting different approaches to parallel programming. We first present a brief description of these two programming tools (see also e.g. Martins [31] for a more thorough description of parallel programming tools for distributed memory environments); then we give an overview of the implementations, where we emphasize the differences resulting from the use of message-passing versus shared-memory paradigms.

5.1 Parallel Programming Tools

PVM (Parallel Virtual Machine) is a programming environment which was developed with the goal of supporting the use of networks of workstations as parallel machines. Communication between concurrent PVM processes is based on message passing. PVM was initially developed for TCP/IP based networks of heterogeneous UNIX workstations. It is currently available for many other platforms. Because of its free availability for a wide range of platforms, ease of installation, and high quality documentation [24], PVM has become highly popular, as can be witnessed by the number of reported experiments which use it (see e.g. [23, 36]).

The main parts of the PVM environment are the PVM daemon, *pvmd*, and a library of PVM routines, available for C and FORTRAN. The daemon runs on each of the machines which make up the virtual parallel machine, providing support for communication and process control. The library contains routines for message-passing, dynamic process creation, process synchronization, and control of the virtual machine (addition and removal of hosts). When the first instance of *pvmd* is started by a user, it creates a virtual machine with the physical machines described in a configuration file. Other users may simultaneously be controlling overlapping virtual machines. The virtual machine can be dynamically modified by a process through calls to the PVM library.

Message-passing routines in PVM follow a conventional send-receive paradigm, with synchronous and asynchronous variants. Before a message is sent, its contents must be explicitly *packed* into a buffer. The program may specify whether data should be converted to a machine-independent format when copied to the transmission buffer; this is analogous to the *marshalling* procedure in remote procedure call [7], but here it is carried out explicitly. When a message is received, it must in turn be explicitly *unpacked*. It is the programmer's responsibility to correctly interpret the data in a received message, i.e., to unpack the right data types from the buffer. Messages may carry a *tag*, which may be used by the destination process for message selection.

Besides the synchronization facilities which may be built on top of synchronous message receipt, PVM provides some pre-defined synchronization primitives, such as barriers. Functions are also available for group management and data reduction. The library also provides some support for fault handling. A process may register its interest in events such as a machine crash or process termination. This information can be used to implement fault tolerant features in an application, for instance by adding some new machine to the virtual machine when a workstation crashes.

Linda was developed as a machine-independent model for the construction of parallel programs. It was initially proposed for parallel architectures and only later implemented on clusters of workstations. Descriptions of work based on Linda for clusters of workstations usually refer to *Network Linda*, Scientific Computing Associates' implementation [38], which is also the one used in this work.

In the Linda model, processes communicate and synchronize through a globally shared associative memory. The motivation for this choice is the belief that it is more natural for programmers to have their programs communicate through shared memory [9] than through message passing primitives. The global memory is organized in *tuples*. Tuples may have any number of fields of any type, and are retrieved from the tuple space by providing a tuple structure (number and type of fields) and some of the field values. Existing Linda implementations provide Linda operations for C and FORTRAN, creating augmented versions of both languages which are referred to as C-Linda and FORTRAN-Linda. The operations **out** and **in** are used to place tuples in and remove tuples from the tuple space, respectively. Implementations must guarantee the atomicity of these operations, thereby creating a basis for process synchronization.

Processes can be created dynamically through the use of the operation `eval`, which creates a new process to evaluate the given tuple. When the new process terminates, its results are placed in the tuple space. For example, the call `eval("foo", 2*2)` creates a new process to evaluate the expression `2*2`. When this process terminates, the tuple `("foo",4)` will be placed in the tuple space. Processes created by `eval`, like any other Linda processes, may communicate through the tuple space.

In Network Linda, similarly to PVM, a virtual machine is controlled by a runtime system. However, the set of participating physical machines can only be defined through a configuration file; there is no explicit dynamic control of the virtual machine.

Both tools are quite easy to use and have very good support. One very interesting feature offered by Linda is a tool called Tuplescope, which is part of the Linda Code Development System and simulates parallel program execution in a uniprocessor environment. Many errors in early versions of the Linda implementation described below were found with the use of this tool.

5.2 Parallel Implementation in PVM and Linda

In both implementations of the parallel tabu search procedure described in Sec. 4, each tabu search thread is conducted by an independent process, called a *searcher*. These processes communicate basically for exchanging good solutions. In both implementations, a *master* process controls the current set of elite solutions, called the solution pool.

As described in Sec. 4, the searchers perform search cycles until some stopping criterion is achieved. The communication between them is done through the pool of elite solutions. From time to time, each of the searchers visits the solution pool. Between these visits, each searcher executes asynchronously either until it finds a new elite solution or until it exhausts the maximum number of iterations without finding an improving solution. In the PVM model, processes do not have access to any shared memory. In this case, the master process maintains the solution pool in its address space. All accesses to the solution pool are carried out through requests to the master. In Linda, as explained above, processes share a tuple space. This allowed us to implement the solution pool as shared information. The advantage of this approach is that it allows for a higher level of concurrency, since the searchers can do part of the processing which is required in a visit to the solution pool.

If a searcher visits the pool because it has found an elite solution, this solution must be compared to those currently in the pool. As explained in Sec. 4, instead of requiring that the new solution be better than all the current ones, some tolerance is introduced by comparing the cost of the new solution to a threshold based on the cost of the best one in the pool. If the new solution passes this test, it is placed in the pool if the latter is not full, or replaces the worst solution currently in the pool otherwise. In the PVM implementation, it is the master process which is responsible for this comparison. In the Linda implementation, the searcher itself can conduct the comparison directly. The substitution of a pool solution, however, is more complex in this case, because of the possibility of inconsistencies generated by concurrent accesses. When a searcher verifies that its proposed solution should be included in the pool, it generates a new tuple in the tuple space, containing the new solution, the identity of the searcher, and the keyword "newsolution". The master continuosly checks for the presence of this kind of tuple in the tuple space. If it finds one, it blocks the pool from further accesses and places the new solution in the pool.

If a searcher visits the pool because it has exhausted the maximum number of iterations without finding an improving one, it will try to retrieve a solution from the pool in order to use it as a new starting point. The program must avoid giving one same solution more than once to the same worker. To this end, lists associating forbidden pairs of searchers and solutions are kept in both implementations. In the PVM implementation, lists with the forbidden searchers associated with each solution are kept in the master's address space. The searcher asks the master for a new solution, and the master verifies the solutions which can be sent to this searcher. In the Linda implementation, the forbidden solution list associated with each searcher is a tuple (one per searcher) kept in the tuple space. This allows a searcher to directly query its own list. If the pool contains solutions which are not in its forbidden list, the searcher blocks access to the tuple space while it retrieves a randomly chosen solution and adds it to its own list of forbidden solutions. This is necessary because of inconsistencies which could result if the searcher was allowed to manipulate the pool and its forbidden solutions list while the master was executing a solution substitution.

When a solution is substituted or a new solution is inserted in the pool, the lists associating forbidden pairs of searchers and solutions must be updated, to reflect the fact that the new solution is available to all but the contributing searcher. This is done by the master in both implementations.

The end of the program is controlled by the master. When a searcher is ready-to-stop or interrupts its search, it indicates this event to the master. In PVM, it sends a message to the master, while in Linda, it includes a new tuple in the tuple space, indicating the status of the process. When the master detects that the number of ready-to-stop and interrupted processes is equal to the number of total searchers, it advises the ready-to-stop searchers to interrupt their searches. In PVM, the master sends a message to the ready-to-stop processes, while in Linda, it puts a tuple in the pool indicating the end of processing. The master then packs all solutions in the pool, selects the best one, and finalizes all processing.

In the discussion above, the need for blocking the pool was mentioned several times for the implementation using Linda. This is done through a technique commonly used in Linda programs. A special *access tuple* represents access control. When a process desires to block access to the tuple space, it asks to remove this tuple from the tuple space. When it finishes executing critical code, it returns the access tuple to the tuple space. Because of Linda's semantic, a process which tries to remove the access tuple will be blocked until it is possible to do so; also, the atomicity of operations in and out guarantees that this scheme works.

6 Computational Experiments

In this section we report computational results obtained with the application of the cooperative multi-thread parallel algorithm presented in Sections 4 and 5 to nine benchmark ISCAS combinational circuits from Berglez and Fujiwara [5]. We give in Table 1 the basic description of each circuit: the number of inputs, gates, outputs and links, as well as the maximum in-degree d_{max}^- and the maximum out-degree d_{max}^+ among all gates in the circuit. Two values have been taken for the testability parameter: $L = 15$ and $L = 20$ (see Sec. 2).

Table 1. ISCAS benchmark circuits

Circuits	Size	Inputs	Gates	Outputs	Links	d_{max}^-	d_{max}^+
C-1	small	36	153	7	432	9	9
C-2	small	41	170	32	499	5	12
C-3	small	60	357	26	880	4	8
C-4	medium	41	514	32	1355	5	12
C-5	medium	33	855	25	1908	8	16
C-6	medium	157	1129	140	2670	5	11
C-7	medium	50	1647	22	3540	8	16
C-8	large	32	2384	32	6288	2	16
C-9	large	207	3405	108	7552	5	15

6.1 PVM Implementation of the Parallel Tabu Search

The first part of the computational experiments was devoted to the PVM implementation. These experiments have been performed on a *multi-user* IBM SP-2 parallel machine with sixteen processors, each of which having 256 Mbytes of RAM, 2.0 Gbytes of disk space, and running AIX version 3.2.5. The processors are connected through an Ethernet network using the IP protocol and through a high-speed switch using the User Space (US) proprietary protocol. The programs were executed using the IP protocol over the Ethernet network. Codes in C were compiled with the XL C compiler version 1.3.0.5, with the option -O for code optimization, and version 3.3.9 of PVM was used.

Ten processors have been used, nine of each are searchers using each of the nine possible different combinations of the initial solution algorithm and the type of move attribute. The pool size was defined to be equal to 27 (three times the number of search threads).

We first investigate the value of the threshold for the acceptance of elite solutions in the pool. As we have already seen, whenever a searcher finds a new elite solution, it first sends the cost of this solution to the master. Only after this cost has been verified to be less than $(1 + \alpha)$ times the cost of the best elite solution in the pool, the master notifies the searcher to send the full solution itself to be inserted in the pool. The main goal of using this threshold is to reduce the rate of solutions effectively sent to the pool, avoiding that the master be overcharged with respect to its maximum capacity of receiving and handling new solutions. Different values of α were tested: 2%, 10%, 50%, and 200%. To illustrate the impact of the value of α in the restrictiveness of the acceptance criterion, we present in Table 2 the number of solutions effectively inserted in the pool and the number of solutions which were not accepted, for each value of α and for one run of the parallel tabu search for each benchmark circuit. Further computational results reported in Aiex [1] have shown that:

- Solution quality does not seem to be very much affected by the acceptance criterion, as far as the best elite solutions (which are the ones most likely to lead to the best final solutions) are always accepted, independently of the acceptance criterion. However, it seems to be useful to slightly restrict the insertion of solutions in the pool, to avoid that very bad solutions be used as initial solutions by the searchers.
- Computational times increase with α. Less restrictive acceptance criteria lead to more time spent by the master with insertions in the pool, also degrading the time it takes to reply to a searcher's request for a solution.
- The execution of the parallel tabu search in a multi-user environment leads to different memory availabilities and computational times at different runs, even if the same data and parameters are used. Each *pvmd* daemon of the virtual machine dynamically allocates memory for message storage. When the machine is lightly loaded, more memory can be allocated by the master's *pvmd* and, consequently, more messages can be dealt with by the master. We have observed that when the machine is more heavily loaded, it becomes imperative to use a restrictive acceptance criterion by the pool, in order to reduce large queues which could lead to slow behavior or even machine crashes due to a small memory availability.

The above observations lead to the use of $\alpha = 50\%$ as the threshold for the acceptance criterion in our PVM implementation. The results obtained by the application of the cooperative multi-thread parallel tabu search procedure to the nine circuits are presented in Table 3. Each search thread runs a copy of the basic sequential tabu search algorithm with the same original criteria, parameters, and objective function, except for the adaptations reported in Sec. 4. For each circuit and for both values of the partitioning parameter $L = 15$ and $L = 20$, we report the best solution (number of circuits and number of cuts) found by the parallel tabu search procedure, side by side with the best solution found by the **TS-CPP** sequential tabu search algorithm (extracted from Andreatta and Ribeiro [2]).

The above results show that the parallel tabu search procedure is able to improve the good results already obtained by the sequential tabu search. In most cases the best solution was found after the first search cycle, illustrating the effectiveness of the use of elite solutions found by other processors as reinitialization points. Significant reductions of up to 30% in the number of circuits and up to 45 % in the number of cuts are observed for the largest circuits C-8 and C-9. To further illustrate the effectiveness of the parallel procedure proposed in this work, with respect to the most recent (and, also, most effective) heuristics in the literature (**cep** and **TS-CPP**), we show in Figures 4 and 5, respectively,

Table 2. Effect of the threshold in the acceptance criterion

Circuit	$\alpha = 2\%$	$\alpha = 10\%$	$\alpha = 50\%$	$\alpha = 200\%$	Solutions...
C-1	43	51	25	285	inserted
	240	263	258	0	not accepted
C-2	35	25	31	24	inserted
	276	286	279	69	not accepted
C-3	21	25	20	253	inserted
	384	344	342	106	not accepted
C-4	46	36	39	233	inserted
	476	471	475	263	not accepted
C-5	52	72	81	493	inserted
	502	476	488	66	not accepted
C-6	100	86	88	493	inserted
	396	402	401	1	not accepted
C-7	191	163	181	564	inserted
	365	359	350	0	not accepted
C-8	49	33	32	253	inserted
	217	223	215	2	not accepted
C-9	267	260	242	654	inserted
	620	619	619	206	not accepted

the number of subcircuits and the number of cuts in the best solutions found for the medium- and large-size circuits by the three algorithms for $L = 20$.

6.2 A Quantitative Comparison of Linda vs. PVM

The second part of the computational experiments concerns the comparison between the efficiency of the implementations in Linda and PVM. For this new set of experiments, we have used a 10 node IBM SP-2 parallel machine. Two THIN nodes are RS6000 mod. 390 processors with 256 Mbytes of RAM and 4.4 Gbytes of disk space each. The other eight WIDE nodes are RS6000 mod. 590 processors with 1 Gbytes of RAM and 9 Gbytes of disk space each. Also in this environment, the processors are connected through an Ethernet network using the IP protocol and through a high-speed switch using both IP and US protocols. All processors run AIX 4.1.4. Both Linda and PVM implementations were executed in *exclusive mode* using the US protocol over the switch. Version 3.3.11 of PVM was used. Codes in C of the PVM implementation were compiled with mpcc, a shell script of POE 2.1.0.12 to use xlC 3.1.3.3. Linda programs were compiled with the Linda compiler clc 4.1.

Since only eight WIDE processors were available, we reduced the number of searchers from nine to eight, with respect to the PVM implementation discussed in Sec.6.1. To this effect, the combination of the initial solution algorithm **cir** with the source subcircuit move attribute was discarded. The pool size was taken as 24, three times the number of searchers. The master runs in one of the THIN processors. In order to have a global stopping criterion not depending on local behavior of the searchers and to enforce that they perform approximately

Table 3. Best solutions for benchmark circuits

Circuit	L	Parallel Procedure		Sequential Algorithm	
		Circuits	Cuts	Circuits	Cuts
C-1	$L = 15$	7	61	7	61
	$L = 20$	5	45	5	55
C-2	$L = 15$	7	53	8	70
	$L = 20$	5	45	5	55
C-3	$L = 15$	11	88	11	102
	$L = 20$	6	57	8	86
C-4	$L = 15$	10	97	10	106
	$L = 20$	6	72	8	99
C-5	$L = 15$	12	144	14	158
	$L = 20$	8	109	9	121
C-6	$L = 15$	24	173	26	208
	$L = 20$	15	118	19	177
C-7	$L = 15$	35	445	38	480
	$L = 20$	21	325	22	370
C-8	$L = 15$	29	349	41	561
	$L = 20$	16	252	23	403
C-9	$L = 15$	42	365	60	657
	$L = 20$	27	279	38	506

the same work in both implementations, the condition defining when a searcher requests a new solution to the pool and the stopping criteria described, respectively, in Sections 4.1 and 4.3 have been slightly modified. Each searcher performs $max\_iterations = 3 \cdot \lceil |P|^{(1.2)} \rceil$ iterations within each search cycle before requesting an elite solution from the pool. The maximum number of iterations used as the stopping criterion for each searcher is set as $total\_iterations = 18 \cdot \lceil |P|^{(1.2)} \rceil$. All other strategies and parameters remain as described before. Both PVM and Linda implementations use the same parameter settings.

We present in Table 4 the results obtained by PVM and Linda implementations, for each circuit and for both values of the partitioning parameter $L = 15$ and $L = 20$. For each implementation we present the best solution found (number of circuits and number of cuts), together with the elapsed time in seconds (runs in exclusive mode, i.e. with no other users). For the PVM implementation, we also report the search cycle in which the best solution was found. As already commented in Sec.6.1, the fact that the best solution was never found in the first search cycle illustrates the effectiveness of the use of elite solutions as reinitialization points by the cooperative multi-thread parallel procedure.

The differences on the best solutions found by the two implementations are not relevant, since they are due to different access chronologies to the pool. However, the differences on elapsed times in exclusive mode are significant: on the average and under the same stopping criteria, elapsed times under PVM are 19.7% smaller than those observed with Linda. However, the new version of Linda used in these experiments offers facilities, such as "keep alive facility" and "tuple broadcast optimization", which were not investigated. The use of

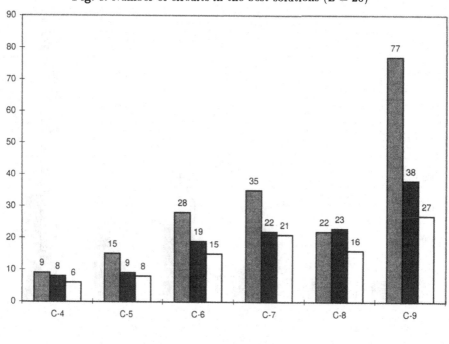

Fig. 4. Number of circuits in the best solutions ($L = 20$)

☐ cep ■ sequential tabu search ☐ parallel tabu search

these facilities could possibly lead to better performance results for the Linda implementation than those reported here.

7 Concluding Remarks

We have proposed in this work a cooperative multi-thread parallel tabu search algorithm for the circuit partitioning problem. Each search thread implements a different variant of a sequential tabu search algorithm from the literature, using a different combination of initial solution algorithm and move attribute definition. These threads communicate by exchanging elite solutions, in order to improve the quality of the solutions that each of them would be able to find by itself working on a stand alone basis. The parallel tabu search procedure was implemented in PVM and Linda, two widely used development tools reflecting different paradigms to parallel programming in distributed memory environments: message-passing and shared-memory.

The numerical results reported for a set of ISCAS benchmark circuits illustrate the effectiveness of the parallel tabu search procedure. Improved solutions are found for all test problems with respect to the original sequential tabu search algorithm, as shown in Table 3. Significant reductions of up to 30% in the number of circuits and up to 45 % in the number of cuts are observed for the largest

Fig. 5. Number of cuts in the best solutions ($L = 20$)

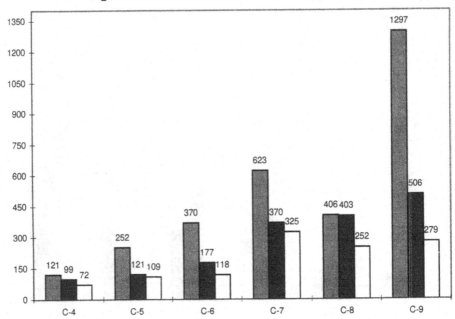

circuits, as further illustrated in Figures 4 and 5. As shown in Table 4, the best solution was never found in one of the first search cycles, illustrating the effectiveness of the cooperative multi-thread procedure, due to the use of elite solutions found by other processors as reinitialization points.

When the two parallel programming tools used for the implementation of the cooperative parallel tabu search procedure are compared, the differences on elapsed times in exclusive mode are significant: on the average and under the same stopping criteria, elapsed times under PVM are approximately 20% smaller than those observed with Linda.

Acknowledgements. The authors acknowledge Scientific Computing Associates for the license to use Linda on the IBM SP-2.

References

1. R.M. AIEX, *Asynchronous Parallel Tabu Search Strategies with an Application to Graph Partitioning* (in Portuguese), M.Sc. Dissertation, Department of Computer Science, Catholic University of Rio de Janeiro, 1996.
2. A.A. ANDREATTA AND C.C. RIBEIRO, "A Graph Partitioning Heuristic for the Parallel Pseudo-Exhaustive Logical Test of VLSI Combinational Circuits", *Annals of Operations Research* 50 (1994), 1–36.

Table 4. PVM vs. Linda

Circuit	L	PVM				Linda		
		Circuits	Cuts	Sec's	Cycle	Circuits	Cuts	Sec's
C-1	$L = 15$	7	61	15.9	2	7	61	20.4
	$L = 20$	5	45	16.0	3	5	44	19.5
C-2	$L = 15$	7	55	21.7	2	7	57	41.4
	$L = 20$	5	41	19.9	2	5	43	25.7
C-3	$L = 15$	10	82	56.4	5	11	89	67.8
	$L = 20$	6	59	53.2	3	6	59	104.2
C-4	$L = 15$	10	98	112.3	6	10	99	126.0
	$L = 20$	7	87	108.5	4	7	81	128.3
C-5	$L = 15$	13	142	282.6	4	10	115	325.4
	$L = 20$	7	91	270.0	3	8	103	315.8
C-6	$L = 15$	23	174	444.4	3	23	174	570.4
	$L = 20$	15	119	403.3	6	15	129	458.8
C-7	$L = 15$	35	448	1342.5	6	35	439	1581.1
	$L = 20$	22	348	1267.6	5	21	311	1408.3
C-8	$L = 15$	29	349	2302.3	6	31	366	2329.7
	$L = 20$	16	255	2096.9	6	16	255	2161.3
C-9	$L = 15$	46	416	4450.3	5	45	414	4809.5
	$L = 20$	28	281	4075.4	5	28	292	4392.2

3. E.C. ARCHAMBEAU AND E.J. MCCLUSKEY, "Fault Coverage of Pseudo-Exhaustive Testing", *Digest of Papers of the 14th International Conference on Fault-Tolerant Computing*, 141–145, IEEE, 1984.

4. P. BADEAU, F. GUERTIN, J.-Y. POTVIN, M.GENDREAU, AND E.D. TAILLARD, "A Parallel Tabu Search Heuristic for the Vehicle Routing Problem with Time Windows", *Transportation Research C* 5 (1997), 109–122.

5. F. BERGLEZ AND H. FUJIWARA, "A Neutral Netlist of 10 Combinational Benchmark Circuits and a Target Translator in Fortran", Special session on "ATPG and Fault Simulation", *IEEE International Symposium on Circuits and Systems*, Kyoto, 1985.

6. S.N. BHATT, F.R.K. CHUNG, AND A.L. ROSENBERG, "Partitioning Circuits for Improved Testability", *Proceedings of the Fourth MIR Conference: Advanced Research in VLSI*, 91–106, The MIT Press, Cambridge, 1986.

7. A. BIRRELL AND B. NELSON, "Implementing remote procedure calls", *ACM Transactions on Computer Systems* 2 (1984), 39–59.

8. S. BOZORGUI-NESBAT AND E.J. MCCLUSKEY, "Structured Design for Testability to Eliminate Test Pattern Generation", *Digest of Papers of the 10th International Symposium on Fault-Tolerant Computing*, 158–163, IEEE, 1980.

9. N. CARRIERO AND D. GELERNTER, "How to Write Parallel Programs: A Guide to the Perplexed", *ACM Computing Surveys* 21 (1989), 323–357.

10. N. CARRIERO, D. GELERNTER, AND T. MATTSON, "Linda in Context", *Communications of the ACM* 32 (1989), 444–458.

11. P. CIGNONI, D. LAFORENZA, R. PEREGO, R. SCOPIGNO, AND C. MONTANI, "Evaluation of Parallelization Strategies for an Incremental Delaunay Triangulator in E^3", *Concurrency: Practice and Experience* 7 (1995), 61–80.

12. T.G. CRAINIC AND M. GENDREAU, "A Cooperative Parallel Tabu Search for Capacited Network Design", Research report, Centre de Recherche sur les Transports, Université de Montréal, 1997.

13. T.G. CRAINIC AND M. TOULOUSE, "Parallel Metaheuristics", Research report, Centre de Recherche sur les Transports, Université de Montréal, 1997.

14. T.G. CRAINIC, M. TOULOUSE, AND M. GENDREAU, "Parallel Asynchronous Tabu Search for Multicommodity Location-Allocation with Balancing Requirements", Publication 935, Centre de Recherche sur les Transports, Université de Montréal, 1993.

15. T.G. CRAINIC, M. TOULOUSE, AND M. GENDREAU, "Towards a Taxonomy of Parallel Tabu Search", INFORMS Journal on Computing 9 (1997), 61–72.

16. T.G. CRAINIC, M. TOULOUSE, AND M. GENDREAU, "Synchronous Tabu Search Parallelization Strategies for Multicommodity Location-Allocation with Balancing Requirements", OR Spektrum 17 (1995), 113–123.

17. M. TOULOUSE, T.G. CRAINIC, AND M. GENDREAU, "Communication Issues in Designing Cooperative Multi-Thread Parallel Searches", in Meta-Heuristics: Theory and Applications (I.H. Osman and J.P. Kelly, editors), 501–522, Kluwer, 1996.

18. M. DAVIS-MORADKHAN, Problèmes de Partitionnement dans la Technologie des VLSI, Doctorate thesis, Université Paris VI, 1993.

19. M. DAVIS-MORADKHAN AND C. ROUCAIROL, "Comparison of Two Heuristics for Partitioning Combinational Circuits for Parallel Pseudo-Exhaustive Testing", Rapport MASI 92.25, Laboratoire MASI, Université Paris VI, 1992.

20. M. DAVIS-MORADKHAN AND C. ROUCAIROL, "Graph Partitioning Applied to the Problem of Logic Testing of VLSI Combinational Circuits", Rapport MASI 92.41, Laboratoire MASI, Université Paris VI, 1992.

21. I. DE FALCO, R. DEL BALIO, E. TARANTINO, AND R. VACARO, "Improving Search by Incorporating Evolution Principles in Parallel Tabu Search", Proceedings of the International Conference on Machine Learning, 823–828, 1994.

22. I. DE FALCO, R. DEL BALIO, AND E. TARANTINO, "Solving the Mapping Problem by Parallel Tabu Search", Research report, Instituto per la Recerca sui Sistemi Informatici Paralleli - CRN, 1995.

23. M. FRANKLIN AND V. GOVINDAN, "A General Matrix Iterative Model for Dynamic Load Balancing", Parallel Computing 22 (1996), 969–989.

24. A. GEIST, A. BEGUELIN, J. DONGARRA, W. JIANG, R. MANCHEK, AND V. SUNDERMAN, PVM: Parallel Virtual Machine - A User's Guide and Tutorial for Networked Parallel Computing, The MIT Press, 1994.

25. M.GENDREAU, P. BADEAU, F. GUERTIN, J.-Y. POTVIN, AND E.D. TAILLARD, "A Solution Procedure for Real-Time Routing and Dispatching of Commercial Vehicles", Publication CRT-96-24, Centre de Recherche sur les Transports, Université de Montréal, 1996.

26. F. GLOVER, "Tabu Search - Part I", ORSA Journal on Computing 1 (1989), 190–206.

27. F. GLOVER, "Tabu Search - Part II", ORSA Journal on Computing 2 (1990), 4–32.

28. F. GLOVER AND M. LAGUNA, "Tabu Search", in Modern Heuristic Techniques for Combinatorial Problems (C.R. Reeves, editor), 70–150, Blackwell, 1993, Londres.

29. F. GLOVER AND M. LAGUNA, Tabu Search, Kluwer, 1997, Boston.

30. A.H. KARP, "Some Experiences with Network Linda", International Journal of High Speed Computing 6 (1994), 55–80.

31. S.L. MARTINS, C.C. RIBEIRO, AND N.R. RODRIGUEZ, "Parallel Programming Tools for Distributed Memory Environments" (in Portuguese), Investigación Operativa 5 (1996), 67–98.

32. A. MATRONE, P. SCHIANO, AND V. PUOTTI, "LINDA and PVM: A Comparison between Two Environments for Parallel Programming", *Parallel Computing* 19 (1993), 949–957.

33. Y. MIN AND Z. LI, "Pseudo-Exhaustive Testing Strategy for Large Combinational Circuits", *Computer Systems Science and Engineering* 1 (1986), 213–220.

34. O. PATASHNIK, *Optimal Circuit Segmentation for Pseudo-Exhaustive Testing*, Doctorate thesis, Stanford University, Department of Computer Science, 1990.

35. A. PETRIE AND R. KERR, "A Qualitative Comparison of Network Linda and PVM", Parallel Processing Memorandum PPM/017, Department of Computing Science, University of Newcastle upon Tyne, 1994.

36. S.C. PORTO AND C.C. RIBEIRO, "Parallel Tabu Search Message-Passing Synchronous Strategies for Task Scheduling under Precedence Constraints", *Journal of Heuristics* 1 (1995), 207–223.

37. M.W. ROBERTS AND P.K. LALA, "An Algorithm for the Partitioning of Logic Circuits", *IEE Proceedings-G* 131 (1984), 113–118.

38. SCIENTIFIC COMPUTING ASSOCIATES, *Linda's User's Guide and Reference Manual*, version 4.0.1 – SP2/POE.

39. F. SUKUP, "Efficiency Evaluation of Some Parallelization Tools on a Workstation Cluster Using the NAS Parallel Benchmarks", Research report, Vienna University of Technology, Computing Center, 1994.

40. V. SUNDERMAN, "PVM: A Framework for Parallel Distributed Computing", *Concurrency: Practice and Experience* 2 (1990), 315–339.

41. E.D. TAILLARD, P. BADEAU, M.GENDREAU, F. GUERTIN, AND J.-Y. POTVIN, "A Tabu Search Heuristic for the Vehicle Routing Problem with Soft Time Windows", *Transportation Science* 31 (1997), 170–186.

Experiments with mpC:
Efficient Solving Regular Problems
on Heterogeneous Networks of Computers
via Irregularization

Dmitry Arapov, Alexey Kalinov, Alexey Lastovetsky, and Ilya Ledovskih

Institute for System Programming, Russian Academy of Sciences
25, Bolshaya Kommunisticheskaya str., Moscow 109004, Russia
mpc@ispras.ru

Abstract. mpC is a medium-level parallel language for programming heterogeneous networks of computers. It allows to write libraries of parallel routines adaptable to peculiarities of any particular executing multiprocessor system to ensure efficient running. The adaptable routines distribute data and computations in accordance with performances of participating processors. In this case even the problems traditionally considered regular, become irregular. Advantages of mpC for efficient solving of regular problems on heterogeneous networks of computers are demonstrated with an mpC routine implementing Cholesky factorization, with efficiency of the mpC routine being compared with ScaLAPACK one.

1 Introduction

A heterogeneous network of computers, being the most common parallel architecture available to common users, can be used for high-performance computing. Taking into account that in the 1990s network capacity increases surpassed processor speed increases [1], pp.6–7, one can predict increasing their importance as a low-cost platform for parallel high-performance computations. Efficient programming heterogeneous networks of computers has some difficulties which do not arise when programming traditional homogeneous supercomputers. The point is that to use the full performance potential of a heterogenous network of computers, it is necessary to distribute data and computations among processors in accordance with their performances. That heterogeneity of data distribution leads to considering irregular such problems as, for example, dense linear algebra problems, traditionally considered regular when solving on homogeneous multiprocessor computing systems.

One can ascertain absence of suitable and handy tools for parallel programming irregular applications for heterogeneous networks of computers. Both high-level parallel languages like HPF [2] and low-level message-passing packages like MPI [3] are not suitable for the purpose, since they do not have facilities to detect performance characteristics of a particular executing multiprocessor hardware and distribute data and computations among its processors in accordance with results of the detection.

The situation has induced us to develop mpC – a medium-level parallel language for programming heterogeneous networks of computers, that, like the C language, combines assembler (MPI) flexibility and efficiency with high-level programming language convenience. The mpC language is an ANSI C extension allowing to write libraries of parallel routines adaptable to peculiarities of any particular executing parallel computer system to ensure efficient running.

We will demonstrate advantages of mpC with such a generally-known challenge in parallel computations as Cholesky factorization. We use almost the same parallel algorithm, that is used in ScaLAPACK [4] in case of 1-D processor grid, and use LAPACK [5] and BLAS [6] for local computations. The main difference between our parallel algorithm and the ScaLAPACK one is data distribution. We consider the Cholesky factorization to be an irregular problem and distribute data among processors of an executing parallel machine in accordance with their relative performances. ScaLAPACK considers the problem regular and distributes data in accordance with homogeneous block-cyclic distribution. Of course, we make no pretensions to solve all problems but only want to demonstrate how mpC allows slightly to modify a good parallel algorithm to obtain an adaptable routine for heterogeneous networks of computers.

Our parallel mpC function will hide its parallel nature from a caller. It corresponds to an easy-to-use style of parallel programming, when all parallel computations are encapsulated in library functions. This approach makes transition from sequential programming in C to parallel programming in mpC very simple.

Section 2 introduces a parallel algorithm of Cholesky factorization. Section 3 describes implementation of the algorithm in mpC, introducing all necessary details of the mpC language. Section 4 compares the mpC and ScaLAPACK Cholesky factorization routines running on networks of workstations.

More about mpC can be found in [7–10] as well as at mpC home page (http://www.ispras.ru/~mpc), where additionally free mpC software is available.

2 Algorithm of Cholesky Factorization in mpC

Cholesky factorization of a real symmetric positive definite matrix is an extremely important computation, arising in a variety of scientific and engineering applications. It is a well-known challenge for efficient and scalable parallel implementation because of large volumes of interprocessor communications.

Cholesky factorization factors an $n \times n$, symmetric, positive-definite matrix A into a product of a lower triangular matrix L and its transpose, i.e., $A = LL^{\mathrm{T}}$. One can partition the $n \times n$ matrices A, L and L^{T} and write the system as

$$\begin{bmatrix} A_{11} & A_{21}^{\mathrm{T}} \\ A_{21} & A_2 2 \end{bmatrix} = \begin{bmatrix} L_{11} & 0 \\ L_{21} & L_2 2 \end{bmatrix} \cdot \begin{bmatrix} L_{11}^{\mathrm{T}} & L_{21}^{\mathrm{T}} \\ 0 & L_2^{\mathrm{T}} 2 \end{bmatrix} = \begin{bmatrix} L_{11}L_{11}^{\mathrm{T}} & L_{11}L_{21}^{\mathrm{T}} \\ L_{21}L_{11}^{\mathrm{T}} & L_{21}L_{21}^{\mathrm{T}} + L_{22}L_{22}^{\mathrm{T}} \end{bmatrix},$$

where blocks $A11$ and $L11$ are $n_{b1} \times n_{b1}$, A_{21}, L_{21} are $(n - n_{b1}) \times n_{b1}$, $A22$, $L22$ are $(n - n_{b1}) \times (n - n_{b1})$. $L11$ and $L22$ is lower triangular, n_{b1} is the size of the first block. Assuming that L_{11}, the lower triangular Cholesky factor of A_{11}, is known, one can rearrange the block equations

$$L_{21} \leftarrow A_{21} \left(L_{11}^{\mathrm{T}}\right)^{-1},$$
$$\bar{A}_{22} \leftarrow \left(A_{22} - L_{21} L_{21}^{\mathrm{T}} = L_{22} L_{22}^{\mathrm{T}}\right).$$

The factorization can be done by recursively applying the step outlined above to the updated matrix A_{22}. The parallel implementation of the corresponding ScaLAPACK routine PDPOTRF [11] is based on the above scheme and a block cyclic distribution of matrix A over a $P \times Q$ process grid with a block size of $n_b \times n_b$. The routine assumes that the lower (upper) triangular portion of A is stored in the lower (upper) triangle of a two-dimensional array and that the computed elements of L overwrite the given elements of A (here and henceforth when speaking of an array we mean a Fortran array, that is, that column elements of an 2-D array are allocated contiguously).

Our mpC Cholesky factorization routine implements almost the same algorithm that is implemented by the ScaLAPACK one in the case of 1-D process grid (P=1) and the lower triangular portion of A used to compute L. Namely, to compute above steps it involves the following operations:

1. process Pr, which has L_{11}, L_{21}, calls LAPACK function dpotf2 to compute Cholesky factor L_{11} and sets a flag if A_{11} is not positively defined;
2. process Pr calls BLAS function dtrsm to compute Cholesky factor L_{21} if A_{11} is positively defined;
3. process Pr broadcasts the column panel, L_{11} and L_{21}, as well as the flag to all other processes and stops the computation if A_{11} is not positively defined;
4. all processes stop the computation if A_{11} is not positively defined or otherwise update matrix A_{22} in parallel, that involves calls to BLAS functions dsyrk and dgemm by each process updating its local portions of matrix A_{22}.

The main difference between the mpC and the ScaLAPACK routines lies in data distribution. In fact, in our case (P=1) the ScaLAPACK routine divides matrix A into a number of column panels with just the same width n_b and distributes them cyclically over Q processes (see Fig. 1) where Q, nb are input parameters of the routine.

The mpC routine distribution is almost the same. The only difference is that column panel with width n_{bi}, calculated as follows

$$n_{bi} = Q \cdot n_b \frac{p_i}{\sum_{j=1}^{Q} p_j} , \tag{1}$$

is placed to i-th process, where Q, n_b are input parameters of the routine, and p_j are the relative speed of j-th process ($j = 1, \ldots, Q$).

Suppose we have a 18×18 matrix, $P = 1$, $Q = 2$, $n_b = 3$ and the underlying network of computers consists of two processors, each running one process and the second being twice faster. Figure 1 shows the ScaLAPACK data distribution and figure 2 shows the mpC data distribution. In this case $n_{b1} = 2$, $n_{b2} = 4$.

Fig. 1. ScaLAPACK(left) and mpC(right) data distributions when the underlying process grid consists of two processors, the second being twice as fast. Black columns belong to the first process and white columns belong to the second one

3   Implementation of the Cholesky Factorization in mpC

The mpC language is an ANSI C superset allowing the user to specify the topology of and to define the so-called network objects (in particular, dynamically) as well as to distribute data and computations over the network objects. The mpC programming environment uses this information to map (in run time) the mpC network objects to any underlying heterogeneous network in such a way that to ensures efficient running of the application on the network.

In mpC, a programmer deals with a new kind of resource – *computing space* – a set of virtual processors represented in run-time by actual processes. The resource can be managed with allocating and discarding regions of computing space called *network objects* (or simply *networks*). Allocating network objects in the computing space and discarding them is performed in similar fashion to allocating and discarding data objects in the storage in the C language. A network object may be used to distribute data, to compute expressions, and to execute statements. Every network has a *parent* – the virtual processor initiated its allocation and belonging to both the newly created network and one of networks created before. The only virtual processor defined from the beginning of program execution till its termination is the pre-defined virtual *host-processor*.

So, in our mpC application we, first of all, define a network object over which we want to distribute data and computations. Every network object, declared in an mpC program, has a type. The *network type* specifies the number and performances of virtual processors, links between these processors, as well as separates the parent. For our purpose we declare the family of network types, named HeteroNet,

```
/*1.1*/   nettype HeteroNet(n, p[n]) {
/*1.2*/       coord I=n;
/*1.3*/       node { I>=0: p[I]; };
/*1.4*/   };
```

parametrized with integer parameter n and vector parameter p consisting of n integers. The family of network types corresponds to network objects consisting of n virtual processors. The virtual processors are related to the coordinate system with coordinate variable I ranging from 0 to n-1. The relative performance of the virtual processor with coordinate I being characterized by the value of p[I]. By default the network parent has coordinate 0.

The high-level mpC function Ch, implementing the Cholesky factorization, looks as follows

```
/* 2.1*/ void [*]Ch(double *[host]A,
/* 2.2*/              int [host]N,
/* 2.3*/              int *[host]INFO) {
/* 2.4*/    repl int nprocs, * ipowers;
/* 2.5*/    repl double * powers;
/* 2.6*/    MPC_Processors_static_info(&nprocs, &powers);
/* 2.7*/    IntPowers(nprocs,powers,&ipowers);
/* 2.8*/    {
/* 2.9*/       net HeteroNet(nprocs,ipowers) w;
/*2.10*/       double* [w]da;
/*2.11*/       int [w]info=0;
/*2.12*/       repl int [w]n, * [w]map, [w]source[1]= {0};
/*2.13*/       n=N;
/*2.14*/       if(I coordof da == 0)
/*2.15*/          [host]da=A;
/*2.16*/       else
/*2.17*/          da=[w]malloc( n*n*[w](sizeof(double)));
/*2.18*/       map=[w]malloc(n*[w]sizeof(int));
/*2.19*/       [w]Distr(n,[w]nprocs,[w]powers,map);
/*2.20*/       ([([w]nprocs)w]) ChScatter(da,n,map);
/*2.21*/       info=([([w]nprocs)w])ParCh(da,n,map );
/*2.22*/       *INFO=[host]info[+];
/*2.23*/       if(I coordof da != 0) [w]free(da);
/*2.24*/       [w]free(map);
/*2.25*/    }
/*2.26*/    free(ipowers);
/*2.27*/ }
```

The function is a so-called *basic* mpC function with three arguments belonging to the virtual host-processor: pointer A to the source matrix, dimension N of the matrix, and pointer INFO to an indicator of the termination status. There are three kinds of function in mpC: basic, network, and nodal functions. *Basic function* is called and executed on the entire computing space. Only in basic functions networks may be defined. *Network function* is called and executed on a network object. *Nodal function* can be executed completely by any one virtual processor. Any C function is considered a nodal function in mpC.

Line 2.4 defines integer variable nprocs and pointer ipowers to integer. Both variable nprocs and data object, that ipowers points to, are declared *replicated* over the entire computing space. By definition, data object *distributed* over a region of the computing space (in particular, over the entire computing space) comprises a set of components of any one type so that each virtual processor of the region holds one component. By definition, a distributed data object is *replicated* if all its components is equal to each other.

Line 2.5 defines pointer **powers** distributed over the entire computing space and specifies that it points to a replicated data object.

Line 2.6 calls library nodal function MPC_Processors_static_info on the entire computing space returning the number of actual processors and their relative performances. So, after this call replicated variable nprocs will hold the number of actual processors, and replicated array powers will hold their relative performances. Note, that the possibility to detect in run time the detailed information about characteristics of an executing actual parallel machine is an important peculiarity of the mpC language making it a suitable tool for efficient programming heterogeneous networks of computers.

Line 2.7 calls nodal function IntPower which allocates and initializes replicated *integer* array ipowers holding relative processor performances.

At the point, we have obtained enough information about characteristics of the executing multiprocessor to define properly a network object to perform our parallel Cholesky factorization. Line 2.9 defines automatic network w, the type of which, being an instance of the corresponding family of network types, is defined only in run time, and which executes the most of the rest of computations and communications. It consists of nprocs virtual processors, the relative performance of the i-th virtual processor being characterized by the value of ipowers[i]. The definition of the network causes its allocation (or creation) in run time. The mpC programming environment will ensure the optimal mapping of virtual processors of the network w into a set of actual processes representing the entire computing space. So, just one process from processes running on each of actual processors will be involved in the Cholesky factorization, and the more powerful is the virtual processor, the more powerful actual processor will execute the corresponding process.

Note, that the possibility to define a network type (or requirements to virtual processors) in run time and to map the virtual processors into actual processes in accordance with the network type requirements is a key advantage of the mpC programming environment making it a suitable tool for programming heterogeneous networks.

Having defined the network, we can distribute data and computations over it. Construct [w] in the definition of pointer da in line 2.10 just says that the pointer is distributed over network w. By default, if the distribution of a variable is not specified, it means that the variable is distributed over:

– the entire computing space if defined in a basic function;
– the corresponding network if defined in a network function.

Construct [*] specifies that the data object is distributed over the entire computing space, and construct [host] specifies that the data object belongs to the host.

The assignment in line 2.13 broadcasts the value of N to all components of distributed variable n. In general, the simple assignment is extended in mpC to express data transfer between virtual processors of the same network object.

The if-else statement in lines 2.14–2.17 sets a value of distributed pointer da. On the virtual host-processor its component points to matrix A, and on other virtual processors of the network w its components point to an allocated array. Unlike the previous statement, the execution of this statement does not need any communications between virtual processors constituting network w. In fact, this statement is divided into a set of independent undistributed statements each of which is executed by the corresponding virtual processor using the corresponding data components. Such statement is called an *asynchronous* statement.

The control expression in the if-else statement contains unusual binary operator coordof. Its result is an integer value distributed over w, each component of which is equal to the value of coordinate variable I of the virtual processor to which the component belongs. The right operand of the operator coordof is not evaluated and used only to specify a region of the computing space. Note, that coordinate variable I is treated as an integer variable distributed over the region.

Line 2.19 calls nodal function Distr on network w. The execution of this function call just on network w is provided with prefix unary network cast operator [w], cutting from values of Distr, nprocs, and powers just the components, belonging to w, and resulting in all operands of the function call are distributed over network w. A compiler uses the information about distribution of operands of the expression to determine the region of the computing space where the expression is evaluated and the statement is executed.

The function Distr allocates array map and calculates its elements defining the distribution of columns of the matrix over virtual processors of network w. The calculation is based on formula (1). So, after this call, map[i] holds the number of the virtual processor to which the i-th column belongs, and the more powerful is the virtual processor, the more columns are assigned to it.

Line 2.20 calls network function ChScatter on network w to scatter the matrix A in accordance with mapping provided by the array map. The function ChScatter has the following function prototype:

```
/*3.1*/ int [net SimpleNet(p)v]ChScatter(double* da,
/*3.2*/                                   const repl int n,
/*3.3*/                                   const repl int* map);
```

In general, a *network function* is called and executed on some network, and its value is also distributed over the same network. The function ChScatter has two special formal parameters – so-called *network parameter* v, representing the network on which the function is executed, and parameter p, treated in the function as a replicated over network v integer variable. The family of network types, named SimpleNet, is the simplest one introducing only a coordinate system and declared in standard header file mpc.h as follows:

```
/*4.1*/ nettype SimpleNet(n) { coord I=n; };
```

Except the network parameter, no network can be used or declared in the network function. Only data objects belonging to the network parameter may be defined in its body. In addition, the corresponding components of an externally-defined distributed data object can be used. Unlike basic functions, network functions (as well as nodal functions) can be called in parallel. Any network of a relevant type can be used as an actual network parameter in a network-function call. In our case, the network w is such a network argument, nprocs being other special argument.

Line 2.21 calls network function ParCh (described below) to compute in parallel the Cholesky factor of the matrix. It returns the value of flag info detecting how computation is terminated on each of virtual processors of network w.

Line 2.22 calculates the sum of all components of info and assigns the result to *INFO. The result of postfix unary operator [+] is distributed over w. All its components are equal to the sum of all components of its operand info. Here, the result of prefix unary network cast operator [host] is the component of its operand belonging to the virtual host-processor. So, the statement assigns the sum of all components of info to *INFO on the virtual host-processor.

Line 2.23 frees the memory allocated for the matrix on all virtual processors of w different from the host-virtual processor.

Network w is discarded when execution of the block in lines 2.8–2.25 ends.

Note, that we do not gather result to the host. The algorithm implemented by ParCh ensures that Cholesky factor of the matrix will appear on each of virtual processors of network w.

The network function strictly computed Cholesky factor of the matrix is the following:

```
/* 5.1*/  int [net SimpleNet(p)w]ParCh(double* da,
/* 5.2*/                        const repl int n,
/* 5.3*/                        const repl int* map) {
/* 5.4*/    repl int ngroups,*displs,*ncols,group,k;
/* 5.5*/    repl int coor,displsg,ncolsg;
/* 5.6*/    int i,j,dim,info=0,displ,coor_1,displsk,ncolsk;
/* 5.7*/    double one=1.0, minus_one=-1.0;
/* 5.8*/    MakeGroups(n, map, &ngroups, &ncols, &displs);
/* 5.9*/    for (group=0;group<ngroups;group++) {
/*5.10*/      displsg=displs[group];
/*5.11*/      ncolsg=ncols[group];
/*5.12*/      coor=map[displsg];
/*5.13*/      if ( coor == (I coordof da)) {
/*5.14*/        /* calculate L11,                      */
/*5.15*/        /* call of the LAPACK function dpotf2 */
/*5.16*/        dpotf2_("L",&ncolsg,da+displsg*(n+1),&n,&info);
/*5.17*/        if(!info && group < ngroups-1) {
/*5.18*/          /* calculate L21,                      */
/*5.19*/          /* call of the BLAS lev.3 function dtrsm */
/*5.20*/          dim=n-displsg-ncolsg;
```

```
/*5.21*/              dtrsm_("R","L","T","N",
/*5.22*/                   &dim,&ncolsg,&one,
/*5.23*/                   da+displsg*(n+1),&n,
/*5.24*/                   da+displsg*(n+1)+ncolsg,&n);
/*5.25*/          }
/*5.26*/        }
/*5.27*/        ([([w]p)w])ChBcast
/*5.28*/                   (&info,coor,group,da,displs,ncols,n);
/*5.29*/        if((repl int)info) return info;
/*5.30*/        for(k=group+1; k<ngroups; k++) {
/*5.31*/          int dim_n;
/*5.32*/          displsk=displs[k];
/*5.33*/          ncolsk=ncols[k];
/*5.34*/          coor_l=map[displsk];
/*5.35*/          if ( coor_l == (I coordof da)) {
/*5.36*/            /* update the triangle part of the group */
/*5.37*/            /* call the BLAS level 3 function dsyrk */
/*5.38*/            dsyrk_("L","N",&ncolsk,&ncolsg,
/*5.39*/                   &minus_one,da+displsg*n+displsk,&n,
/*5.40*/                   &one,da+displsk*(n+1),&n);
/*5.41*/            /* update the rectangle part of the group */
/*5.42*/            /* call the BLAS level 3 function dgemm */
/*5.43*/            dim_n=n-displsk-ncolsk;
/*5.44*/            if(dim_n != 0) {
/*5.45*/              dgemm_("N","T",&dim_n,&ncolsk,&ncolsg,
/*5.46*/                   &minus_one,da+displsg*n+displsk+ncolsk,
/*5.47*/                   &n,da+displsg*n+displsk,&n,
/*5.48*/                   &one,da+displsk*(n+1)+ncolsk,&n);
/*5.49*/            }
/*5.50*/          }
/*5.51*/        }
/*5.52*/      }
/*5.53*/    return 0;
/*5.54*/ }
```

Line 5.8 calls nodal function MakeGroups to collect columns in groups. After the call, variable ngroups holds the total number of groups, ncols[i] holds the number of columns in i-th group, and displs[i] holds its displacement. In the above example (presented in Fig. 2), there are 6 groups: each of three groups, belonging to the slower virtual processor, consists of two columns, and each of three groups, belonging to the faster virtual processor, consists of four columns.

Lines 5.9–5.52 are the main loop of the algorithm.

Lines 5.10–5.12 calculate the number of columns in the current group as well as the coordinate of the virtual processor holding it.

The statement in lines 5.13–5.26 computes matrix L_{11} by call to LAPACK function dpotf2 and, if it has been successfully computed, and it is not the latest group, compute matrix L_{21} by call to BLAS level 3 function `dtrsm`. The computations are performed sequentially by the virtual processor holding the corresponding group.

Lines 5.27–5.28 calls network function `ChBcast` to broadcast flag `info` and matrix L_{21} over network `w`.

Line 5.29 returns control and the value of `info` to a caller, if matrix L_{11} has not been successfully computed. The value of control expression must be replicated, since all virtual processors must either return or not return control to the caller coordinately, and it is strictly checked by the compiler. This requirement is an example of the programming style, supported by mpC, which allows to avoid many errors which may arise in parallel programming. In this case, if we do not say to compiler, that we guarantee replication of the value, it will detect an error.

The loop in lines 5.30–5.51 updates matrix A_{22} in parallel, the triangle part of the group is updated by BLAS level 3 function `dsyrk`, and the rectangle part of the group is updated by BLAS level 3 function `dgemm`.

We have presented the most interesting part of about 200 lines of mpC code implementing Cholesky factorization. One can see that it is not very difficult to implement such a complex application in mpC, and as we demonstrate below, our implementation is good enough. It consists of basic library function `Ch` and a small set of network and nodal functions. We have demonstrated two levels of modularity in mpC. The first level of modularity is provided by basic function `Ch`, which creates a network and calls the network function `ParCh` on it, providing the second level of modularity. One can see that our network function does not depend on data distribution (1-D processor grid is assumed only) provided by array `map`, and can be called from different basic functions on different networks as well as from different network functions.

4 Experimental Results

We compared the running time of our `ParCh` mpC function and its ScaLA-PACK counterpart `PDPOTRF`. We used SPARCstation 20 (hostname `alpha`), three SPARCstations 5 (hostnames `gamma`, `beta`, and `delta`), and SPARCclassic (`omega`), with relative performances 180, 160, 160, 160, and 77 correspondingly connected via 10Mbits Ethernet. We used the following networks: network `gbo` consisting of `gamma`, `beta`, and `omega`; network `gbd` consisting of `gamma`, `beta`, and `delta`, and so on. Note, that performances of processors in these networks are detected automatically with a command of the mpC programming environment.

We used MPICH version 1.0.13 as a communication platform, GNU C compiler with optimization option -O2, and GNU fortran 77 compiler with optimization option -O4. We started one process per workstation for both programs and tuned n_b to provide the best performance for the mpC and ScaLAPACK routines.

Tables 1 and 2 demonstrate speedups computed relative to the LAPACK routine `dpotf2` executing sequential Cholesky factorization on `gamma`. One can see that the mpC function `ParCh` and the ScaLAPACK routine `PDPOTRF` take approximately the same time when running on homogeneous networks **gb**, **gbd**, and practically homogeneous **gbda** (Table 1). After we enhance these networks with low-performance **omega**, the mpC program allows to utilize the parallel potential of performance-heterogeneous **gbo**, **gbdo**, and **gbdao** speeding up the Cholesky factorization. At the same time, its ScaLAPACK counterpart does not allow this, slowing down the Cholesky factorization (Table 2).

Table 1. Speedups on homogeneous networks

n	gb		gbd		gbda	
	mpC	ScaL	mpC	ScaL	mpc	ScaL
300	1.03	1.03	1.25	1.27	1.26	1.25
400	1.13	1.10	1.47	1.42	1.57	1.60
500	1.18	1.17	1.56	1.54	1.76	1.73
600	1.27	1.24	1.69	1.65	1.91	1.90
700	1.29	1.26	1.75	1.75	2.01	2.03
800	1.33	1.32	1.82	1.84	2.11	2.12

Table 2. Speedups on heterogeneous networks

n	gbo		gbdo		gbdao	
	mpC	ScaL	mpC	ScaL	mpc	ScaL
300	1.15	0.85	1.25	1.02	1.33	1.11
400	1.29	0.98	1.50	1.17	1.62	1.37
500	1.38	1.04	1.64	1.29	1.84	1.43
600	1.48	1.09	1.76	1.36	1.97	1.60
700	1.53	1.13	1.85	1.41	2.11	1.69
800	1.57	1.12	1.90	1.49	2.20	1.77

5 Summary

Efficient programming heterogeneous networks of computers implies distributing data and computations over processors in accordance with their performances. It makes even regular problems be considered irregular. We have demonstrated how mpC can be used for solving such irregular problems. The key facilities making the mpC language and its programming environment unique tools for programming such irregular applications are:

- facilities to detect the number and performances of processors of the executing network of computers;

- convenient facilities to formulate requirements on performances of virtual processors constituting the abstract parallel machine (network object) executing computations and communications (facilities to specify a network type in run time);
- convenient and natural facilities to allocate and discard network objects as well as distribute data and computations over them.

References

[1] El-Rewini, H., and Lewis, T.: Introduction To Distributed Computing. IEEE Computer Society Press, Los Alamitos, CA, 1997.

[2] High Performance Fortran Forum, High Performance Fortran Language Specification, version 1.1. Rice University, Houston TX, November 10, 1994

[3] Message Passing Interface Forum, MPI: A Message-passing Interface Standard, version 1.1, June 1995.

[4] Choi, J., Demmel, J., Dhillon, I., Dongarra, J., Ostrouchov, S., Petitet, A., Stanley, K., Walker, D., and Whaley, D.: ScaLAPACK: A Portable Linear Algebra Library for Distributed Memory Computers – Design Issues and Performance. UT, CS-95-283, March 1995.

[5] Anderson, E., Bai, Z., Bischof, C., Demmel, J., Dongarra, J., Du Croz, J., Greenbaum, A., Hammarling, S., McKenney, S., Octrouchov, S., and Sorensen, D.: LAPACK Users' Guide, Second Edition. SIAM, Philadelphia, PA, 1995.

[6] Dongarra, J., Du Croz, J., Duff, I., and Hammarling, S.: A Set of Level 3 Basic Linear Algebra Subprograms. ASM Trans. Math. Soft., 16, 1, pp.1–17, March 1990

[7] Lastovetsky, A.: The mpC Programming Language Specification. Technical Report, ISPRAS, Moscow, December 1994.

[8] Arapov, D., Kalinov, A., and Lastovetsky, A.: Managing the Computing Space in the mpC Compiler. Proceedings of the 1996 Parallel Architectures and Compilation Techniques (PACT'96) conference, IEEE CS Press, Boston, MA, Oct. 1996, pp.150–155.

[9] Arapov, D., Kalinov, A., and Lastovetsky, A.: Resource Management in the mpC Programming Environment. Proceedings of the 30th Hawaii International Conference on System Sciences (HICSS'30), IEEE CS Press, Maui, HI, January 1997.

[10] Arapov, D., Kalinov, A., Lastovetsky, A., Ledovskih, I., and Lewis, T.: A Programming Environment for Heterogeneous Distributed Memory Machines. Proceedings of the 1997 Heterogeneous Computing Workshop (HCW'97) of the 11th International Parallel Processing Symposium (IPPS'97), IEEE CS Press, Geneva, Switzerland, April 1997, pp.32–45.

[11] Choi, J., Dongarra, J., Ostrouchov, S., Petitet, A., Walker, D., and Whaley, R.C.: The Design and Implementation of the ScaLAPACK LU, QR, and Cholesky Factorization Routines. UT, CS-94-246, September, 1994.

Balancing the Load in Large-Scale Distributed Entity-Level Simulations

Sharon Brunett

California Institute of Technology, Pasadena, California, USA 91125
sharon@cacr.caltech.edu

Abstract. A distributed, parallel implementation of the widely used Modular Semi-Automated Forces (ModSAF) Distributed Interactive Simulation (DIS) is presented, using networked high-performance resources to simulate large-scale entity-level exercises. Processing, communication and I/O demands increase dramatically as the simulation grows in terms of size or complexity. A general framework for functional decomposition and scalable communications architecture is presented. An analysis of the communications load within a single computer and between computers is presented. Ongoing activities to address more dynamically communication limitations and processing load using Globus are discussed.

Large-Scale Distributed Simulation Background

An ongoing area of interest within the DIS community is increasing synthetic simulation capabilities. Improving scale and complexity for virtual representations of real operational environments yield cost-effective methods for training, system modeling, evaluation and experimentation. ModSAF is a widely used DIS application, designed to provide realistic entity-level computer generated forces interacting with real-time manned simulators. Elements of the synthetic simulation environment include visualization stations, high-fidelity individually modeled entities (i.e., various types of vehicles, troops, and weapons), complex terrain, and support for man-in-the-loop interactions. ModSAF instantiations generally run across a collection of heterogeneous network-linked workstations. Individual simulators (workstations) host small numbers of entities, 50–150 depending on exercise complexity and workstation capabilities. Connectivity between multiple simulators is based on the premise that each simulator has a common representation of a fixed world environment.

As the size of the simulation increases or the complexity of an exercise grows, the load on a collection of ModSAF workstations climbs dramatically. Each simulator must process additional high-priority tasks (route planning, obstacle avoidance, line-of-sight detection of nearby vehicles ...) associated with realistically modeling each local vehicle. Moreover, every simulator regularly sends, receives and processes entity state updates (information about dynamic activities and events) through the exchange of standard DIS Protocol Data Units (PDUs)[1].

The Synthetic Forces (SF) Express project [2] addresses ModSAF's communication bottlenecks and simulator performance issues by providing a scalable communications scheme, proven to support the computational, communication and I/O demands for very large-scale entity-level simulations. In this paper, an approach for functionally decomposing the ModSAF v2.1 application across resources in a high-performance computing environment is presented. Interest-based communication rules enforcing restricted communication are described. The supporting communications architecture, enabling efficient movement of relevant simulator data, is examined. Results from various large-scale runs are discussed. The paper concludes with a description for future plans.

SF Express Overview

Functional Decomposition

Increasing the entity count for ModSAF requires adding more workstations or increasing the number of entities on existing resources. For ModSAF run on networked workstations, both options fail to scale for two main reasons:

1. As the number of simulated entities increases, the total (broadcast) message traffic saturates the communications network.
2. Independent of effectively moving these messages, the processing requirements for the incoming data overwhelm the capabilities of the individual workstations (simulator engines).

Failure to processes an assigned workload for a real-time simulation causes a window of time where the realism of the simulation may be at risk. Within ModSAF, distributed simulators do not wait for overloaded simulation engines to "catch up", as is typical for many other applications. Instead, dead reckoning algorithms are used to extrapolate remote object behavior when state updates from overburdened simulators do not appear in a timely fashion. This "best-guess" approach makes assessing when to rearrange workload within the collection of resources somewhat complex. Even with a metric for determining acceptable simulator load, reassigning the work domain between distributed resources can come at a high communications cost.

SF Express addresses ModSAF simulator performance and communication limitations with a scalable architecture utilizing distributed supercomputers, connected by high-speed networks. High-performance computers offer benefits of substantial computational power, access to fast I/O devices, and reliable high-speed dataflow among processors. The basic strategy used to move ModSAF into the supercomputing environment is hetergenous assignment of tasks to processors. The three key components to the system are as follows:

Simulators: Processors responsible for running the vanilla ModSAF simulation engine. Most processors in the supercomputer are assigned simulator tasks.

Routers: Processors responsible for moving data among simulators. Each simulator uses a router processor to send out information packets describing the activities of its own locally simulated vehicles and receive messages describing the state and activities of entities hosted on other simulators.

Data Servers: Processes responsible for managing access to disk and cached data (i.e., terrain) needed by the simulators.

Interest-Filtered Communications Architecture

In addition to functionally decomposing the simulation's computational, communication, and I/O elements, additional logic must be added to limit the incoming message traffic seen by each simulator. In a very large-scale simulation, much of the network data available to an individual simulator is irrelevant. For example, tanks with geographic separations exceeding a few kilometers hosted by different simulators need not communicate superfluous data which will be processed and ultimately discarded by the receiving simulator. The SF Express scalable communications scheme can be divided into three parts:

Interest Specification Procedures: PDUs can be associated with specific interest class indices. Procedures exist for evaluating the total interest state of vehicles local to a simulator.

Intra-SPP Communications: Within a computer, a collection of processors are assigned designated message routing responsibilities. The number of processors can be adjusted at run-time, according to exercise interaction expectations and prior experiences. These router processors are in charge of receiving and storing interest declarations from the simulator processors and moving simulation data packets according to the interest declarations.

Inter-SPP Interest-Constrained Communications: Additional interest - restricted data exchange procedures provide necessary support for SF Express execution in a distributed high-performance computing environment.

Operational Model

The basic model for interopating SF Express components, as illustrated in Fig. (1), is straightforward. Simulators process tasks required for accurately modeling their locally simulated vehicles, with respect to the entire distributed exercise. Typical tasks might include route planning and intervisibility calculations possibly requiring terrain data accessed via data server processors). Regularly, each simulator sends and receives state updates and event reports, in order to update local representations of remotely simulated vehicles.

Event messages can often be "tagged" or mapped onto indices within a grid (see Fig. (2)). For example, state update messages containing information about a vehicle's orientation or position upon the terrain can easily be translated into cell numbers within a position grid. The cell values associated with each vehicle determine an "interest map", or area of geographical interest, which affects the collection of locally simulated vehicles. Each simulator periodically recomputes its interest map and sends this information to an assigned router processor. Interest map declarations are the simulator's method for subscribing to relevant

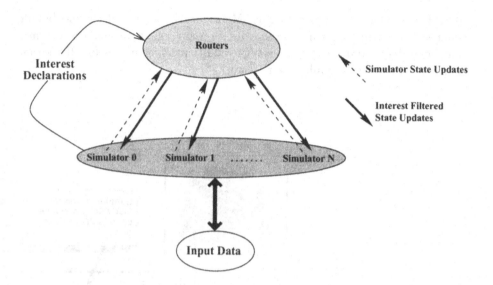

Fig. 1. Functional components of ModSAF interoperating via interest-restricted communication links

subsets of the exercise wide flow of PDU traffic. Simulators receive PDUs of interest by sending a request to the router processors.

Router processors are responsible for the following main tasks:

1. As simulators regularly send data messages describing the state of their locally simulated vehicles, store this information in a large circular buffer.
2. Maintain a client list which corresponds to each simulator's current interest declaration. Also, manage pointers to pending outbound PDU messages located in the circular buffer.
3. Assess each incoming message's interest tag, and determine which simulators have subscribed to, or declared interest in, this value. Messages are delivered to the appropriate client simulators when the router receives a probe for PDUs matching declared interest.

Router Network Discussion

The network of router processors is built around a basic model of fixed communication channels among specific subsets of processors within a computer. As discussed above, simulators communicate with a specified router processor, called the "Primary" router. Primary routers forward all simulator generated state updates to a top layer of routers and also sends its collective interest maps (the union of the Simulators' interest maps) to a middle layer of routers. The result is a three tiered router architecture, responsible for distributing data of interest to simulators, in a flow controlled efficient manner. Top layer routers provide a distributed repository for active messages within the simulation, enabling easy data logging and statistics gathering. Middle level routers collect and bundle interest-filtered messages in parallel with Primary router ⇔simulator communications. This parallelism minimizes the additional time delays for PDUs that must travel through the full router network.

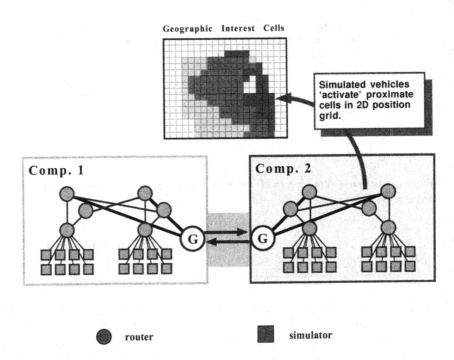

Fig. 2. Two computers hosting SF Express, using Gateway nodes to transfer overlapping areas of interest and associated simulator data simulators

Single Computer Performance Highlights Bottom level Primary routers can comfortably manage the communication requirements for client simulator processors hosting 1K-2K total vehicles. Simulator processors typically send out 20–40 PDUs/sec (a typical PDU has a size of 200 bytes) while receiving 100–200 PDUs/sec from the communication network. Sampled message rates of 4,000 PDUs/sec (about 1Mbyte/sec) are typical in Primary routers. Regular IBM SP runs show Primary routers spend 80% of their time waiting for simulator requests. This suggests more simulators (causing additional message traffic) could easily be assigned to a Primary router.

Detailed studies of the SF Express communications architecture are contained in Refs.[4],[5]. Highlights of these analyses are as follows:

1. The router network approach has been run successfully on a variety of high-performance computers, including the Intel Paragon, IBM SP, Hewlett Packard Exemplar, Silicon Graphics Origin 2000, and "Beowulf" PC Cluster [6].
2. Runs conducted on a single Paragon have included simulations involving up to 18,000 vehicles.
3. The scaling behavior of the router network as problem sizes increase is well-understood, with "theoretical" expectations validated by measured performance results.
4. The effective inter-processor communications within a computer reduce PDU communication overhead significantly for individual simulators (relative to standard ModSAF performance on a LAN/WAN network).

SF Express Communications Architecture Extended for Multiple Computers

For large-scale simulations requiring more computational power or diverse resources than a single high-performance computer can deliver, expanding the SF Express architecture over distributed heterogenous resources is a powerful solution. Just as entity state information exchanged between simulators within a single computer needs filtering by the router network, interest-restricted communications among multiple computers needs to be enforced. Required router network extentions enabling balanced inter-computer message transfer are relatively simple. Dedicated Gateway processors are attached to the router network, and behave as communication servers for two classes of clients:

Local Clients: Router processors on the same computer as the Gateway hold continually changing PDUs and interest maps representing the local computer's state. Local Clients send interest declarations and simulation data to the Gateway for subsequent delivery to remote resources.

External Clients: Processors on remote machines receive and process interest declarations and PDU bundles generated locally. An external client could be a standard ModSAF workstation or GUI. For inter-computer links, an External Client is essentially a mirror image of a Local Client that resides on the external computer.

Gateways manage interest-selected bi-directional data flow by using four basic operations:

1. The local computer's collective interest state is sent out to each of the external computers.
2. Corresponding interest maps are received from remote computers, defining client interests. The union of these external interest states defines the collective (external) gateway interest, which is sent up to the attached local routers.

3. The Gateway receives interest-screened PDUs from local routers in the usual manner, and forwards these to the appropriate external hosts.
4. The Gateway receives relevant event updates from external computers and forwards it to the attached local routers for subsequent distribution.

Fig. (2) shows the router network architecture, spanning multiple computers. Note each computer's Gateway processor exchanges an "interest map" obtained from local middle layer routers, and associated PDU state updates collected from upper layer routers. Only data relevant to simulators residing on different computers is transferred, thus controlling the inter-computer communications load.

Metacomputing Performance Highlights

The performance issues for the metacomputing model in Fig. (2) center on data movement through the Gateway processors. Results presented in this section demonstrate that communications levels were easily managed, with one, essentially expected, exception.

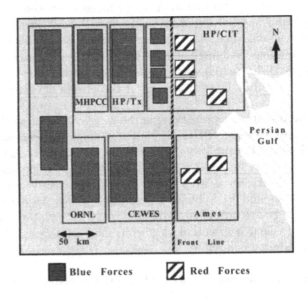

Fig. 3. Scenario and assignments of force groups to specific computers

Fig. (3) depicts a 50K vehicle, nine computer SF Express distributed simulation. Vehicles are assigned to reside on particular computers. Various command and control orders are given to the vehicles, during force group creation, to create a militarily relevant exercise. The evolution of the exercise over time is simple:

all Blue forces move east and attack while the Red forces sit and defend. Intense interaction along the dashed "Front Line" arises. High levels of Gateway traffic are expected where force groups on distributed computers are geographically close. Many of the Gateway⇔Gateway communications links have no appreciable activity, i.e., ORNL⇔ HP/Tx, due to effective inter-computer interest management. Table 1 contains a detailed look at three of the more active inter-computer links:

HP/Tx ⇔ MHPCC: Successful, moderately high bandwidth communications between machines on a Wide-Area Network.

MHPCC ⇔ ORNL: Saturated/Failed communications between machines on a Wide-Area Network.

HP/C1 ⇔ HP/C2: Successful communications between machines on a Local-Area Network.

Table 1. Details of Gateway Performance on Three Busy Links of the 50K SF Express Run

Local SPP Remote SPP	HP/Tx MHPCC	MHPCC HP/Tx	ORNL MHPCC	MHPCC ORNL	HP/C1 HP/C2	HP/C2 HP/C1
Local PDU Busy	0.168	0.074	0.041	0.050	0.023	0.014
Remote PDU Busy	0.147	0.080	0.926	0.032	0.031	0.030
Read Time [msec]	0.60	0.63	22.26	0.77	0.61	0.60
Write Time [msec]	0.97	0.28	27.52	0.26	0.45	0.34

The "PDU Busy" rows list the fraction of (wall clock) time spent in PDU communications within a computer and through the Gateway to a remote computer. The last two rows give the mean times for PDU bundle communications across the network. The UDP-ethernet reads and writes on the ORNL Paragon are about 30 times slower than on the other platforms, leading to an overwhelmed Gateway and the significant data loss. Note the lopsided data exchange between ORNL and MHPCC:

> MHPCC Sends 63.9 Kbytes/sec to ORNL
> ORNL Receives 7.9 Kbytes/sec from MHPCC

Clearly, the Gateway processors connecting MHPCC⇔ORNL need to use a more efficient transport mechanism than UDP/IP ethernet. Otherwise, the dropped packet rate for state messages between computers was well within the tolerable range for distributed ModSAF.

Lessons Learned and Planned Improvements

Even with a communication architecture able to restrict uninteresting events from overloading simulators, at least two areas for concern remain:

1. As an exercise evolves, some simulators might be under utilized (all their vehicles have been destroyed) while other simulators might require more processing or I/O power to keep up with increased demands from vehicle interactions. How should we dynamically respond to the load imbalance?
2. The Gateway processors which permit inter-computer communication are not cognizant of optimal transport mechanisms. In addition, the current implementation dedicates a Gateway processor for each remote resource. For metacomputing runs involving many computers, many Gateways sit idle or underutilized, rather than serving other useful purposes (i.e., hosting vehicles).

Integrating services from the Globus [7] toolkit will allow a "resource aware" implementation of Gateway nodes and adaptive resource configurations. Globus provides a Metacomputing Directory Service (MDS) [9] which provides a uniform method for accessing timely information about the underlying networked collection of resources. Configuration details including CPU speed, types and numbers of network interfaces allow for a much better initial functional decomposition of SF Express across the allocated resources. Instantaneous performance information, such as CPU load, available network bandwith, and point-to-point latency can be obtained to more precisely determine how and when to reconfigure the simulation according to the existing load. Application program structures and run-specific information may also be stored in and retrieved from the MDS. Such information would allow successive runs to learn from one another, and enable better application performance.

The Globus communication library providing models cognizant of network quality of service parameters is called Nexus [8]. From the application level, interfaces are provided for rule-based selction regarding transport mechanisms and parameter values (e.g., "use TCP over ATM", "use packet size Y"). Resource property inquiry is also available to the application, via the MDS. Note how locating Gateway processors on available HiPPI or ATM nodes of the ORNL Paragon would have dramatically helped the inter-computer communications performance. Such optimizations are possible with Nexus integration. Notification mechanisms allow an application to specify constraints on quality of service and call-back functions to be invoked when constraints are violated. Such a service would be extremely useful when a communication link or processor becomes overloaded, permitting the simulation to determine when and how to switch between networks or adjusting the processing load.

Integrating Nexus into SF Express can be done in two phases. First, the Gateway code would be re-written. Gateways currently communicate with each other via socket calls and with their local router network using MPI. These socket calls would be replaced with a straightforward communication link mechanism supported by Nexus. A communication link consists of startpoints (SP) and endpoints (EP). Gateway processors on a local computer would be startpoints, endpoint would represent remote Gateways. Multiple endpoints can be bound to a single startpoint (multicast communication) and vice versa. Such dynamic communication structures will greatly reduce the number of Gateways

SF Express currently sets up at start-up time in order to accomodate possibly loaded inter-computer communication channels. Fig. 4 depicts Gateway processes implemented using Nexus communications mechanisms.

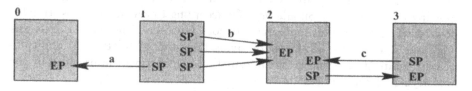

Fig. 4. Nexus Communication Links

Link "a" represents typical Gateway to logging device communication. Link "b" is an example of a Gateway forwarding simulation events into a multiplexer. Such a model would be very useful for allowing one visualization station to selectively process incoming PDU streams from many computers.

Different communication methods can be associated with different communication links, using rule-based selection to determine which method is used for new links. Such versitility is exactly what was needed when the inter-computer link between ORNL and MHPCC showed tremendous data loss. Using the MDS to locate available HiPPI or ATM Paragon nodes, in addition to swapping communication links when connectivity was poor would have dramatically helped inter-computer communications performance and overall system load. Fig. 4 Link "c" represents splitting a typical bi-directional Gateway link into two links, each exercising optimal transport mechanisms.

Communication links support Remote Service Request (RSR) communication operations. RSRs are one-sided asynchronous remote procedure calls which transfer data from a startpoint to the associated endpoint(s) and then integrate the data into the process(es) containing the endpoint(s) by invoking a remote function in the process(es). The SF Express router network is primarily concerned with remote enqueing of simulation events (passing relevant PDUs to the requesting simulators), hence can naturally be expressed as Nexus RSRs instead of MPI calls.

Only the socket-based Gateway code need be replaced with Nexus to make substantial progress producing a more dynamic, resource efficient SF Express inter-computer communication implementation. Overloaded simulator nodes within a computer can also be addressed by reworking the MPI based router network code with Nexus RSRs. For shared-memory computers, Nexus could be much more efficient than MPI due to multithreaded execution. A lightweight thread of control is created for each RSR, enabling remote operations to execute in parallel with each other and with other computation. This multithreaded approach is perfectly suited for the simulator engine's task scheduler mode of operation.

Using the MDS to provide information about which simulators are currently overloaded (gleaned from the MDS' application specific data structure dumps),

RSRs could be generated to appropriately spawn new simulator processes. Processors would no longer be assigned one and only one function throughout the course of the simulation. The static links connecting simulators⇔routers and routers within the router network shown in Fig. 2 would be replaced by RSR invocations. Reclaiming idle resources (data server nodes) or assigning more resources (i.e., Gateway, Simulator, I/O) according to load demands is straightfoward using RSRs and appropriate remote functions. The integrated Nexus approach also has the advantage of giving the entire simulation the ability to react and adjust according to the information-rich environment provided by the underlying Globus services.

Acknowledgements

Support for this research was provided by the Information Technology Office, DARPA, with contract and technical monitoring via Naval Research and Development Laboratory (NRaD).

Access to various computational facilities and significant system support were essential for this work. The 1024 node Intel Paragon was made available by the Oak Ridge Center for Computational Sciences. The smaller Intel Paragon and 256 processor Hewlett Packard Exemplar were made available by Caltech/CACR. The IBM SPs were provided by the Maui High Performance Computing Center (MHPCC), U.S. Army Corps of Engineers Waterways Experiment Station Information Technology Laboratory (CEWES) , and the Numerical Aerodynamic Simulation Systems Division at NASA Ames Research Center. Indiana University and NASA Ames provided access to the Silicon Graphics machines. A 128 CPU Hewlett Packard Exemplar was provided by Hewlett Packard Division Headquarters, Richardson Texas. We thank the system administrators and support staff at all these sites.

Globus experiments and integration were made possible by Argonne National Laboratory and USC Information Science Institute. Key technical guidance and code development was given by Ian Foster, Steve Tuecke, Carl Kesselman, Steve Fitzgerald, Karl Czajkowski, Mei-Hui Sue, and Marcus Thiebaux.

Caltech's SF Express design and implementation are greatly due to Paul Messina's direction and Thomas Gottschalk's technical expertise and dedication.

An eternal debt of gratitude to Caltech's Terri Canzian, Tina Mihaly Pauna, and Heather Young for their help preparing SF Express documents and presentations.

References

1. J. S. Dahmann and D. C. Wood, 'Scanning the Special Issue on Distributed Interactive Simulations', Proceedings of the IEEE, Volume 83 (1995) 1111, and references therein.

2. S. Brunett, D. Davis, T. Gottschalk, P. Messina, and C. Kesselman, 'Implementing Distributed Synthetic Forces Simulations in Metacomputing Environments', IPPS/SPDP '98 Heterogeneous Computing Workshop, 1998.

3. C. M. Keune and D. Coppock, 'Synthetic Theater of War-Europe (STOW-E) Technical Analysis', NRaD Technical Report 1703 (1995).

4. S. Brunett and T. Gottschalk, ' Scalable ModSAF Simulations with more than 50,000 Vehicles using Multiple Scalable Parallel Processors', Technical Report CACR - 156

5. S. Brunett and T. Gottschalk, 'Large-Scale Metacomputing Framework for Mod-SAF', Technical Report CACR-152, January 1998.

6. http://cesdis.gsfc.nasa.gov/beowulf, the Beowulf WWW site, and references/links therein.

7. I. Foster and C. Kesselman, 'Globus: A Metacomputing Infrastructure Toolkit', to be published in *International Journal of Supercomputer Applications.*

8. I. Foster, C. Kesselman, S. Tuecke, 'The Nexus Approach to Integrating Multithreading and Communication', J. Parallel and Distributed Computing, 37:70–82, 1996.

9. S. Fitzgerald, I. Foster, C. Kesselman, G. von Laszewski, W. Smith, and S. Tuecke, 'A Directory Service for Configuring High-Performance Distributed Computations', Proc. 6th IEEE Symp. on High-Performance Distributed Computing 1997.

Modeling Dynamic Load Balancing in Molecular Dynamics to Achieve Scalable Parallel Execution *

Lars Nyland,[1] Jan Prins,[1] Ru Huai Yun,[2] Jan Hermans,[2]
Hye-Chung Kum,[1] and Lei Wang[1]

[1] Computer Science Department,
Univ. of North Carolina, Chapel Hill, NC 27599
http://www.cs.unc.edu/
[2] Biochemistry and Biophysics Department,
Univ. of North Carolina, Chapel Hill, NC 27599
http://femto.med.unc.edu/

Abstract. To achieve scalable parallel performance in Molecular Dynamics Simulation, we have modeled and implemented several dynamic spatial domain decomposition algorithms. The modeling is based upon Valiant's Bulk Synchronous Parallel architecture model (BSP), which describes supersteps of computation, communication, and synchronization. We have developed prototypes that estimate the differing costs of several spatial decomposition algorithms using the BSP model.

Our parallel MD implementation is not bound to the limitations of the BSP model, allowing us to extend the spatial decomposition algorithm. For an initial decomposition, we use one of the successful decomposition strategies from the BSP study, and then subsequently use performance data to adjust the decomposition, dynamically improving the load balance. We report our results here.

1 Introduction

A driving goal of our research group is to develop a high performance MD simulator to support biochemists in their research. Our goals are to study large timescale behavior of molecules and to facilitate *interactive* simulations [7]. Two main characteristics of the problem impede our goal: first is the large number of interactions in solvated biomolecules, and second is the small timestep that is required to adequately capture high frequency motions. To meet our goal, we must develop a parallel implementation that scales well even on small problem sizes. Because the communication cost, memory reference and load balance across processors trade in a complex fashion, we have analyzed candidate implementations using the BSP [10, 1] model. The most promising implementation was implemented and improved outside of the constraints of the BSP model.

---

* This work has been supported in part by the National Institutes of Health's National Center for Research Resources (grant RR08102 to the UNC/Duke/NYU Computational Structural Biology Resource).

At each step in MD simulation, the sum of all forces on each atom is calculated and used to update the positions and velocities of each atom. The bonded forces seek to maintain bond lengths, bond angles, and dihedral angles on single bonds, two-bond chains and three-bond chains, respectively. Non-bonded forces are comprised of the electrostatic forces and the Van der Waals forces.

A cutoff radius is introduced to limit non-bonded atom interactions to pairs closer than a preset radius, R_c. This is still the most time-consuming portion of each step, even though the cutoff radius reduces the $O(n^2)$ work to $O(n)$. The remaining longer-range forces are calculated by some other method [2, 5], calculated less frequently, or completely ignored.

Good opportunities for parallelization in MD exist; all of the forces on each of the atoms are independent, so they can be computed in parallel. Once computed, the non-bonded forces are summed and applied in parallel. Good efficiency depends on good load-balancing, low overhead, and low communication requirements.

Using a spatial decomposition increases data coherence, reducing communication costs. Two nearby atoms interact with all atoms that are within R_c of both, providing two opportunities for reduced communication. First, the atoms are near each other, thus accessing the data for many nearby atoms data will not require interprocessor communication. Second, for those interactions that require data from neighboring regions, atomic data can be fetched once and then reused many times, due to the similarities of interactions of nearby atoms.

The use of a spatial decomposition for MD has become widespread in recent years. It is used by AMBER [3, 9], Charmm [6], Gromos [4] and NAMD [8], all of which run in a message-passing paradigm, as opposed to our shared-memory implementation. In general, each of these implementations found good scaling properties, but it is difficult to compare overall performance of the parallel MD simulators, as machine speeds have improved significantly since publication of the cited reports. Compared with these other implementations, our shared-memory implementation allows very precise load-balancing on small systems.

2 Modeling Parallel Computation with the BSP Model

The Bulk Synchronous Parallel (BSP) model has been proposed by Valiant [10] as a model for general-purpose parallel computation. It was further modified in [1] to provide a *normalized cost* of parallel algorithms, enabling uniform comparison of algorithms. The BSP model is both simple enough to quickly understand and use, but realistic enough to achieve meaningful results for many parallel computers.

A parallel computer that is consistent with BSP architecture has a set of processor-memory pairs, a communication network that transmits values in a point-to-point manner, and a mechanism for efficient barrier synchronization of the processors. Parallel computers are parameterized with 4 values:

1. The number of processors, P.
2. The processor speed, s, measured in floating-point operations per second.
3. The latency, L, which reflects the minimum latency to send a packet through the network, which also defines the minimum time to perform global synchronization.

4. The gap, g, reflecting the network communication bandwidth on a per-processor basis, measured in floating-point operation cycles taken per floating-point value sent.

An algorithm for the BSP is written in terms of S supersteps, where a single superstep consists of some local computation, external communication, and global synchronization. The values communicated are not available for use until after the synchronization. The cost of the ith superstep is $C_i = w_i + gh_i + L$ where w_i is the maximum number of local operations executed by any processor and h_i is the maximum number of values sent or received by any processor. The total cost of executing a program of S steps is then:

$$C_{tot} = \sum_{i=1}^{S} C_i = W + Hg + SL, \text{ where } W = \sum_{i=1}^{S} w_i \text{ and } H = \sum_{i=1}^{S} h_i$$

The normalized cost is the ratio between the BSP cost using P processors and the optimal work perfectly distributed over P processors. The optimal work, W_{opt}, is defined by the best known sequential algorithm. The normalized time is expressed as

$$C(P) = \frac{P \cdot C_{tot}}{W_{opt}}$$

The normalized cost can be reformulated as $C(P) = a + bg + cL$, where $a = P \cdot W/(W_{opt}), b = P \cdot H/(W_{opt})$, and $c = P \cdot S/(W_{opt})$.

When the triplet $(a, b, c) = (1, 0, 0)$, the parallelization is optimal. Values where $a > 1$ indicate extra work is introduced in the parallelization and/or load imbalance among the processors. Values of $b > 1/g$ or $c > 1/L$ indicate that the algorithm is communication bound, for the architecture described by particular values of g and L.

```
for t = 1 to T by k {
    if Processor == 0 {
        distribute atoms to processors
        calculate local pairlist
    }
    for s = t to t+k - 1 {
        get remote atom information
        synchronize
        calculate forces on local atoms
        apply forces to update
        local positions/velocities
    }
}
```

Fig. 1. A high-level, multiple timestep algorithm for performing parallel molecular dynamics computations (k small timesteps per large timestep).

3 BSP Modeling of Parallel MD

In this section, we model several domain decompositions for MD simulation. We describe a simplified MD algorithm, the domain decompositions, and show the results of modeling.

3.1 A Simplified MD Algorithm

The most time-consuming step of MD simulations is the calculation of the non-bonded forces, typically exceeding 90% of the execution time, thus we limit our modeling study to this aspect. Our simplified algorithm for computing the non-bonded forces is shown in figure 1. It consists of an outer loop that updates the pairlist every k steps, with an inner loop to perform the force computations and application. The value k ranges from 10 to 50 steps, and is often referred to as the *pairlist calculation frequency*.

In our modeling of MD, we examine the cost of executing k steps to amortize the cost of pairlist calculation. Computing the cost of k steps is adequate as the cost of subsequent steps is roughly the same.

Molecular Input Data					
Name	Atoms	Name	Atoms	Name	Atoms
Alanine	66	Water	798	Eglin	7065
Dipeptide (wet)	231	Argon	1728	Water	8640
SS Corin	439	SS Corin (wet)	3913	Polio (segment)	49144

Fig. 2. The input dataset names and number of atoms used for measuring different decompositions

The outer loop distributes the atoms to processors. We modeled it with three super-steps that distribute the data, send perimeter atoms to neighboring processors, and build local pairlists.

There is only one superstep in the inner loop. It consists of distributing positions of perimeter atoms to nearby processors; a synchronization barrier to ensure all computation is using data from the same iteration; followed by a force calculation and application. The computations performed by the inner loop are the same for all decompositions.

3.2 Modeling Experiment

The goal of the experiment is to find values of a, b, and c for each combination of four data decompositions using nine molecular data sets (summarized in figure 2) with varying numbers of processors (normalized execution costs can be computed by choosing values for g and L). The values of a, b, and c show how work and communication affect parallel performance, and are computed for the inner and outer loops using

$$a = \frac{P(w_{outer} + k \cdot w_{inner})}{W_{opt}}, b = \frac{P(h_{outer} + k \cdot h_{inner})}{W_{opt}}, c = \frac{P(S_{outer} + k \cdot S_{inner})}{W_{opt}}$$

to compute $C(P) = a + bg + cL$.

The four decomposition strategies in this study are:

- *Uniform Geometric Decomposition*. This decomposition simply splits the simulation space (or sub-space) equally in half along each dimension until the number of subspaces equals the number of processors.
- *Orthogonal Recursive Bisection Decomposition (ORB)*. ORB recursively splits the longest dimension by placing a planar boundary such that half the atoms are on one side, and half are on the other. This yields an assignment of atoms to processors that varies by at most 1.
- *Pairlist Decomposition*. This decomposition yields perfect load-balance by evenly decomposing the pairlist among the processors. A drawback is that it does not have spatial locality, and is included it to better understand this aspect.
- *Spatial Pairlist Decomposition*. We also consider a spatial decomposition that is based upon the number of entries in the pairlist assigned to each processor, placing spatial boundaries based on pairlist length.

3.3 Results

Normalized coefficients a, b, and c were computed for every combination of decomposition, input data set, and processor count P in $\{2, 4, 8, 16, 32\}$, resulting in a table of triplets with 180 entries. An example where the decomposition is uniform geometric, the data set is polio, and the number of processors is 32 has the entry $(a, b, c) = (3.59, 0.0032, 3.42 \times 10^{-7})$. The performance improves by changing the decomposition to ORB leading to the values $(a, b, c) = (1.64, 0.005, 3.42 \times 10^{-7})$ (note the reduced a-value, representing load-balance improvement). These values are typical (as can be seen in figure 3 which shows all the data), in that the values for a are in the range 1.0 to 10, the values of b are typically less than $1/100$ the value of a, and the values of c are typically less than $1/10^5$ of a.

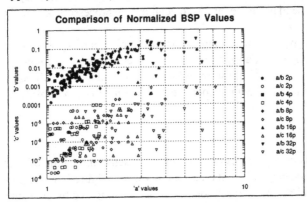

Fig. 3. A comparison showing the magnitude of difference between a and b, and a and c for all (a, b, c) triplets in our experiment. The values of a are plotted along the x-axis. The b values are plotted with solid markers against a. Similarly, the c values are plotted along the y-axis with hollow markers against a.

A general conclusion is that for modern parallel computers with values of g in $1 - 100$ and L in $25 - 10000$, the parallel overhead plus load-imbalance (amount that $a > 1$) far outweighs the cost of communication and synchronization on virtually all of the results in the study. Any decomposition that seeks a more evenly balanced load (reduction of a) will improve performance far more than solutions that seek reduced communication (lower b) or reduced synchronization (lower c).

Thus, even parallel computers with the slowest communications hardware will execute well-balanced, spatially-decomposed MD simulations with good efficiency.

The graph in figure 4 shows the effect of communication speed on the overall performance of the different decompositions. In this dataset, P is set to 32, and two different machine classes are examined. The first is a uniform memory access machine (UMA), with $(g, L) = (1, 128)$, representing machines such as the Cray vector processors that can supply values to processors at processor speed once an initial latency has been charged. The second is a non-uniform memory access machine (NUMA), much like the SGI parallel computers and the Convex SPP.

There are two interesting conclusions to be drawn from figure 4. The first is that executing MD on a machine with extremely high communication bandwidth (UMA) performs, in normalized terms, almost identically with machines with moderate communications bandwidth. This is seen in the small difference between the same data using the same decomposition, where the normalized execution cost for both architectures is nearly the same. The second interesting point in figure 4 is that decomposition matters much more than communications bandwidth. The decompositions that attempt to balance work and locality (ORB and spatial pairlist) have a bigger impact on performance than parallel computers with extremely high performance communications. This is a good indication that either the ORB or spatial pairlist decomposition should be used for a parallel implementation on any parallel computing hardware.

The most significant conclusion drawn from this study is that load-balancing is by far the most important aspect of parallelizing non-bonded MD computations. This can be seen in the significantly larger values of a when compared to values of b and c, as well as the results in figure 4 that show the improvement gained in using load-balanced decompositions. The spatial decomposition using pairlist-length as a measure shows the advantage that is achieved by increasing locality over the non-spatial pairlist decomposition. These results are important not only in our work implementing simulators, but to others as well, guiding them in the choices of their parallel algorithms.

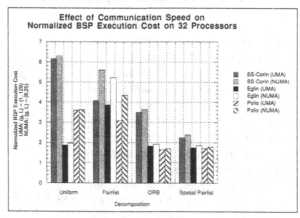

Fig. 4. This graph shows the normalized execution cost on 32 processors, comparing different decomposition strategies on machines with differing communication performance. For the UMA architecture, $(g, L) = (1, 128)$; for NUMA, $(g, L) = (8, 25)$. Note that the normalized cost of a program on a machine with very high performance communication is only marginally better than machines with substantially lower communication performance (except for pairlist decomposition).

4 Implementation of Dynamic Load Balancing in Molecular Dynamics

The results of the previous section are a stepping stone in the pursuit of our overall goal. In this section, we describe the parallelization using spatial decomposition for shared-memory computers of our Sigma MD simulator. The performance results in this section show that the modeling provides a good starting point, but good scaling is difficult to achieve without dynamic load-balancing.

Optimizations in programs often hamper parallelization, as they usually reduce work in a non-uniform manner. There are (at least) two optimizations that hamper the success of an ORB decomposition in Sigma. The first is the optimized treatment of water. Any decomposition based on atom count will have less work assigned when the percentage of atoms from water molecules is higher.

The second is the creation of *atom groups*, where between 1 and 4 related atoms are treated as a group for non-bonded interactions, so the amount of work per group can vary by a factor of 4. This optimization has the benefit of reducing the pairlist by a factor of 9, since the average population of a group is about 3.

4.1 A Dynamic Domain Decomposition Strategy

One troubling characteristic of our static parallel implementation was the consistency of the imbalance in the load over a long period. Typically, one processor had a heavier load than the others, and it was this processor's arrival at the synchronization point that determined the overall parallel performance, convincing us that an adaptive decomposition was necessary.

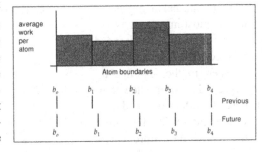

To achieve an evenly balanced decomposition in our MD simulations, we use past performance as a prediction of future work requirements. One reason this is viable is that the system of molecules, while undergoing some motion, is not moving all that much. The combination of this aspect of MD with the accurate performance information leads to a dynamic spatial decomposition that provides improved performance and is quick to compute.

Fig. 5. Unbalanced work loads on a set of processors. If the boundaries are moved as shown, then the work will be more in balance.

To perform dynamic load balancing, we rely on built-in hardware registers that record detailed performance quantities about a program. The data in these registers provide a cost-free measure of the work performed by a program on a processor-by-processor basis, and as such, are useful in determining an equitable load balance.

Some definitions are needed to describe our work-based decomposition strategy.

- The dynamics work, w_i, performed by each processor since the last load-balancing operation (does not include communication and synchronization costs)
- The total work, $W = \sum_{i=1}^{P} w_i$, since the last load-balancing operation
- The estimated ideal (average) work, $\overline{w} = W/P$, to be performed by each processor for future steps
- The average amount of work, $a_i = w_i/n_i$, performed on behalf of each atom group on processor i (with n_i atom groups on processor i)
- The number of decompositions, d_x, d_y, d_z, in the x, y and z dimensions

4.2 Spatial Adaptation

We place the boundaries one dimension at a time (as is done in the ORB decomposition) with a straightforward $O(P)$ algorithm. Figure 5 shows a single dimension split into n subdivisions, with $n-1$ movable boundaries (b_0 and b_n are naturally at the beginning and end of the space being divided). In Sigma, we first divide the space along the

x-dimension into d_x balanced parts, then each of those into d_y parts, and finally, each of the "shafts" into d_z parts, using the following description.

Consider the repartitioning in a single dimension as shown in figure 5. Along the x-axis, the region boundaries separate atoms based on their position (atoms are sorted by x-position). The height of a partition represents the average work per atom in a partition, which as stated earlier, is not constant due to density changes in the data and optimizations that have been introduced. Thus, the area of the box for each partition is w_i, and the sum of the areas is W. The goal is to place b_i far enough from b_{i-1} such that the work (represented by area) is as close to $\overline{w}$ as possible. This placement of the boundaries can be computed in $O(n)$ time

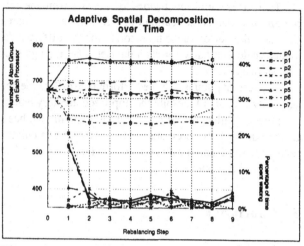

Fig. 6. This graph shows two views of the adaptive decomposition working over time using 8 processors. The upper traces show the number of T4-Lysozyme atom groups assigned to each processor. The lower traces show the percentage of time spent waiting in barriers by each process since the previous balancing step. At step 0, an equal number of atom groups is assigned to each processor, since nothing is known about the computational costs. From then on, the decomposition is adjusted based on the work performed.

for n boundaries. While this does not lead to an exact solution, a few iterations of work followed by balancing yield very good solutions where the boundaries settle down.

Figure 6 shows the boundary motion in Sigma as the simulation progresses. Initially, space is decomposed as if each atom group causes the same amount of work. This decomposes space using ORB such that all processors have the same number of atom groups. As the simulation progresses, boundaries are moved to equalize the load based on historical work information. This makes the more heavily loaded spaces smaller, adding more volume (and therefore atoms) to the lightly loaded spaces. As the simulation progresses, the number of atom groups shifted to/from a processor is reduced, but still changing due to the dynamic nature of the simulation and inexact balance.

4.3 Results

We have tested the implementation on several different parallel machines, including SGI Origin2000, SGI Power Challenge and KSR-1 computers. Figure 7 shows the performance of several different molecular systems being simulated on varying numbers of processors. The y-axis shows the number of simulation steps executed per second, which is indeed the metric of most concern to the scientists using the simulator. We ran tests using decompositions where we set $P = (d_x \cdot d_y \cdot d_z)$ to 1, 2, 4, 6, 8, 9, 12, and 16.

There are several conclusions to be drawn from the performance graph, the most important of which is the scaling of performance with increasing processors. The similar slopes of the performance trajectories for the different datasets shows that the performance scales similarly for each dataset. The average speedup on 8 processors for the data shown is 7.59.

The second point is that the performance difference between the two architectures is generally very small, despite the improved memory bandwidth of the Origin 2000 over the Power Challenge. Our conjecture to explain this, based on this experiment and the BSP modeling in the previous section, is that the calculation of non-bonded interactions involves a small enough dataset such that most, if not all, atom data can remain in cache once it has been fetched.

5 Conclusions

We are excited to achieve performance that enables interactive molecular dynamics on systems of molecules relevant to biochemists. Our performance results also enable rapid execution of large timescale simulations, allowing many experiments to be run in a timely manner. The methodology described shows the use of high-level modeling to understand what the critical impediments to high-performance are, followed by detailed implementations where optimizations (including model violations) can take place to achieve even better performance.

Prior to our BSP modeling study, we could only conjecture that load-balancing was the most important aspect of parallelism to explore for high performance parallel MD using a spatial decomposition. Our BSP modeling supports this claim, and also leads us to the conclusion that the use of 2 or 4 workstations using ethernet communications should provide good per-

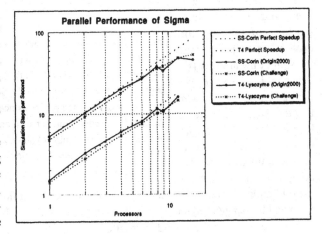

Fig. 7. Parallel Performance of Sigma. This graph shows the number of simulations steps per second achieved with several molecular systems, T4-Lysozyme (13642 atoms), and SS-Corin (3948 atoms). The data plotted represent the performance of the last 200fs (100 steps) of a 600fs simulation, which allowed the dynamic decomposition to stabilize prior to measurement. A typical simulation would carry on from this point, running for a total of 10^6 fs (500,000 simulation steps) in simulated time, at roughly these performance levels.

formance improvements, despite the relatively slow communications medium. Unfortunately, we have not yet demonstrated this, as our implementation is based upon a shared-memory model, and will require further effort to accommodate this model.

Our BSP study also shows that, for MD, processor speed is far more important than communication speed, so that paying for a high-speed communications system is not necessary for high performance MD simulations. This provides economic information for the acquisition of parallel hardware, since systems with faster communication usually cost substantially more.

And finally, we've shown that good parallelization strategies that rely on information from the underlying hardware or operating system can be economically obtained and effectively used to create scalable parallel performance. Much to our disappointment, we have not been able to test our method on machines with large numbers of processors, as the trend with shared-memory parallel computers is to use small numbers of very fast processors.

We gratefully acknowledge the support of NCSA with their "friendly user account" program in support of this work.

References

1. R. H. Bisseling and W. F. McColl. Scientific computing on bulk synchronous parallel architectures. Technical report, Department of Mathematics, Utrecht University, April 27 1994.
2. John A. Board, Jr., Ziyad S. Hakura, William D. Elliott, and William T. Rankin. Scalable variants of multipole-accelerated algorithms for molecular dynamics applications. Technical Report TR94-006, Electrical Engineering, Duke University, 1994.
3. David A. Case, Jerry P. Greenberg, Wayne Pfeiffer, and Jack Rogers. AMBER – molecular dynamics. Technical report, Scripps Research Institute, see [9] for additional information, 1995.
4. Terry W. Clark, Reinhard v. Hanxleden, J. Andrew McCammon, and L. Ridgway Scott. Parallelizing molecular dynamics using spatial decomposition. In *Proceedings of the Scalable High Performance Computing Conference*, Knoxville, TN, May 1994. Also available from ftp://softlib.rice.edu/pub/CRPC-TRs/reports/CRPC-TR93356-S.
5. Tom Darden, Darrin York, and Lee Pedersen. Particle mesh ewald: An $n\log(n)$ method for ewald sums in large systems. *J. Chem. Phys.*, 98(12):10089–10092, June 1993.
6. Yuan-Shin Hwang, Raja Das, Joel H. Saltz, Milan Hodošček, and Bernard Brooks. Parallelizing molecular dynamics programs for distributed memory machines: An application of the chaos runtime support library. In *Proceedings of the Meeting of the American Chemical Society*, August 21–22 1994.
7. Jonathan Leech, Jan F. Prins, and Jan Hermans. SMD: Visual steering of molecular dynamics for protein design. *IEEE Computional Science & Engineering*, 3(4):38–45, Winter 1996.
8. Mark Nelson, William Humphrey, Attila Gursoy, Andrew Dalke, Laxmikant Kale, Robert D. Skeel, and Klaus Schulten. NAMD - a parallel, object-oriented molecular dynamics program. *Journal of Supercomputing Applications and High Performance Computing*, In press.
9. D.A. Pearlman, D.A. Case, J.W. Caldwell, W.R. Ross, T.E. Cheatham III, S. DeBolt, D. Ferguson, G. Seibel, and P. Kollman. AMBER, a computer program for applying molecular mechanics, normal mode analysis, molecular dynamics and free energy calculations to elucidate the structures and energies of molecules. *Computer Physics Communications*, 91:1–41, 1995.
10. L. F. Valiant. A bridging model for parallel computation. *CACM*, 33:103–111, 1990.

Relaxed Implementation of Spectral Methods for Graph Partitioning

Chengzhong Xu and Yibing Nie

Department of Electrical and Computer Engineering
Wayne State University, Detroit, MI 48020
{czxu,yibing}@ece.eng.wayne.edu

Abstract. This paper presents a fast implementation of the recursive spectral bisection method for p-way partitioning. It is known that recursive bisections for p-way partitioning using optimal strategies at each step may not lead to a good overall solution. The relaxed implementation accelerates the partitioning process by relaxing the accuracy requirement of spectral bisection (SB) method. Considering the solution quality of a SB method on a graph is primarily determined by the accuracy of its Fiedler vector, we propose to set a tight iteration number bound and a loose residual tolerance for Lanczos algorithms to compute the Fiedler vector. The relaxed SB was tested on eight representative meshes from different applications. Experimental results show that the relaxed SB on six meshes produces approximately equivalent quality solutions as the Chaco SB while gaining 10% to 35% improvements in execution time. On the other two meshes, the relaxed SB approaches the Chaco SB in quality as p increases and reduces the execution time by 40% to 70%. Coupled with the Kernighan-Lin local refinement algorithm, the relaxed SB is able to yield high quality solutions in all test cases as p goes beyond 32. Multilevel and spectral quadrisection algorithms benefit from relaxed implementations, as well.

1 Introduction

Efficient use of a parallel computer requires that the computational load of processors be balanced in such a way that the cost of interprocessor communication is kept small. For bulk synchronous applications where computation alternates with communication, representing computational tasks as vertices of a graph and inter-task communication as edges reduces the load balancing problem to graph partitioning [2, 15]. For example, in computational fluid dynamics, numerical calculations are carried out at grid points of a tessellated physical domain. Parallel simulations require splitting the grid into equal-sized partitions so as to minimize the number of edges crossing partitions. Graph partitioning arises in many other applications, including molecular dynamics, VLSI design, and DNA sequencing.

In general, finding an optimal decomposition of unstructured graphs is NP-hard. Due to its practical importance, many heuristic algorithms were developed in the past. The most general approach is physical optimization, including simulated annealing based on crystal annealing in statistical physics and genetic algorithm based on evolution in biology. Both the simulated annealing and the genetic algorithm are able to generate very high quality partitions as long as they proceed at a slow rate. At the other extreme are the algorithms based on geometric information of vertices. Recursive coordinate bisection sorts the vertices according to their coordinates in the direction of the longest spatial extent of the domain. Accordingly, the domain is divided into two partitions. For p-way partitioning, the process continues recursively until p partitions are generated. A more elaborated geometric approach is inertial method [7, 14]. It employs a physical analogy in which the grid points are treated as point masses and the grid is cut with a plane orthogonal to the principal inertial axis of the mass distribution. Even though geometric partitioners are simple and fast, they tend to produce moderate solution qualities because they do not consider the connectivity information in decision-making.

Between the two extreme are a class of less intuitive methods based on spectral theory. A representative is spectral bisection(SB) [13]. It partitions a graph into two pieces according to the graph's Fiedler vector. The vector is actually the eigenvector corresponding to the second smallest eigenvalue of the graph's Laplacian matrix. For p-way partitioning, the spectral bisection method is applied recursively on subgraphs. Recursive spectral bisection is commonly regarded as one of the best partitioners due to its generality and high quality.

Computation of the spectral method is dominated by the calculation of the Fiedler vector. For large graphs, it is a time-consuming process. Many researches were directed towards accelerating the process. Barnard and Simon [1], Hendrickson and Leland [5], and Karypis and Kumar [6] proposed multilevel approaches to coarsen big graphs into small ones for fast calculation of the Fiedler vector. Hendrickson and Leland [3] improved recursive spectral bisection by using more than one eigenvectors to allow for division of a graph into four or eight parts at each recursive step. Most recently, Simon, Sohn, and Biswas made an effort on integrating the spectral bisection with the inertial method for a better tradeoff between partition quality and time [11].

Note that recursive bisection for p-way partitioning using optimal strategies at each step may not lead to a good overall solution due to the greedy nature of and the lack of global information [12]. In other words, having a high quality bisection via the expensive spectral method may not be a good idea for p-way partitioning. It is known that the quality of the spectral method is primarily determined by the accuracy of Fiedler vector. This paper presents a relaxed implementation of Lanczos algorithm, by using an iteration number bound of $2\sqrt{n}$ and a residual tolerance of 10^{-2}, for the Fiedler vector. We test the relaxed implementation on eight representative meshes from the Harwell-Boeing and RIACS collections. Experimental results show that the relaxed SB on six meshes achieves up to 35% improvement in execution time over the SB in Chaco library [4] while without sacrificing solution qualities. On the other two meshes, the relaxed SB approaches the Chaco SB in quality as p increases and reduces the execution time by 40% to 70%. Coupled with the Kernighan-Lin local refinement algorithm, the relaxed SB is able to yield high quality solutions in all test cases as p goes beyond 32. Benefits from the relaxed implementation are also demonstrated in Chaco's multilevel and spectral quadrisection algorithms.

The rest of the paper is organized as follows. Section 2 formulates the graph partitioning problem and relates the spectral bisection method on a graph to the calculation of the Fiedler vector of the graph. Section 3 presents Lanczos algorithms for computing the Fiedler vector and the relaxed accuracy conditions in implementation. Relative performances of the relaxed implementation are reported in Section 4. Section 5 summarizes the results.

2 Spectral Bisection Method

Given an undirected graph $G = (V, E)$, the spectral bisection uses its Fiedler vector to find two disjoint subsets V_1 and V_2 in such a way that $V = V_1 \cup V_2$, $|V_1| = |V_2|$, and the number of edges cross between sets (edge cut) is minimized. Let v_i denote the vertex with index i and $e_{i,j}$ be the edge between v_i and v_j.

Assume there are n vertices in the graph. Then, an equal-sized bisection can be represented by a vector X satisfying $X_i = \pm 1$, $i = 1, 2, \ldots, n$, and $\sum_{i=1}^{n} X_i = 0$; and the number of edge cuts as a function:

$$f(X) = \tfrac{1}{4} \sum_{e_{i,j} \in E} (X_i - X_j)^2. \tag{1}$$

Let A be the adjacency matrix of the graph G and D be a diagonal matrix with $D_{i,i} = diag(d_i)$, where d_i is the number of edges incident upon v_i. Then, Eq. (1) can be rewritten as

$$f(X) = \tfrac{1}{4} X^t L X, \tag{2}$$

where $L = D - A$ and is called *Laplacian matrix* of G. Then, the bisection problem can be reduced to finding a *separator* vector X that

$$\begin{aligned} \text{Minimize} \quad & f(X) = \tfrac{1}{4} X^t L X, \\ \text{Subject to} \quad & X^t \mathbf{1} = 0, \text{ and } X_i = \pm 1 \end{aligned} \tag{3}$$

This discrete optimization problem is NP-hard. However, we can have an approximate X by relaxing the discreteness of constraints on X to that $X_i \in \mathbb{R}$ and $X^t X = n$. Let $u_1, u_2, \ldots, u_n$ be the normalized eigenvectors of L with corresponding eigenvalues $\lambda_1, \lambda_2, \ldots, \lambda_n$. Hendrickson and Leland showed that the u_i are pairwise orthogonal and λ_1 is the only zero eigenvalue of L if G is connected [3]. Expressing X as a linear combination of λ_i, we obtain that $f(X) \geq n\lambda_2/4$. It follows that $f(X)$ is minimized by choosing $X = \sqrt{n}u_2$. The second smallest eigenvector u_2 is often referred to as *Fiedler vector*. Computing the Fiedler vector in $\mathbb{R}^n$ space is then followed by a simple operation that converts the Fiedler vector to a partitioning separator.

3 Relaxed Implementation

Most implementations of the SB employ Lanczos algorithm or its variants to compute the Fiedler vector [13, 4, 6]. The Lanczos method expands the eigenvector spectrum of a symmetric matrix in an iterative fashion. It is so far the most efficient algorithm for computing a few extreme eigenpairs of large, symmetric sparse matrices.

3.1 Lanczos Algorithm

The Lanczos method computes eigenvectors of L as it reduces L to a orthogonally congruent tridiagonal matrix T_n:

$$T_n = Q_n^t L Q_n, \tag{4}$$

where $Q_n = (q_1, q_2, \ldots, q_n)$, $Q_n^t Q_n = I$, and T_n is in the form of

$$\begin{bmatrix} \alpha_1 & \beta_1 & & & \\ \beta_1 & \alpha_2 & \beta_2 & & \\ & \cdots & & \cdots & \\ & & \beta_{j-2} & \alpha_{j-1} & \beta_{j-1} \\ & & & \beta_{j-1} & \alpha_j \end{bmatrix}.$$

By equating the first j columns of Eq. (4), we obtain

$$LQ_j = Q_j T_j + r_j e_j^t, \quad j = 1, 2, \ldots, n, \tag{5}$$

where $r_j = q_{j+1}\beta_j$, $e_j^t = (0, \ldots, 0, 1)$, $\alpha_i = q_i^t L q_i$, and $\beta_i = \|r_i\|$.

The Lanczos algorithm builds up Q_j and T_j one column per step. Suppose that the Lanczos algorithm pauses to compute eigenpairs of T_j at the jth step. Since $r_j = q_{j+1}\beta_j = Lq_j - \alpha_j q_j - \beta_{j-1}q_{j-1}$, this leads to Lanczos recursion—at each step, a new Lanczos vector is generated and an additional row and column is added to the tridiagonal matrix T_j. The eigenvalues of the growing matrix T_j converge towards the eigenvalues of L.

3.2 Accuracy Relaxation

Assume $\theta_1, \theta_2, \ldots, \theta_j$ are the eigenvalues of T_j, associated with eigenvectors $g_1, g_2, \ldots, g_j$. Define Ritz vectors $\{y_i\}$ as $y_i = Q_j g_i$, for $i = 1, 2, \ldots, j$. Accuracy of θ_i as approximations of L's eigenvalues is determined by norm of the *residuals*

$$s_i = Ly_i - \theta_i y_i, \quad i = 1, 2, \ldots, j. \tag{6}$$

Given an initial vector q_1, the Lanczos algorithm generates Lanczos vectors until $\beta_j = 0$. This is bound to happen for some $j < n$ in theory. However, in practice, when using finite arithmetic, the process will continue for ever producing more and more copies of the Ritz vectors once a small residual has occurred. and the orthogonality between q_i is lost. Hence, the Lanczos algorithm is usually supplemented by periodic re-orthogonalizations against those Lanczos vectors to which the orthogonality is lost [13].

With periodic pauses for re-orthogonalization, the Lanczos procedure for graph partitioning could be terminated safely by setting an appropriate *tolerance of residuals* $\|s_i\|$ and a loose *iteration number bound*. Selecting an appropriate residual tolerance is an ad hoc experience. In [10], Poth, Simon, and Liou showed that a tolerance of 10^{-3} is small enough to get the best Fiedler vector on all meshes they tested and a tolerance of 10^{-2} will fail in some cases. For generality, the Chaco library set the tolerance to 10^{-3}. In this paper, we show that the tolerance of 10^{-2} can yield the same high quality solutions for p-way partitioning in all test test cases when it is complemented by a local refinement operation.

In addition to setting a residual tolerance of 10^{-2}, we relax the accuracy of the Lanczos algorithm further by setting a tight iteration number bound. The iteration number bound guarantees the Lanczos algorithm terminates when using finite arithmetic. The Chaco library sets it to $2n$ and the MeTis [6] sets it to n. Both of them was shown to be safe to generate best Fiedler vectors. However, they are too conservative for a sub-optimal Fielder vector. In theory, it was observed that the Lanczos procedure *often* stop at certain iteration j as small as $2\sqrt{n}$ and that the ends of the spectrum usually converge faster than the interior [9, 8]. In light of these, the relaxed implementation sets the bound to $2\sqrt{n}$.

The remaining questions include whether this bound is robust for partitioning any kind of graphs and whether the bound together with a loose residual tolerance of 10^{-2} will generate high quality solutions for p-way partitioning.

4 Results

We tested the relaxed SB on eight meshes from the Boeing-Harwell and RIACS collections. The meshes arise in flow simulation, structural mechanics, and sparse matrix factorization and vary in size from thousands to hundred thousands of vertices. Their sizes and characteristics are shown in Table 1.

We compared the relaxed SB and spectral quadrisection (SQ) methods to the related methods in Chaco library in terms of edge cut and runtime. We define

Table 1. Characteristics of Test Meshes

Meshes	VertexNum	EdgeNum	Characteristics
3ELT	4720	13722	2D finite element mesh
BIG	15606	45878	2D Finite element mesh
BRACK2	62631	366559	3D finite element mesh
WAVE	156317	1059331	3D finite element mesh
BCSSTK31	35588	572914	3D Stiffness matrix
BCSSTK32	44609	985046	3D Stiffness matrix
NASA2146	2146	35052	2D Sparse matrix with 72250 nonzero entries
NASA4704	4704	50026	2D Sparse matrix with 104756 nonzero entries

quality loss as

$$QualityLoss = \frac{\text{EDGECUT}_{Relaxed} - \text{EDGECUT}_{Chaco}}{\text{EDGECUT}_{Chaco}} \times 100\%,$$

and *speedup* as

$$Speedup = \frac{\text{TIME}_{Chaco} - \text{TIME}_{Relaxed}}{\text{TIME}_{Chaco}} \times 100\%.$$

Negative quality loss means quality gain due to relaxed implementation.

Unless otherwise specified, the tolerance of residuals is set to 10^{-2} and the iteration number bound is set to $2\sqrt{n}$. All experiments were performed on an SUN Ultra Enterprise 4000 with 512 MBytes of memory and 233 MHz UltraSPARC processor. All times reported are in seconds.

Relaxed Spectral Bisection

Fig. 1 reports relative performances of the relaxed SB. Since the method is applicable to all test cases, it is conceivable that setting an iteration bound of $2\sqrt{n}$ in the Lanczos process would only sacrifice the accuracy of the Fiedler vector and would not destroy the orthogonality of the Lanczos vectors.

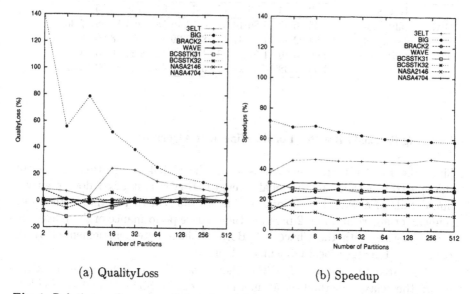

(a) QualityLoss   (b) Speedup

Fig. 1. Relative performance of the relaxed SB in comparison with the Chaco SB.

From Fig.1(a) and (b), it can be seen that the relaxed SB produces nearly the same quality solutions as the Chaco SB while reducing the execution time by 10 to 35 percentages on all meshes except BIG and 3ELT. The relaxed SB can generate even higher quality solutions in certain cases. because an accurate Fiedler vector may not necessarily lead to a good separator vector.

On the BIG and 3ELT meshes, the relaxed SB generates poorer quality solutions for small p due to big losses of the accuracy of their Fiedler vectors. However, their quality loss curves drop quickly as p increases. The curves agree with the results shown by Simon and Teng [12] that recursive optimal bisection may not necessary produce optimal partitions. Since the relaxed SB terminates the Lanczos process prematurely, it obtains an improvement of 40% to 70% speedups on the two meshes. Since the improvement is insensitive to p, the relaxed SB makes a better tradeoff between quality and runtime for large p-way partitioning.

The reason for the relaxed SB's inferior quality for small p on the BIG and 3ELT meshes is that the residual tolerance of 10^{-2} is too loose to be a criteria for approximating the Fiedler vector on the meshes. Setting the tolerance to 10^{-3} (the same as the Chaco SB) and keeping $2\sqrt{n}$ as the iteration bound yield results in Table 2. The table shows that relaxation by setting a tighter iteration bound alone can accelerate the recursive spectral bisection at a sacrifice of a few percentages of solution qualities on the two meshes.

Table 2. Relative performance of the relaxed SB on the BIG and 3ELT meshes, where the tolerance is set to 10^{-3} and the iteration number bound is set to $2\sqrt{n}$, in comparison with the Chaco SB.

Meshes		2	4	8	16	32	64	128	256	512
BIG	QualityLoss(%)	1.05	0.39	13.86	14.05	16.52	16.15	10.3	11.55	7.30
	Speedup(%)	22.54	22.55	31.71	47.45	28.98	29.37	29.01	28.89	29.30
3ELT	QualityLoss(%)	0.00	3.10	1.07	5.09	1.80	7.46	8.24	6.50	4.89
	Speedup(%)	3.58	16.38	17.68	24.75	25.55	29.60	29.39	33.02	33.42

Coupling Relaxed SB with a Local Refinement Algorithm

Since the Fiedler vector characteristics of the global information of a graph, the spectral method shows global strength. However, it often does poorly in fine details of a partition. In practice, it is usually used together with a local refinement algorithm. Fig. 2 presents the relative performance of the relaxed SB in comparison with the Chaco SB. Both of them were complemented by the Kernighan-Lin (KL) local refinement algorithm.

Compared to Fig. 1(a), Fig. 2(a) shows that the KL refinement algorithm favors the relaxed method on all meshes, except BCSSTK31 and BCSSTK32. Due to the KL refinement, the quality-loss curve of the BIG mesh drops quickly and approaches zero as p goes beyond 32. The relaxed SB/KL on the 3ELT yields superior quality solutions for small p. The claim that the KL algorithm improves more on the relaxed SB is also supported by the results shown in Fig. 2(b). Since the KL algorithm spends more time in refining the partitions generated by the relaxed SB, the relaxed SB/KL gains relatively smaller speedups in most cases in comparison to Fig. 1(b).

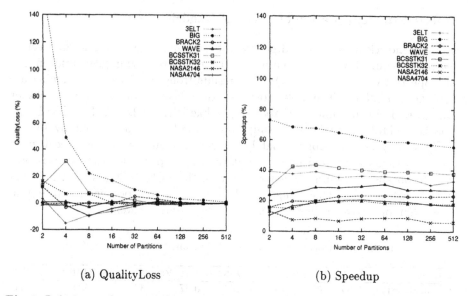

(a) QualityLoss   (b) Speedup

Fig. 2. Relative performance of relaxed SB, coupled with KL refinements, in comparison with the Chaco SB/KL for p-way partitioning.

On the BCSSTK31 and BCSSTK32 meshes, the Chaco SB/KL outperforms the relaxed SB/KL for small p. Recall from Fig. 1(b) that we knew that the relaxed SB without KL is superior to the Chaco SB. That is, the partitions resulted from the Chaco SB have more potentials to be refined by the subsequent KL algorithm. This is why the speedups of BCSSTK31 increase from 25% or so, as shown in Fig. 1(b), to about 40%.

Relaxed SB in Multilevel Algorithm

Compared with geometric partitioners, spectral methods generate premium quality solutions at a higher cost. The multilevel (ML) approach provides a good compromise between solution quality and partition time. It first coarsens a mesh down to a smaller one. Partitions of the small mesh resulted from the spectral method are then projected back to the original mesh. Table 3 presents the relative performance of the relaxed ML method for 256-way partitioning, by setting the cutoff for graph contraction to 100. The table shows that the relaxed ML accelerates the Chaco ML method significantly for small graphs, while producing the same quality partitions. The ML algorithm benefited few from the relaxed implementation for large graphs, like WAVE, because its computation becomes dominated by the coarsening step.

Table 3. Performance of the relaxed ML, relative to the Chaco ML.

Meshes	3ELT	BIG	BRACK2	WAVE	BCSSTK31	BCSSTK32	NASA2146	NASA4704
QualityLoss(%)	0.21	-0.20	0.00	0.07	-0.40	0.27	0.01	-1.62
Speedup(%)	21.92	16.09	4.59	0.53	0.46	7.18	8.66	9.21

Relaxed Implementation of Spectral Quadrisection

In [3], Hendrickson and Leland proposed to use more than one Lanczos vectors for partitioning at a time. We employed the idea of relaxed implementation into the Spectral Quadrisection (SQ). The results from a 256-way partitioning are tabulated in Table 4. The table shows that the relaxed implementation works for SQ, as well. For medium and large graphs like BRACK2, WAVE, BCSSTK31, and BCSSTK32, the relaxed setting reduces the SQ execution time by 16% to 25%, while producing approximately the same quality partitions. Recall from Fig. 1(a) that the relaxed implementation is appropriate for the Lanczos process to obtain an accurate Fiedler vector on the BCSSTK31, NASA2146, and NASA4704 meshes. Table 4 suggests that the setting is good enough to obtain more than one extreme eigenvectors on these meshes, as well.

Table 4. Performance of relaxed SQ, relative to the Chaco SQ.

Meshes	3ELT	BIG	BRACK2	WAVE	BCSSTK31	BCSSTK32	NASA2146	NASA4704
QualityLoss(%)	6.33	7.09	1.44	1.12	0.67	3.27	0.40	0.41
Speedup(%)	20.57	43.56	25.11	16.96	25.06	19.46	4.45	12.23

5 Summary

It is known that recursive bisections for graph partitioning using optimal strategies at each step may not lead to a good overall solution. That is, having a high quality bisection via an expensive spectral bisection may not be a good idea for p-way partitioning. In this paper, we have proposed an implementation of the spectral bisection with a relaxed accuracy requirement to accelerate the algorithm as well as attain high solution qualities for p-way partitioning.

The dominant computation in the spectral method is a Fiedler vector calculated by Lanczos algorithm. It is an iterative process, expanding the eigenvector spectrum of the graph's Laplacian matrix recursively. The relaxed SB uses an iteration bound of $2\sqrt{n}$ and a residual tolerance of 10^{-2} to control required accuracies of the Fiedler vector. We have evaluated the relaxed SB on eight representative meshes arising from different applications with different characteristics and compared to the Chaco SB in terms of runtime and cut edges. Experimental results show that the relaxed SB on six meshes yield almost the same quality solutions while achieving an improvement of up to 35% speedups. On the other two meshes, the relaxed SB approaches the Chaco SB in quality as p increases and reduces the execution time by 40We have also shown that the Kernighan-Lin local refinement algorithm favors the relaxed implementation. Coupled with the Kernighan-Lin local refinement algorithm, the relaxed SB is able to yield high quality solutions in all test cases as p goes beyond 32. Benefits from the relaxed implementation have also been demonstrated in Chaco multilevel and spectral quadrisection algorithms.

Acknowledgments

The authors thank Bruce Hendrickson and Robert Leland for providing the Chaco library, Ralf Diekmann and Robert Preis for providing some of the example meshes. Special thanks go to Jiwen Bao for her help in the collection of experimental data.

References

1. S. T. Barnard and H. D. Simon. Fast multilevel implementation of recursive spectral bisection for partitioning unstructured problems. *Concurrency: Practice and Experience*, 6(2):101–117, 1994.
2. R. Diekmann, B. Monien, and R. Preis. Load balancing strategies for distributed memory machines. Technical Report TR-RSFB-97-050, Department of Computer Science, University of Paderborn, 1997.
3. B. Hendrickson and R. Leland. An improved spectral graph partitioning algorithm for mapping parallel graphs. Technical Report SAND 92-1460, Sandia National Lab., USA, 1992.
4. B. Hendrickson and R. Leland. The Chaco user's guide. Technical Report SAND 93-2339, Sandia National Lab., USA, 1993.
5. B. Hendrickson and R. Leland. A multilevel algorithm for partitioning graphs. Technical Report Tech. Rep. SAND 93-11301, Sandia National Lab., USA, 1993.
6. G. Karypis and V. Kumar. A fast and high quality multilevel scheme for partitioning irregular graphs. Technical Report 95-035, Department of Computer Science, University of Minnesota, 1995.
7. J. De Keyser and D. Roose. Grid partitioning by inertial recursive bisection. Technical Report TW 174, Katholieke Universiteit leuven, Belgium, 1992.
8. B. N. Parlett. *The Symmetric Eigenvalue Problem*. Prentice-Hall, Englewood Cliffs, NJ., 1980.
9. B. N. Parlett and D. S. Scott. The Lanczos algorithm with selective orthogonalization. *Mathematics of Computation*, 33(145):217–238, 1979.
10. A. Pothen, H. D. Simon, and K. Liou. Partitioning sparse matrices with eigenvectors of graphs. *SIAM J. Matrix Analysis and Applications*, 11(3):430–452, July 1990.
11. H. D. Simon and A. Sohn. HARP: A fast spectral partitioner. In *Proc. of the 9th ACM Symp. on Parallel Algorithms and Architectues*, June 1997.
12. H. D. Simon and S.-H. Teng. How good is recursive bisection. Technical Report RNR-93-12, NASA Ames Report, 1993.
13. H. D. Siomn. Partitioning of unstructured problems for parallel processing. *Computing Systems in Engineering*, 2:135–148, 1991.
14. R. D. Williams. Unification of spectral and inertial bisection. Technical report, CalTech, 1994. Available at http://www.cacr.caltech.edu/ roy/papers.
15. C. Xu and F. Lau. *Load Balancing in Parallel Computers: Theory and Practice*. Kluwer Academic Publishers, November 1996.

S-HARP: A Parallel Dynamic Spectral Partitioner
(A short summary)

Andrew Sohn[1]
Computer Science Dept.
New Jersey Institute of Technology
Newark, NJ 07102; sohn@cis.njit.edu

Horst Simon[2]
NERSC
Lawrence Berkeley National Laboratory
Berkeley, CA 94720; simon@nersc.gov

Abstract. Computational science problems with adaptive meshes involve dynamic load balancing when implemented on parallel machines. This dynamic load balancing requires fast partitioning of computational meshes at run time. We present in this report a scalable parallel partitioner, called S-HARP. The underlying principles of S-HARP are the fast feature of inertial partitioning and the quality feature of spectral partitioning. S-HARP partitions a graph from scratch, requiring no partition information from previous iterations. Two types of parallelism have been exploited in S-HARP, fine-grain loop-level parallelism and coarse-grain recursive parallelism. The parallel partitioner has been implemented in Message Passing Interface on Cray T3E and IBM SP2 for portability. Experimental results indicate that S-HARP can partition a mesh of over 100,000 vertices into 256 partitions in 0.2 seconds on a 64-processor Cray T3E. S-HARP is much more scalable than other dynamic partitioners, giving over 15-fold speedup on 64 processors while ParaMeTiS1.0 gives a few-fold speedup. Experimental results demonstrate that S-HARP is three to 10 times faster than the dynamic partitioners ParaMeTiS and Jostle on six computational meshes of size over 100,000 vertices.

1 Introduction

Computational science problems are often modeled using adaptive meshes. The meshes are partitioned at compile time and distributed across processors for parallel computation. These meshes change at runtime to accurately reflect the computational behavior of the underlying problems. When these large number of processors are used, load balancing, however, becomes a critical issue. Some processors will have a lot of work to do while others may have little [14]. It is imperative that the computational loads be balanced at runtime to improve the efficiency and throughput. Runtime load balancing of computational science problems typically involves graph partitioning. Fast runtime mesh partitioning is therefore critical to the success of computational science problems on large-scale distributed-memory multiprocessors.

Runtime graph partitioning needs to satisfy several criteria, including execution time, imbalance factor, and edge cuts. Among the dynamic partitioning methods are Jostle [15], ParaMeTiS [9], and HARP [11]. These methods can be classified into two categories based on the use of initial partitions or runtime partitions between successive iterations. The first group simply partitions from scratch, requiring no runtime or initial partitions. HARP belongs to the first group. The second group relies on the initial or runtime partitions. Jostle and ParaMeTiS belong to this second group.

---

1. This work is supported in part by the NASA JOVE Program, by travel support from USRA RIACS, and by summer support from MRJ, NASA Ames Research Center.
2. This work was supported by the Director, Office of Computational Sciences of the U.S. Department of Energy under contract number DE-AC03-76SF00098.

Jostle and ParaMeTiS employ a multilevel method which was introduced by Barnard and Simon in MRSB [1] and by Hendrickson and Leland in Chaco [5]. The idea behind the multilevel method is to successively shrink (coarsen) the original mesh in such a way that the information of the original mesh is preserved in the course of shrinking. When it is sufficiently small, the original graph will be partitioned into a desired number of sets. The partitioned sets are then uncoarsened to bring the partitions to the original graph. Jostle introduces a diffusion scheme to give a global view of the original mesh. At the coarsest mesh, vertices are moved (diffused) to neighboring partitions to balance the coarse graph. ParaMeTiS performs similar multilevel operations as Jostle.

The traditional argument against partitioning from scratch is that it is computationally expensive and involves an unnecessarily large amount of vertex movement at runtime. However, this argument has been investigated in the JOVE global load balancing framework [12] and the fast spectral partitioner HARP [11]. The sequential partitioner HARP partitioned from scratch a graph of over 100,000 vertices into 256 sets in less than four seconds on a single processor SP2. This fast execution time is three to five times faster than the multilevel partitioner MeTiS [6]. The idea behind HARP is spectral partitioning [4] and inertial partitioning [7]. Spectral partitioning first introduced by Hall in 1970 [4] was later popularized by Simon [10].

This report introduces S-HARP, a scalable parallel version of HARP. S-HARP can run on any parallel machine which supports Message Passing Interface. It can partition a graph into any number of partitions using any number of processors. There is no restriction such as one partition per processor. No local optimization is employed in the partitioner. S-HARP requires no initialization, except to find the eigenvectors of the graph under consideration for runtime partitioning. The imbalance factor of S-HARP is guaranteed to be at most one vertex each time a graph is divided into two sets, provided all vertices are equally weighted.

2 The Ideas of S-HARP

2.1 The spectral basis

S-HARP combines Laplacian-based spectral method and coordinate-based inertial method. The basic idea of S-HARP was reported in 1994 as a part of the dynamic load balancing framework JOVE [12]. A similar method was reported in 1995 [2]. The spectral method constructs a d-dimensional eigen space to embed a graph of n dimensions, where d is much smaller than n. Inertial method finds a $d-1$ dimensional hyperplane to separate the embedded graph into half. Laplacian-based method was first introduced by Hall in 1970 [4]. The method precisely defines the relationship between the connectivity of vertices and the eigenvectors of a Laplacian matrix. The Laplacian matrix $\mathbf{L}$ is defined where the diagonal entry d_{ii} is the connectivity of vertex v_i while other entry d_{ij} is -1 if there is a connection between two vertices v_i and v_j; $d_{ij} = 0$ otherwise. Eigenvectors of the Laplacian matrix $\mathbf{L}$ help transform a graph represented in a cartesian coordinate into a graph represented in an eigen space spanned by eigenvectors.

For a graph $G = (v_0, v_1, ..., v_{n-1})$ of n vertices, S-HARP constructs a d-dimensional eigen space spanned by d eigenvectors $E_0, ..., E_{d-1}$. The dimension of eigen space d is much smaller than the number of vertices n, typically a few to tens for practical purposes. An eigenvector E_i, $i = 0 .. d-1$, has n entries $E_{i,0},...,E_{i,n-1}$, where $E_{i,j}$ corresponds to vertex v_j, for $j = 0 ... n-1$. Each eigenvector of dimension n contains information of all the n vertices in the original graph but to a limited extent. Together with other $d-1$ remaining eigenvectors, a vertex can be located in the d-dimensional eigen space. To be more precise, the location of a vertex v_j can be accurately defined by d eigen entries $E_{0,j},...,E_{d-1,j}$ each of which is derived from a different eigen vector. The location of a vertex can be more accurately defined as d increases.

The traditional argument against spectral partitioning is that there exists a class of graphs for which spectral partitioners are not suitable. Figure 1 shows a "roach" mesh with 16 vertices used by Guattery and Miller in [3]. The argument is that spectral partitioners will partition the graph vertically, resulting in four edge cuts while it can do so horizontally, resulting in the minimal of four edge cuts. We find that the argument presents a serious technical problem. Partitioning with the dotted line will indeed gives the number of edge cuts. However, it does so with *three* components! The top partition will have two *disconnected* components while the bottom partition will have a single connected component of eight vertices. This type of partitioning is highly undesirable and can defeat the purpose of partitioning because the neighboring vertices are not placed into the same partition. The disconnected components can cause a serious communication problem when they are mapped to a multiprocessor. We continue to use the roach mesh to illustrate the internal working of S-HARP.

Figure 1(b) shows the Laplacian matrix **D** for the mesh shown in Figure 1(a). Figure 1(c) shows the first three eigen vectors of the matrix **D**. The first eigenvector E_0 which is an orthonormal vector is discarded since it has no directional information. The second and third vectors, E_1 and E_2, have directional information and are used for partitioning. The plot shown in Figure 1(d) is exactly based on these two eigenvectors. The 16 vertices of the roach mesh are plotted in the two-dimensional space spanned by the two eigenvectors. The x-axis shows the second eigenvector E_1 while the y-axis shows the third eigenvector E_2.

Given the eigen plot shown in Figure 1(d), partitioning the roach graph is straightforward. The simplest way is to partition along the first eigenvector E_1 (x-axis); The eight vertices, 0..7, of left half will be partition 0 while the eight vertices, 8..15, of right half will be partition 1. The vertical axis passing through $x=0$ and orthogonal to the x-axis is all it needs to partition the graph. The resulting partition gives the minimal number of edge cuts while all the vertices in each partition are *connected*. The locations of vertices 7 and 8 are not very clear because of the plotting precision. They are numerically different. Should there be ties after the graph is partitioned into half using E_1, spectral partitioners will spend little time resolving such cases. The second eigenvector E_2 (y-axis) will be used to break the tie(s). Should there still be ties after using E_1 and E_2, the fourth eigenvector E_3 will be used as a tie-breaker. The example graph has no ties after the first partition using E_1 and hence, it is not necessary to use E_2 to break ties.

2.2 The inertial basis

For small meshes like the one shown in the previous section, it is clear that a single eigenvector can clearly partition a graph into two connected components with the minimal edge cuts. For meshes with many ties in the eigen space, a few additional eigenvectors are necessary to break ties that often result from numerical precision. For large complex graphs, however, it is often not obvious if partitioning using one to several eigenvectors is effective. This is where the inertial feature of S-HARP is used to further clarify the partitioning process.

(a) The Roach mesh [3]   (b) The Laplacian matrix for the roach mesh

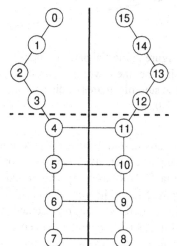

	0	1	2	3	4	5	6	7	8	9	10	11	12	13	14	15
0	1	-1	0	0	0	0	0	0	0	0	0	0	0	0	0	0
1	-1	2	-1	0	0	0	0	0	0	0	0	0	0	0	0	0
2	0	-1	2	-1	0	0	0	0	0	0	0	0	0	0	0	0
3	0	0	-1	2	-1	0	0	0	0	0	0	0	0	0	0	0
4	0	0	0	-1	3	-1	0	0	0	0	0	-1	0	0	0	0
5	0	0	0	0	-1	3	-1	0	0	0	-1	0	0	0	0	0
6	0	0	0	0	0	-1	3	-1	0	-1	0	0	0	0	0	0
7	0	0	0	0	0	0	-1	2	-1	0	0	0	0	0	0	0
8	0	0	0	0	0	0	0	-1	2	-1	0	0	0	0	0	0
9	0	0	0	0	0	0	-1	0	-1	3	-1	0	0	0	0	0
10	0	0	0	0	0	-1	0	0	0	-1	3	-1	0	0	0	0
11	0	0	0	0	-1	0	0	0	0	0	-1	3	-1	0	0	0
12	0	0	0	0	0	0	0	0	0	0	0	-1	2	-1	0	0
13	0	0	0	0	0	0	0	0	0	0	0	0	-1	2	-1	0
14	0	0	0	0	0	0	0	0	0	0	0	0	0	-1	2	-1
15	0	0	0	0	0	0	0	0	0	0	0	0	0	0	-1	1

(c) The first three eigenvectors E0, E1, E2 and 16 eigenvalues

	0	1	2	3	4	5	6	7	8	9	10	11	12	13	14	15
E0	0.25	0.25	0.25	0.25	0.25	0.25	0.25	0.25	0.25	0.25	0.25	0.25	0.25	0.25	0.25	0.25
E1	0.45	0.40	0.31	0.19	0.05	0.02	.004	.001	-.001	-.004	-0.02	-0.05	-0.19	-0.31	-0.40	-0.45
E2	0.35	0.29	0.20	0.07	-0.07	-0.20	-0.29	-0.35	-0.35	-0.29	-0.20	-0.07	0.07	0.20	0.29	0.35

Eigenvalues

0.00	0.10	0.15	0.59	0.87	1.24	2.00	2.00	2.24	2.77	3.10	3.41	3.54	3.80	4.56	5.60

(d) The roach mesh in the eigen space spanned by E1 and E2

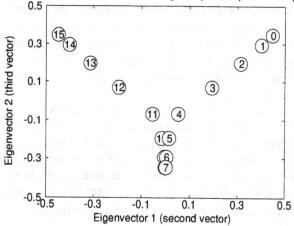

Figure 1: The spectral basis of S-HARP. S-HARP separates the roach graph into two connected components using the *vertical* line. The resulting partitions give the minimal edge cuts while keeping all the vertices in each partition connected.

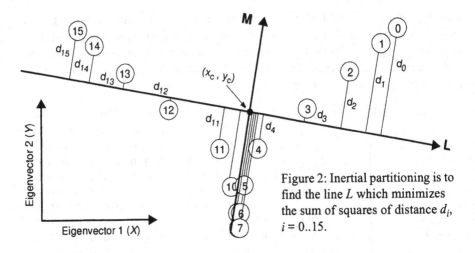

Figure 2: Inertial partitioning is to find the line L which minimizes the sum of squares of distance d_i, $i = 0..15$.

Inertial partitioning finds the center of inertia of vertices and decides the partition of each vertex according to the inertial center. Inertial partitioners use *geometric* coordinate information of a mesh to find the center of inertia. S-HARP uses *eigen* coordinate information. We continue to use the roach mesh. Figure 2 shows how the center of inertia of the 16 vertices helps partition the vertices. Given the 16 vertices represented in the 2-d space spanned by E_1 and E_2, S-HARP finds a line L which separates the vertices into half (for three vectors, S-HARP finds a plane to separate the vertices). The line L is computed in a way that the sum of squares of distance d_i, $\Sigma\, d_i^2$, $i=0..15$, is minimized. Once the line is located, the rest is straightforward. The line M passing through the center of inertia and orthogonal to L separates the vertices into half. Those vertices on the right hand side are projected to partition 0 while those on the left hand side to partition 1.

When the vertices are uniformly weighted, the partitioning is as simple as above. However, S-HARP assumes weighted vertices, i.e., vertices can have different computational intensity due to mesh adaption. This adds an extra step of *sorting* to the above procedure. Finding the lines L and M is the same as above. It can be shown that finding L is equivalent to finding the first eigenvector of a 2x2 inertial matrix R, derived from $\Sigma\, d_i^2$, $i=0..15$ [13]. However, projection will be different since weighted vertices will be projected to L according to their weight. A simple splitting is not possible as this weighted projection has changed the center of inertia which was originally designed for a uniform weight. The projected vertices are, therefore, sorted to find the new median. The last step divides the sorted vertices into half based on the new median. Or, the vertices can be divided into any number of partitions.

Figure 3 shows how S-HARP partitions the roach graph. The inertial line L is exactly the same as the x-axis of Figure 1 while the orthogonal line M is exactly the same as the y-axis of Figure 1. The figure shows the vertices projected to L. Sorting is not necessary since all the vertices have the same weight. The line M passing through the center of inertia splits the vertices into half. The right hand side, labeled partition 0, has 8 vertices, $v_0..v_7$ while the left hand side, labeled partition 1, has 8 vertices, $v_8..v_{15}$. The partition gives 4 edge cuts, which is the optimal number for connected partitions.

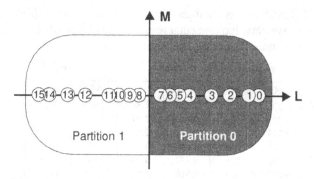

Figure 3: Inertial partitioning: finding the two lines L and M, projection of vertices to L, sorting on the projected values, and splitting with respect to M. Note that projected vertices are scaled for readability.

2.3 Summary of S-HARP in the context of dynamic partitioner

The internal working of S-HARP for partitioning *weighted* graphs is summarized below:

0. Find k eigenvectors of the Laplacian matrix of a graph. For practical purposes, $k = 10$ eigenvectors are typically used [13,11]. For the roach mesh where $k = 2$, a separator of 1-dimensional line was used.
1. To prepare for finding a $(k-1)$-dimensional hyperplane L, compute the inertial distance of the vertices in the eigen space spanned by k eigenvectors.
2. Find a hyperplane L. This is equivalent to finding the first eigenvector of a k x k inertial matrix **R** derived from the minimization of $\Sigma \, d_i^2$, $i=0..V-1$, where d_i is an inertial distance from vertex v_i to L and V is the number of vertices.
3. Project individual weighted vertices to the principal inertial direction L.
4. Sort the vertices according to the projected values.
5. Split the sorted vertices into half by using the orthogonal line M that passes through the median point of the principal inertial direction L.

S-HARP assumes the topology of the underlying mesh remains the same but the computational intensity of each vertex changes at runtime. When a mesh is refined, some vertices will split into several vertices. This refinement can be modeled by adjusting the computational weight of the vertex which is to be refined. The adjustment of computational intensity leads to an important observation that the original eigenvectors which retain the topology of the original mesh remain the same. Since the original mesh remains the same but the vertex weights change at runtime, it is not necessary to recompute the eigenvectors at runtime. Therefore, step 0 is performed only once for the initial mesh. The parallel partitioner will be concerned only for the remaining five steps.

3 Experimental Results and Discussion

S-HARP has been implemented in Message Passing Interface (MPI) on Cray T3E and IBM SP-2. It exploits parallelism both in recursive and function-level. Three of the five steps of S-HARP have been parallelized, which are inertial computation, projection, and sort. The reason the three steps are chosen for parallelization is because they occupy almost 95% of the total computation time. The remaining two steps account for approximately 5% of the total time. The details of S-HARP are presented in [13].

Six test meshes are used in this study: Spiral = (1200, 3191), Labarre = (7959, 22936), Strut = (14504, 57387), Barth5 = (30269, 44929), Hsctl = (31736, 142776), Ford2 = (100196, 222246). Table 1 lists the execution times of HARP1.0 and MeTiS2.0

on a single-processor SP2. MeTiS2.0 results are drawn from the early report [11]. All the HARP1 results used 10 eigenvectors. Experimental results show that HARP1 performs approximately three to five times faster than MeTiS2.0.

# of sets S	STRUT		BARTH5		HSCTL		FORD2	
	HARP1	MeTiS2	HARP1	MeTiS2	HARP1	MeTiS2	HARP1	MeTiS2
8	0.146	0.65	0.305	0.88	0.320	1.84	1.023	3.59
16	0.196	0.92	0.408	1.21	0.426	2.24	1.360	4.78
32	0.249	1.22	0.513	1.59	0.533	2.93	1.698	5.92
64	0.306	1.65	0.623	2.08	0.646	3.76	2.040	7.50
128	0.376	2.17	0.745	2.70	0.772	4.90	2.391	9.23
256	0.465	2.87	0.887	3.29	0.916	5.97	2.761	11.35

Table 1: Comparison of the execution times (seconds) on a single-processor SP2.

Table 2 shows the execution times for Ford2 on T3E. S-HARP can partition a mesh of over 100,000 vertices in 0.2 seconds on a 64-processor T3E. Note the diagonal timing entries. The execution times decrease gradually from the upper left corner to the lower right corner, with the exception of $P = S = 2$. This is a unique feature of S-HARP. These entries will be used to compare with other dynamic partitioners.

P	$S=2$	$S=4$	$S=8$	$S=16$	$S=32$	$S=64$	$S=128$	$S=256$
1	0.358	0.827	1.290	1.740	2.183	2.526	2.833	3.117
2	**0.230**	0.466	0.699	0.927	1.143	1.318	1.472	1.617
4	0.154	**0.290**	0.406	0.520	0.629	0.716	0.792	0.866
8	0.123	0.214	**0.283**	0.340	0.395	0.438	0.476	0.513
16	0.104	0.168	0.212	**0.248**	0.275	0.296	0.313	0.330
32	0.090	0.140	0.168	0.189	**0.201**	0.215	0.223	0.230
64	0.088	0.137	0.159	0.180	0.188	**0.194**	0.198	0.202

Table 2: Partitioning times of Ford2 on T3E. P = # of processors, S = # of partitions

Figure 4 summarizes the scalability of S-HARP. T3E has given over 15-fold speedup on 64 processors. The speedup of 15 may be considered modest considering the parallelization efforts expended. However, graph partitioning is known to be difficult to parallelize especially when the parallel partitioners perform very fast, partitioning 100,000 vertices in 0.2 seconds. In fact, partitioners such as ParaMeTiS1.0 give very modest speedup. For the Spiral mesh, ParaMeTiS1.0 gives the speedup of 0.133/0.099 = **1.3** on 64 processors [8]. For the Hsctl mesh with V=31736, E=142776, ParaMeTiS1.0 gives the speedup of 1.812/0.676 = **2.7**. For the mesh Ford2, ParaMeTiS1.0 performs worse than a single processor, giving the speedup of 0.742/ 0.919 = **0.81**. Comparing the speedup results with ParaMeTiS1.0, S-HARP simply outperforms as it gives over 15-fold speedup. Results for the Jostle dynamic partitioner for the test meshes are not available and could not be compared in this report.

While S-HARP is much more scalable than other dynamic partitioners such as ParaMeTiS1.0, its 16-fold speedup on 64 processors can be still improved. The main reason for the modest speedup is due to the fact that the current version of S-HARP is not crafted for performance. The major source of improvement is the sorting step. The current version takes approximately half the time in sorting. Several solutions are currently being undertaken to reduce the sorting time and the overall partitioning time.

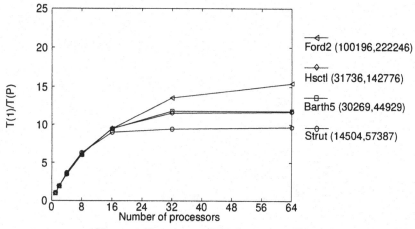

Figure 4: Speedup on T3E for 256 partitions.

4 Comparison of S-HARP with Other Dynamic Partitioners

The performance of S-HARP is compared against two other dynamic partitioners, ParaMeTiS1.0 [9] and Jostle [15]. Jostle comes with two different versions: a multilevel version Jostle-MD and Jostle-D. Comparisons are made in terms of execution time and edge cuts. The execution results of ParaMeTiS and Jostle are provided by L. Oliker of RIACS at NASA Ames [8]. Figure 5 plots the results of S-HARP, ParaMeTiS, Jostle-MD, and Jostle-D. The x-axis indicates the number of processors and the number of partitions. For example, 64 refers that 64 processors partition the mesh into 64 sets. This restriction was due to MeTiS and Jostle since they require the number of partitions to be the same as the number of processors, i.e., one partition per processor. S-HARP does not impose such a restriction. It can partition any number of partitions on any number of processors, as listed in Table 2.

The performance of S-HARP in terms of edge cuts is comparable to the other three. Unlike the other partitioners, S-HARP produces exactly the same solution quality, regardless of the number of processors. The performance of S-HARP in terms of edge cuts is presented in detail in [13]. S-HARP does not perform local optimization as done in ParaMeTiS and Jostle. If local optimization is employed at the expense of slightly increased computation time, the partition quality will improve to the level of MeTiS and Jostle-MD. The reason we stress "slightly" is because S-HARP will not randomly optimize the partitions. S-HARP sorts vertices in terms of their projected locations on the dominant inertial direction in the eigen space spanned by the vertices. Local optimization will be performed where the vertices are divided into two sets since errors come mostly from numerical precision. Hence, local optimization in the area where there are a lot of ties due to numerical precision will likely improve the partition quality. Partition quality can also be improved by using more eigenvectors. The number of eigenvectors used in this report is 10. If the number of eigenvectors is increased to 20, the performance will improve as presented in [11] at the expense of increased execution time. However, we have not performed local optimization since fast execution is more important in dynamically changing computations than the highest possible partition quality.

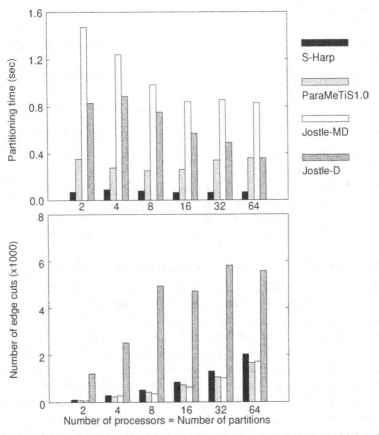

Figure 5: Comparison of dynamic partitioners for Barth 5 (30269,44929) on SP-2.

It is clear from Figure 5 that S-HARP outperformed the other two dynamic partitioners in terms of execution time. For all the test meshes, S-HARP performs three to 10 times faster than ParaMeTiS and Jostle. The performance of S-HARP is consistent throughout different sizes of meshes. The performance of S-HARP is better when the number of processors is large. For 64 processors, S-HARP is five times faster than ParaMeTiS and 10 times faster than Jostle. This clearly suggests that when a large number of processors is used, S-HARP can be the choice among parallel dynamic partitioners. In fact, this can be easily observed by scanning the **bold** diagonal entries of T3E from Table 3 which are listed below again:

- T3E: 0.230, 0.290, 0.283, 0.248, 0.201, 0.194
- SP2: 0.244, 0.302, 0.301, 0.292, 0.275, 0.287

We note that the execution times gradually decrease from 0.290 sec for $P=S=4$ to 0.194 sec for $P=S=64$. This is a unique feature which only S-HARP possesses among the three dynamic partitioners. The fast nature of S-HARP together with the comparable edge-cuts is certainly a viable tool for runtime partitioning of adaptive meshes on large-scale distributed-memory multiprocessors.

Acknowledgments

Andrew Sohn thanks Joseph Oliger of RIACS at NASA Ames for providing travel support and Doug Sakal of MRJ Technology Solutions at NASA Ames for providing summer support. The authors thank Leonid Oliker of RIACS at NASA Ames for providing some of the ParaMeTiS and Jostle data. S-HARP will be made available in the near future at http://www.cs.njit.edu/sohn/harp.

References

1. S. T. Barnard and H. D. Simon, Fast multilevel implementation of recursive spectral bisection for partitioning unstructured problems, *Concurrency: Practice and Experience* 6, 1994, pp.101-117.

2. T. Chan, J. Gilbert, and S. Teng. Geometric spectral partitioning. Xerox PARC Technical Report, January 1995.

3. S. Guattery and G. L. Miller, On the performance of the spectral graph partitioning methods, in *Proc. Sixth Annual ACM-SIAM Symposium on Discrete Algorithms*, 1995, pp.233-242.

4. K. Hall, An *r*-dimensional quadratic placement algorithm, *Management Science 17*, November 1970, pp.219-229.

5. B. Hendrickson and R. Leland, A Multilevel Algorithm for Partitioning Graphs, in *Proc. Supercomputing '95*.

6. G. Karypis and V. Kumar, A fast and high quality multilevel scheme for partitioning irregular graphs, Tech. Report 95-035, University of Minnesota, 1995.

7. B. Nour-Omid, A. Raefsky and G. Lyzenga, Solving Finite Element Equations on Concurrent Computers, in Parallel Computations and their Impact on Mechanics, Ed. A.K.Noor, ASME, New York, 1986, p.209.

8. L. Oliker, Personal communication on the results of ParaMeTiS 1.0 and Jostle on SP2 and T3E, July 30, 1997.

9. K. Schloegel, G. Karypis, and V. Kumar, Parallel Multilevel Diffusion Schemes for Repartitioning of Adaptive Meshes, Tech. Report, Univ. of Minnesota, 1997.

10. H. D. Simon, Partitioning of unstructured problems for parallel processing, *Computing Systems in Engineering*, Vol. 2, 1991, pp. 135-148.

11. H. D. Simon, A. Sohn, and R. Biswas, HARP: A dynamic spectral partitioner, *Journal of Parallel and Distributed Computing 50*, April 1998, pp.88-103.

12. A. Sohn and H. D. Simon, JOVE: A dynamic load balancing framework for adaptive computations on distributed-memory multiprocessors, Technical Report, NJIT CIS 94-60, September 1994. (Also in *Proc. of the ACM Symposium on Parallel Algorithms and Architectures*, June 1996 and *IEEE SPDP*, October 1996.)

13. A. Sohn and H. D. Simon, S-HARP: A parallel dynamic spectral partitioner, Technical Report, NJIT CIS 97-20, September 1997.

14. A. Sohn and R. Biswas, Special Issue on Dynamic Load Balancing, *Journal of Parallel and Distributed Computing 47*, December 1997, pp.99-101.

15. C. Walshaw, M. Cross, and M. Everett. Dynamic mesh partitioning: a unified optimization and load-balancing algorithm. Tech. Rep. 95/IM/06, University of Greenwich, London SE18 6PF, UK, 1995.

Information Filtering Using the Riemannian SVD (R-SVD)

Eric P. Jiang and Michael W. Berry

Department of Computer Science, 107 Ayres Hall, University of Tennessee,
Knoxville, TN, 37996-1301, USA,
jiang@cs.utk.edu, berry@cs.utk.edu

Abstract. The Riemannian SVD (or R-SVD) is a recent nonlinear generalization of the SVD which has been used for specific applications in systems and control. This decomposition can be modified and used to formulate a filtering-based implementation of Latent Semantic Indexing (LSI) for conceptual information retrieval. With LSI, the underlying semantic structure of a collection is represented in k-dimensional space using a rank-k approximation to the corresponding (sparse) term-by-document matrix. Updating LSI models based on user feedback can be accomplished using constraints modeled by the R-SVD of a low-rank approximation to the original term-by-document matrix.

1 Introduction

Traditionally, information is retrieved by literally matching terms in a user's query with terms extracted from documents within a collection. Lexical-based information retrieval techniques, however, can be incomplete and inaccurate due to the variability in word usage. On one side, people typically use different words to describe the same concept (*synonymy*) so that the literal terms in a user's query may be inadequate to retrieve relevant documents. On the other side, many words can have multiple meanings (*polysemy*) so that searching based on terms in a user's query could retrieve documents out of context.

Latent Semantic Indexing (LSI) attempts to circumvent the problems of lexical matching approaches by using statistically derived conceptual indices rather than individual literal terms for retrieval [1, 2]. LSI assumes that an underlying semantic structure of word usage exists in a document collection and uses this to estimate the semantic content of documents in the collection. This estimation is accomplished through a rank-reduced term-by-document space via the singular value decomposition (SVD). Using LSI, a search is based on semantic content rather than the literal terms of documents so that relevant documents which do not share any common terms with the query are retrieved. As a vector space IR model, LSI has achieved significant performance improvements over common lexical searching models [3]. However, the ability to constrain the representation of terms and documents according to users' preferences or knowledge has not been incorporated into previous LSI implementations. Such constraints may define nonlinear perturbations to the existing rank-reduced model.

Recently, an interesting nonlinear generalization of the singular value decomposition (SVD), referred to as the *Riemannian* SVD (R-SVD), has been proposed by De Moor [4] for solving *structured total least squares* problems. Updating LSI models based on user feedback can be accomplished using constraints modeled by the R-SVD of a low-rank approximation to the original term-by-document matrix. In this paper, we present (briefly) the formulation, implementation and performance analysis of a new LSI model (RSVD-LSI) which is equipped with an effective information filtering mechanism.

The remaining sections of this paper are organized as follows: Section 2 provides an overview of the conventional SVD and the Riemannian SVD. In Section 3, the formulation of the LSI-motivated R-SVD for a low-rank matrix is described, and a new iterative algorithm for solving these R-SVD problems is proposed. The RSVD-LSI model for information retrieval and filtering is discussed in Section 4.

2 The Riemannian Singular Value Decomposition

The singular value decomposition (SVD) [5] is a well-known theoretical and numerical tool. It defines a diagonalization of a matrix based on an orthogonal equivalence transformation. Specifically, given a rectangular m by $n\,(m \gg n)$ matrix A of rank r, the SVD of A can be defined as

$$A = U\Sigma V^T, \tag{1}$$

where $U = (u_1, \ldots, u_m)$ and $V = (v_1, \ldots, v_n)$ are unitary matrices and $\Sigma = diag(\sigma_1, \sigma_2, \ldots, \sigma_n), \sigma_1 \geq \sigma_2 \geq \cdots \geq \sigma_r > \sigma_{r+1} = \cdots = \sigma_n = 0$. The σ_i's are called the singular values of A, and the u_i's and v_i's are, respectively, the left and right singular vectors associated with $\sigma_i, i = 1, \ldots, r$, and the ith singular triplet of A is defined as $\{u_i, \sigma_i, v_i\}$. The matrix A can also be expressed as a sum of r rank-one matrices:

$$A = \sum_{i=1}^{r} \sigma_i u_i v_i^T. \tag{2}$$

Further, for any unitarily invariant norm $\|.\|$, the Eckart-Young-Mirsky formula [9]

$$\min_{rank(B) \leq k} \|A - B\| = \|A - A_k\|, \text{ where } A_k = \sum_{i=1}^{k} \sigma_i u_i v_i^T, \tag{3}$$

provides an elegant way to produce the best (in a least squares sense) rank-reduced approximation to A.

The SVD is an important computational tool used in numerous applications ranging from signal processing, control theory to pattern recognition and time-series analysis. The total least squares (TLS) method [6] is a useful data fitting technique for solving an over-determined system of linear equations $Ax \approx b$, and the most common TLS algorithm is based on the SVD of augmented matrices such as $[A \mid b]$. In fact, the TLS problem can be reduced to an equivalent minimization problem of finding a rank-deficient matrix approximation B in the Frobenius norm to A, i.e.,

$$\min_{B \in \mathcal{R}^{m \times n}, y \in \mathcal{R}^n} \|A - B\|_F, \quad \text{subject to } B\,y = 0, \ y^T y = 1. \tag{4}$$

By using the method of Lagrange multipliers it can be shown that the solution of (Eq. 4) satisfies the set of equations

$$Ay = x\tau, \quad x^T x = 1, \tag{5}$$
$$A^T x = y\tau, \quad y^T y = 1,$$

with $\|A - B\|_F = \tau$. Hence, (x, τ, y) is the singular triplet of A corresponding to its smallest singular value.

As an extension of the TLS problem to incorporate the algebraic pattern of errors in the matrix A, *structured* total least squares (STLS) problems have been formulated. Motivated by the derivation from (Eq. 4) to (Eq. 5) and applications in systems and control, De Moor [4] has proposed an interesting nonlinear generalization of SVD, referred to as the *Riemannian* SVD (R-SVD), for solving STLS problems. The STLS has numerous applications in signal processing, system identification and control system design and analysis.

2.1 Riemannian SVD for a Full-Rank Matrix

Let $B(b) = B_0 + b_1 B_1 + \cdots + b_q B_q$ be an affine matrix function of b_i and B_i, where $b = (b_1, b_2, \ldots, b_q)^T \in \mathcal{R}^q$ is a parameter vector and $B_i \in R^{m \times n}$, $i = 0, \ldots, q$, are fixed basis matrices. Further, let $a = (a_1, a_2, \ldots, a_q)^T \in \mathcal{R}^q$ be a data vector, and $w = (w_1, w_2, \ldots, w_q)^T \in \mathcal{R}^q$ be a given weighting vector, then the *structured total least squares* (STLS) problem, that often occurs in systems and control applications, can be formulated as

$$\min_{b \in \mathcal{R}^q, y \in \mathcal{R}^n} [a, b, w]_2^2 \quad \text{subject to } B(b)\,y = 0, \ y^T y = 1. \tag{6}$$

Consider the STLS problem (Eq. 6) with the quadratic criterion $[a, b, w]_2^2 = \sum_{i=1}^{q} (a_i - b_i)^2$, where, for the sake of simplicity, the weights w are ignored (the general case with weights can be treated very similarly). By using the method of Lagrange multipliers, this minimization problem (Eq. 6) can be formulated as an equivalent Riemannian SVD problem for the matrix $A = B_0 + \sum_{i=1}^{q} a_i B_i \in \mathcal{R}^{m \times n}$.

This formulation is addressed in the following theorem[1] from [4].

---

[1] See [7] or [4] for details on the proof of this theorem.

Theorem 1 (STLS to R-SVD). *Consider the STLS problem*

$$\min_{b \in \mathcal{R}^q, y \in \mathcal{R}^n} \sum_{i=1}^{q} (a_i - b_i)^2 \quad subject\ to\ B(b)\,y = 0,\ y^T y = 1, \tag{7}$$

where a_i, $i = 1, \ldots, q$, are the components of the data vector $a \in R^q$, and $B(b) = B_0 + b_1 B_1 + \cdots + b_q B_q$, with $B_i \in R^{m \times n}$, $i = 0, \ldots, q$, a given set of fixed basis matrices. Then the solution to (Eq. 7) is given as follows:

(a) Find the triplet (u, τ, v) corresponding to the minimal τ that satisfies

$$\begin{aligned} Av &= D_v u \tau, & u^T D_v u &= 1, \\ A^T u &= D_u v \tau, & v^T D_u v &= 1, \end{aligned} \tag{8}$$

where $A = B_0 + \displaystyle\sum_{i=1}^{q} a_i B_i$. Here D_u and D_v are defined by

$$\sum_{i=1}^{q} B_i^T (u^T B_i v)\, u = D_u v \quad and \quad \sum_{i=1}^{q} B_i (u^T B_i v)\, v = D_v u,$$

respectively. D_u and D_v are symmetric positive or nonnegative definite matrices [2], and the elements of D_u and D_v are quadratic in the components of the left and right singular vectors u and v, respectively, of the matrix A corresponding to the singular value τ.

(b) The vector y is computed by $y = v/\|v\|$.

(c) The components of b, b_k, are obtained by

$$b_k = a_k - (u^T B_k v)\tau, \quad k = 1, \ldots, q.$$

In order to understand the R-SVD as posed in the above theorem, a few comments are needed. First, Part (c) of the theorem implies that $B = A - f(u, \tau, v)$, where f is a multilinear function of the singular triplet (u, τ, v). Second, it is interesting to notice the similarity between (Eqs. 4 and 5) and (Eqs. 7 and 8). In fact, when D_u and D_v are identity matrices, the Riemannian SVD of the matrix A will reduce to the traditional SVD of A. Lastly, it has been noted that as a nonlinear extension of SVD the R-SVD may not have a complete decomposition with $\min(m, n)$ unique singular triplets, nor the dyadic decomposition as defined in (Eq. 2). Also, the solution to (Eq. 7) may not be unique [4].

2.2 Inverse Iteration

It is known that if both D_u and D_v are constant matrices that are independent of u and v, then the solution of (Eq. 8) can be obtained by computing the minimal eigenvalue (τ) via inverse iteration methods [5]. Based on this observation an

---

[2] The Riemannian SVD of A can be interpreted as a restricted SVD (RSVD) [11] with the Riemannian metrics (D_u and D_v) in the column and row space of A.

inverse iteration algorithm for computing the R-SVD (Eq. 8) of a *full-rank* matrix A has been proposed in [4]. Essentially, for given matrices D_u and D_v, this algorithm performs one step of inverse iteration (using the QR decomposition of A) to obtain new estimates of the singular vectors u and v. These new vectors u and v are then used in updating matrices D_u and D_v. Recent experiments with this method [7], however, have revealed breakdowns associated with singular iteration matrices.

3 The Riemannian SVD for a Low-Rank Matrix

It has been found that, for the LSI approach [2], the polysemy problems in information retrieval can be circumvented by creating a new mechanism to incorporate nonlinear perturbations to the existing rank-reduced model. Such perturbations, which can arise in the form of users' feedback and interactions with LSI model, can be represented by changes in certain single or clustered entries, rows, and columns of the original term-by-document matrix used to define the rank-reduced LSI model. The updated LSI implementation using a modified R-SVD has the ability to filter out irrelevant documents by moving them to the bottom of the retrieved (ranked) document list.

3.1 Riemannian SVD for a Low-Rank Matrix

Assume A is an $m \times n\,(m \gg n)$ term-by-document matrix, and A_k is a corresponding LSI rank-k approximation of A [1]. Suppose A_k has the splitting

$$A_k = A_k^F + A_k^C, \tag{9}$$

where A_k^F reflects the set of terms in A_k whose term-document associations are not allowed to change, and A_k^C is the complement of A_k^F.

Consider the following minimization problem:

$$\min_{B_k \in \mathcal{R}^{m \times n}, y \in \mathcal{R}^n} \|A_k - B_k\|_F^2 \quad \text{subject to } B_k\, y = 0,\ y^T y = 1,\ B_k = A_k^F + B_k^C. \tag{10}$$

Here, the matrix B_k is an approximation to the matrix A_k satisfying the constraints in (Eq. 10), which include preserving the specified elements of A_k (i.e., A_k^F). Using Lagrange multipliers [4], this constrained minimization problem can be formulated as an equivalent R-SVD problem for the rank-k model A_k:

$$A_k v = D_v u \tau, \quad u^T D_v u = 1, \tag{11}$$
$$A_k^T u = D_u v \tau, \quad v^T D_u v = 1,$$

where the metrics D_u and D_v are defined by

$$D_v = \text{diag}\,(P\,\text{diag}(v)\,v)\,,\ D_u = \text{diag}\,(P^T\,\text{diag}(u)\,u)\,, \tag{12}$$

and P is a *perturbation* matrix having zeros at all entries of A_k^F and nonzeros elsewhere.

With regard to LSI applications, the solution of (Eq. 11) generates an updated semantic model B_k from the existing rank-reduced model A_k. In the model B_k, some of the term-document associations, which are represented by the original term-by-document matrix A, are perturbed or changed according to the splitting specified in (Eq. 9). This new model B_k has the same or near rank as A_k but reflects imposed (i.e., filtered) changes on term-document associations.

3.2   The RSVD-LR Algorithm

Using an inverse iteration approach, a new algorithm for solving the R-SVD (Eq. 11) with a low-rank matrix A_k can be constructed [7]. Assume the SVD (Eq. 1) of the matrix A has been computed and written as

$$A = (U_k, U_{m-k})\, \mathrm{diag}(\Sigma_k, \Sigma_{n-k})\, (V_k, V_{n-k})^T,$$

where $U = (U_k, U_{m-k})$ and $V = (V_k, V_{n-k})$. Then, the updated estimates of the singular vectors u and v as well as the singular value τ from (Eq. 11) can be computed by the following iteration:

$$
\begin{aligned}
x &= \Sigma_k^{-1} V_k^T D_u v\, \tau, \\
u^{(new)} &= [U_k - U_{m-k}\, (U_{m-k}^T D_v U_{m-k})^{-1} (U_{m-k}^T D_v U_k)]\, x, \\
s &= \Sigma_k^{-1} U_k^T D_v u^{(new)} \tau, \\
v^{(new)} &= [V_k - V_{n-k}\, (V_{n-k}^T D_u V_{n-k})^{-1} (V_{n-k}^T D_u V_k)]\, s, \\
\tau^{(new)} &= u^{(new)T} A_k v^{(new)}.
\end{aligned}
\tag{13}
$$

Once the vectors $u^{(new)}$ and $v^{(new)}$ have been obtained, the matrices D_u and D_v can be updated by $D_u^{(new)} = \mathrm{diag}\,(P^T \mathrm{diag}\,(u^{(new)})\, u^{(new)})$, and $D_v^{(new)} = \mathrm{diag}\,(P \,\mathrm{diag}\,(v^{(new)})\, v^{(new)})$.

In order to compute the R-SVD (Eq. 11) for information retrieval applications and avoid the singularity problems associated with the matrices $U_{m-k}^T D_v U_{m-k}$ and $V_{n-k}^T D_u V_{n-k}$, an inverse iteration algorithm (RSVD-LR) has been developed. This method (shown on next page) updates the left singular vector u and the right singular vector v by decomposing them into two orthogonal components, and then uses these new vectors to update the matrices D_u and D_v. Specifically, it adopts the same formulas to compute intermediate vectors x and s as presented in (Eq. 13), but uses a different and much simpler way to estimate the orthogonal projection of the vector u onto the null space of A^T and the vector v onto the null space of A, respectively.

One stopping criterion (Step 15) for Algorithm RSVD-LR, which has been successfully used in [7], is $\|B_k v\| + \|B_k^T u\| < tol$, where tol is a user-specified (small) constant. For $n \ll m$, it can be shown that each iteration in Algorithm RSVD-LR2 requires $O(m^2)$ flops. Since only a small number of iterations are typically needed [7] to generate the new perturbed model B_k from the existing model A_k, the overall computational costs in applying RSVD-LR are modest.

Algorithm RSVD-LR

1. Set $i = 0$
2. Initialize $u^{(0)}$, $v^{(0)}$ and $\tau^{(0)}$
3. Compute matrices $D_u^{(0)}$ and $D_v^{(0)}$
4. While (not converged) do
    5. Set $i = i + 1$
    6. $x^{(i)} = \Sigma_k^{-1} V_k^T D_u^{(i-1)} v^{(i-1)} \tau^{(i-1)}$
    7. $y^{(i)} = U_{m-k}^T u^{(i-1)}$
    8. $u^{(i)} = U_k x^{(i)} + U_{m-k} y^{(i)}$
9. $s^{(i)} = \Sigma_k^{-1} U_k^T D_v^{(i-1)} u^{(i)} \tau^{(i-1)}$
10. $t^{(i)} = V_{n-k}^T v^{(i-1)}$
11. $v^{(i)} = V_k s^{(i)} + V_{n-k} t^{(i)}$
12. $\tau^{(i)} = u^{(i)T} A_k v^{(i)}$
13. $D_u^{(i)} = \text{diag}\left(P \,\text{diag}\left(v^{(i)}\right) v^{(i)}\right)$
14. $D_v^{(i)} = \text{diag}\left(P^T \text{diag}\left(u^{(i)}\right) u^{(i)}\right)$
15. Convergence Test

---

4 RSVD-LSI Model

Compared to other information retrieval techniques, LSI is a fully automatic indexing method and performs surprisingly well. In terms of precision (the proportion of returned documents that are relevant) at different levels of recall (the proportion of all relevant documents in a collection that are retrieved), it has achieved up to 30% better retrieval performance over lexical searching techniques [3]. However, in order to improve the viability and versatility of current LSI implementations, a mechanism to incorporate nonlinear perturbations to the existing rank-reduced model has been lacking. Updating LSI models based on user feedback can be accomplished using constraints modeled by the R-SVD of a low-rank approximation to the original term-by-document matrix (see Section 3.2).

4.1 Current LSI Implementations

LSI is implemented using a rank-reduced model of the original term-by-document matrix [2]. Specifically, let A be a constructed $m \times n$ ($m \gg n$) term-by-document matrix $A = [a_{ij}]$, where a_{ij} denotes the frequency in which term i occurs in document j. Usually A is very sparse since each term typically occurs only in a subset of documents. Each of the m terms in the collection is represented as a row in the matrix, while each of the n documents in the collection is represented as a column in the matrix.

In practice, a local and a global weighting may be assigned to each nonzero entry in A to indicate (as an indexing term) its relative importance within documents and its overall importance in the entire document collection, respectively. Therefore, a nonzero weighted entry of the term-by-document matrix A can be expressed as $a_{ij} = l_{ij} * g_i$, where l_{ij} is the local weighting for term i in document j and g_i is the global weighting for term i in the document collection. Various weighting schemes are discussed in [3].

Once the incidence matrix A has been constructed and properly weighted, the SVD (Eq. 1) is used to compute a rank-k approximation ($k \ll \min(m, n)$) to A, defined by $A_k = \sum_{i=1}^{k} \sigma_i u_i v_i^T = U_k \Sigma_k V_k^T$, where U_k and V_k, respectively, contains the first k columns of U and V (Eq. 1). The columns of U_k and V_k

define the singular vectors which correspond to the first (largest) k singular values of A, i.e., diagonal entries of Σ_k. In other words, A_k captures the first k singular triplets of the matrix A. The Eckart-Young-Mirsky formula (Eq. 3) implies, with regard to LSI, that the k-dimensional term-by-document space represented by A_k is the closest rank-k approximation in a least squares sense to the original term-by-document space which is represented by A. It is assumed that the dimension reduction of the term-by-document space helps reveal the underlying semantic structure in the association of terms with documents and removes the obscuring noise in word usage.

In the reduced term-by-document space, individual terms and documents are positioned by the left (U_k) and right (V_k) singular vectors, respectively, and coordinates of the space are often scaled by the corresponding singular values (diagonal elements of Σ_k), allowing clusters of terms and documents to be easily identified. Semantically related terms and documents are assumed to be near each other in this space.

In LSI, a user's query, treated as a *pseudo-document* $\hat{q}$, is usually represented as the sum of the term vectors corresponding to the terms specified in the query scaled by the inverse of the singular values (Σ_k^{-1}) [2,8], $\hat{q} = q^T U_k \Sigma_k^{-1}$, where q is vector containing the weighted term-frequency counts of the terms that appear in the query, and it can be projected into the term-by-document space. The *relevance feedback* mechanism can be incorporated in a typical LSI model. Relevance feedback uses the terms in relevant documents to supplement the user's original query, allowing better retrieval performance [10]. Let $d \in \mathcal{R}^n$ be a vector whose nonzero elements are indices specifying the relevant document vectors. Then, a relevant feedback query can be constructed from the original query by adding the vector sum of the relevant documents identified by the user. The modified query is given by $\hat{q} = q^T U_k \Sigma_k^{-1} + d^T V_k$.

Using one of several similarity measures such as the cosine measure, the pseudo-document vector is then compared against all document vectors in the space. The terms and documents are ranked according to the results of the similarity measure used, and only those semantically related documents that presumably are the *closest* vectors are retrieved.

4.2 New LSI Implementation

This new LSI implementation attempts to overcome problems associated with polysemy by performing appropriate nonlinear perturbations to the existing term-by-document model A_k in the sense that some term-document associations are changed and the corresponding term and document vectors are reconstructed and repositioned in the new perturbed rank-reduced term-by-document space. Based on user feedback or interaction with the system, a perturbation matrix P in (Eq. 12) can be constructed which defines the splitting of A_k as in (Eq. 9). For all the terms and documents in A_k, the matrix P determines the set of terms and the set of documents whose term-document associations are allowed to change (while other associations are fixed). The perturbation matrix P can also be used to have more control in ranking the returned documents (see [7]).

Once the perturbation matrix P has been constructed, the R-SVD problem for the low-rank matrix A_k (Eqs. 11-12) can be formulated. Algorithm RSVD-LR is then applied to compute this R-SVD, and its approximate solution generates a new perturbed term-by-document space B_k which satisfies the minimization problem in (Eq. 10). The new model B_k, having the same or near rank as A_k, reflects the changes in the term-document associations specified by the user in order to better capture the context of the user's query and thereby retrieve more relevant information (documents).

4.3 Retrieval Performance

The RSVD-LSI model has been implemented in MATLAB Version 5, and experiments have shown that the model performs an effective conceptual information retrieval/filtering technique [7]. The performance of this new LSI implementation has been compared with the traditional LSI approach with and without *relevance feedback* using the MEDLINE document collection composed of medical abstracts from the National Library of Medicine [2]. The MEDLINE collection contains 1033 documents, 5831 terms, and a set of 30 queries (see [7]) with *relevance judgments* for each query. The original or current LSI model (C-LSI) [2], and LSI with relevance feedback (RF-LSI) [3] as well as the new enhanced LSI implementation based on R-SVD (RSVD-LSI) were tested with this collection [7].

The performance of information retrieval systems is often measured in terms of two factors: *precision* and *recall*. Given a query, an information retrieval system returns a ranked list of documents. Let r_i be the number of relevant documents among the top i documents in the returned list, and r_q be the number of relevant documents to the query in the collection. The precision for the top i documents, $\mathcal{P}_i$, is then defined as $\mathcal{P}_i = \frac{r_i}{i}$, i.e., the proportion of the top i documents that are relevant. The recall $\mathcal{R}_i$ is then defined as $\mathcal{R}_i = \frac{r_i}{r_q}$, i.e., the proportion of all relevant documents in a collection that are retrieved.

Precision-vs-recall curves for the C-LSI, RF-LSI and RSVD-LSI methods (based on $k = 50$ dimensions) are shown in Figure 1. Precision is plotted as a function of recall for nine levels of recall, from 0.10 through 0.90. These curves represent *average* performance over the 30 queries available with MEDLINE collection. For all levels of recall, precision of the RSVD-LSI method lies well above that obtained with the C-LSI model, and the gap between these two curves is especially significant at recall levels from 0.20 through 0.80. For all but the lowest recall level (0.10), the RSVD-LSI method demonstrates a higher precision than that obtained by the RF-LSI approach. The difference in performance between RSVD-LSI and other LSI implementations is especially impressive at recall levels from 0.30 through 0.70. The results from the MEDLINE experiments suggest that the RSVD-LSI implementation can achieve moare than 20% better performance than the original LSI model, and more than 7% better performance than the RF-LSI approach.

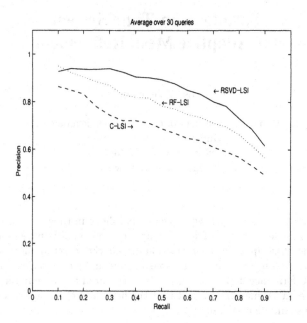

Fig. 1. Precision-recall curves for C-LSI, RF-LSI and RSVD-LSI on the MEDLINE collection.

References

1. M. Berry, S. Dumais and G. O'Brien (1995). Using Linear Algebra for Intelligent Information Retrieval, *SIAM Review*, 37:4:573-595.
2. S. Deerwester, S. Dumais, G. Furnas, T. Landauer and R. Harshman (1990). Indexing by Latent Semantic Analysis, *Journal of the American Society for Information Science*, 41(6):391-409.
3. S. T. Dumais (1991). Improving the Retrieval from External Sources, *Behavior Research Methods, Instruments, & Computers*, 23(2):229-236.
4. B. De Moor (1993). Structured Total Least Squares and L_2 Approximation Problems, *Linear Algebra and its Applications*, 188,189:163-205.
5. G. Golub and C. Van Loan (1996). *Matrix Computations*, John-Hopkins, Baltimore, Third Ed.
6. S. Van Huffel and J. Vandewalle (1991). *The Total Least Squares Problems: Computational Aspects and Analysis*, SIAM, Philadelphia, PA.
7. E. P. Jiang (1998). *Information Retrieval and Filtering Using the Riemannian SVD*, Ph.D. Thesis, Dept. of Computer Science, The University of Tennessee, Knoxville, TN.
8. T. A. Letsche and M. Berry (1997). Large Scale Information Retrieval with Latent Semantic Indexing, *Information Sciences*, 100:105-137.
9. L. Mirsky (1960). Symmetric Gage Function and Unitarily Invariant Norms, *Q. J. Math*, 11:50-59.
10. G. Salton and C. Buckley (1990). Improving Retrieval Performance by Relevance Feedback, *J. Amer. Soc. Info. Sci.*, 41:288-197.
11. H. Zha (1991). The Restricted Singular Value Decomposition of Matrix Triplets, *SIAM J. Matrix Anal. Appl.*, 12(1):172-194.

Parallel Run-Time System
for Adaptive Mesh Refinement

Nikos Chrisochoides*

Department of Computer Science & Engineering
University of Notre Dame
Notre Dame, IN 46556
nikos@cse.nd.edu

Abstract. In this paper we present a parallel run-time system for the efficient implementation of adaptive applications on distributed memory machines. Our approach is application driven; the target applications are characterized by very large variations in time and length scales. Preliminary performance data from parallel unstructured adaptive mesh refinement on an SP machine suggest that the flexibility of our approach does not cause undue overhead.

1 Introduction

In this paper we present a *lean, language-independent and easy to port and maintain* run-time system for implementing adaptive applications on large-scale parallel systems. Our design philosophy is based on the principle of *separation of concerns*. Figure 1 depicts the architecture of the overall system and its layers that address the different requirements (or concerns) of the application.

The first layer, Data-Movement and Control Substrate (DMCS), provides thread-safe one-sided communication. DMCS implements an application program interface (API) very similar to the one proposed by the PORTS consortium [1]. Similar APIs are implemented by Nexus [2] and Tulip [3]. However, DMCS can co-exist with MPI without obscuring access to it. The second layer, Mobile Object Layer (MOL), supports a global addressing scheme in the context of object mobility, as do many other previously developed run-time systems for object-oriented programming languages like Amber [4], COOL [5], and Charm++ [6], to mention a few. However, MOL is simple, easy to understand, port, and maintain on different machines. For data-movement operations, our measurements on an SP2 show that the communication parameters of DMCS and MOL are within 10% and 14 % respectively of the underlying message layer [7].

MOL does not provide support for shared memory management nor implements policies specifying how and when the mobile objects must be moved. Application-driven scalable memory management and load balancing schemes are implemented in the third layer. A shared memory management scheme tries to: (1) minimize false sharing at the SMP level and (2) reduce extra work due

---

* This work was supported by NSF grant #9726388 and JPL award #961097.

to data-migration at the inter-node level. Shared memory models at the hardware or operating system level have been studied extensively during the last 15 years [9] and only recently application-specific memory management strategies were used in parallel applications like n-body [10] and parallel finite difference solvers [11]. To the best of our knowledge this is the first time that application-driven schemes are generalized and uncoupled from the applications so that they can be implemented in the run-time system.

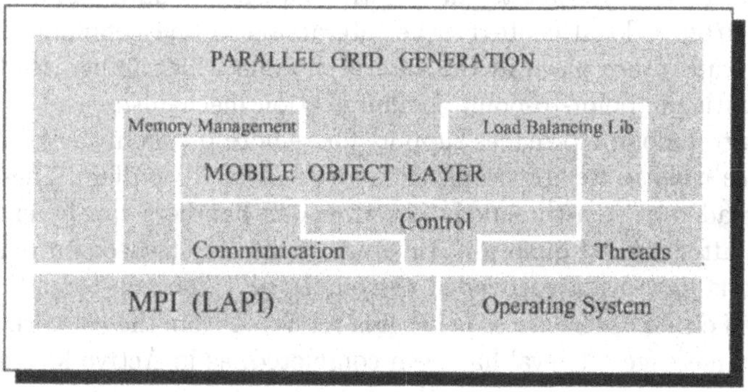

Fig. 1. Parallel run-time system: Architecture.

The rest of the paper is organized as follows: In Section 2 we briefly describe the DMCS layer. In Section 3 we describe in more detail the MOL because of its critical role in the efficient implementation of parallel adaptive applications. Section 4 outlines the basic ideas for the memory management and load balancing libraries. In Section 5 we describe an unstructured mesh refinement method which guarantees grid quality. In Section 6 we present preliminary performance data and in Section 7 we conclude with a summary and future work.

2   Data-Movement and Control Substrate

Data-Movement and Control Substrate [12] is a mid-level substrate that targets traditional homogeneous MPPs like IBM's SP-series and non-traditional architectures like JPL's HTMT design [13]. DMCS implements thread-safe one-sided communication primitives, static global address space, and a threaded model of execution. DMCS consists of three modules: (i) a *threads* module (ii) a *communication* module and (iii) a *control* module. Figure 1 depicts these three modules and their interaction. The threads and communication modules are independent, with clearly defined interface requirements, while the control module is built on top of the point-to-point communication and thread primitives.

The *threads* module supports a subset, PORTS0, of POSIX API defined by the PORTS consortium. This subset is augmented by an additional primitive: *dmcs_thread_create_atonce* which is necessary to minimize the scheduling latency of urgent *remote procedure invocations* required for the load balancing of adaptive applications.

DMCS uses the *global pointer* [14, 15] as a communication abstraction, in addition to processor (or context) *id*. A global pointer essentially consists of a processor (or context) *id* and a pointer to the local address space. The functionality of the communication module for point-to-point data-movement includes routines like *get/put* to initiate the transfer of data from/to a remote context to/from a local context. The interaction of the communication module and threads takes place in the *control* module which integrates the thread scheduler with the point-to-point communication mechanism.

The *control* module provides support for remote procedure invocation or *remote service request*. Remote requests are user defined handlers. These handlers can be threaded or non-threaded. The threaded handlers can be either *urgent* (scheduled after a fixed quanta of time or can be *lazy* (scheduled only after all other threads have been suspended or completed). The non-threaded handlers are executed either as the message is being retrieved from the network interface[1], or after the message retrieval has been completed, as in Active Messages [7].

3 The Mobile Object Layer

The MOL provides tools to build migratable, distributed data structures consisting of mobile objects linked via mobile pointers to these objects (e.g. a distributed graph using a mobile object for each node, and lists of mobile pointers to adjacent nodes). The MOL guarantees [8] that, if a node (or an object) in such a structure were moved from one processor to another, messages sent to the migrated object would be received by forwarding the messages from the sending processors to the object's new location.

3.1 Mobile Pointers and Distributed Directories

The basic building block provided by the MOL is a "mobile pointer," (processor number, index number) pair. The processor number is the object's "home node," the processor where the object was originally allocated. The index number assigned to new mobile objects is unique on that processor. This (home node, index number) pair forms a name for a mobile object that is unique over the whole system. Thus, mobile pointers are valid on all processors, and can be passed as data in messages without extra help from the system.

---

[1] This works only for commutative and associative operations. The overlapping of computation and communication occurs at the instruction level by interleaving the computation and flow control that corresponds to the incoming message and various copy operations needed to retrieve the message from the network interface to users address space.

A directory entry consists of a processor number (the object's "best guess" location), a sequence number (how up to date the best guess is), and a physical pointer to the object, if it's local. A directory is a two dimensional array of directory entries; an entry for any mobile pointer (and thus the best guess location of the corresponding mobile object) can be located by indexing into the directory with the mobile pointer's home node number and index number. The directory grows to accommodate more entries as mobile objects are allocated. Since directories are sparse, it would be possible to store them in a more compact form (such as a hash table), but we have not implemented this yet.

To send a message to a mobile object, the object's best guess location is retrieved from the sending processor's directory (using the object's mobile pointer). If the location is incorrect, the MOL forwards the message from the incorrect processor towards the real location, and updates the sending processor to reflect the change in the object's best guess location. A processor's local directory gets updated only when the processor sends to an object a message which "misses" the object's correct location. Only processors showing explicit interest in an object get informed of the object's current location. This avoids the need to broadcast updates to all processors each time an object moves.

There are three possibilities for the state of an object's location. First, the object may reside on the current processor. In this case, the message can be handled locally. Second, there may be an entry in the directory indicating that an object may reside at another processor; the message is sent to the processor indicated by the directory entry. If the target processor does not contain the object, it will forward the message to the best guess location given by its local directory. Third, the directory may have no entry for the mobile pointer; the mobile pointer's home node entry is used as the default best guess location for the object.

The forwarding mechanism only affects the source and target processors of a message to a mobile object; this "lazy" updating minimizes the communication cost of moving an object. By using the sequence number of the mobile object and the field in the corresponding directory entry, the MOL prevents older updates from overwriting newer ones. This is necessary, since the MOL does not assume causal ordering and it allows network delays to halt message reception for arbitrary lengths of time.

The MOL directory structure and forwarding protocol are similar to those used in ABC++ [16]. However, the protocol in ABC++ would have required communication with the home node each time a message is sent to the object. MOL eliminates additional communication for every message because its directories are automatically updated to keep track of where objects have migrated. Furthermore, MOL updates are not broadcast (as is the case for ABC++) to all processors in the system, but are lazily sent out to individual processors as needed. The MOL uses sequence numbers to distinguish new updates from old updates in a way that appears to be essentially the same as that described by Fowler [17] (which was subsequently used in Amber [4]).

3.2 Message Layer

Our implementation of the MOL builds its own message layer on top of DMCS to handle messages sent to objects. In our implementation we use incoming and outgoing pools of messages for small (64 bytes) to medium (1024 bytes) size messages, similar to those used in the Generic Active Message specification from Berkeley. The MOL supports "processor requests" with *mob_request*, and "object messages" with *mob_message*, to transfer a small amount of data to processors and mobile objects, respectively. Both types require a user–supplied procedure to handle the message when it arrives at its destination.

Three types of handlers are available to process a message initiated by the MOL. First, a function handler, e.g. an Active Messages handler, may be used to process the message. This is the fastest handler type, but neither communication nor context switching is allowed within the handler. Second, the message may be processed from within a delayed handler, which is queued internally by the MOL. The delayed handler is slower, but communication from within the handler (but not context switching) is allowed. Third, a threaded handler can spawn a thread to process the message. This is the most flexible handler type but it is also the slowest.

Although the MOL only directly supports small and medium size messages (store and forwarding of large messages is inefficient), efficient large message protocols can be built using MOL messaging. For instance, a large "get" operation directed to a mobile object can be performed easily. First, create a local buffer to hold incoming data. Then send a mobile object message including the buffer's address to the target object. The remote (delayed or threaded) handler can then performs a "put" memory operation to store the requested data to the buffer in the originating processor. In this case, the object must be "locked" by the programmer to keep the object from moving before the put operation completes, since the MOL does not control the migration of objects.

4 Application Specific Runtime Libraries

Scalable Memory Management Both in industry and academia is not clear yet what is the best approach in supporting scalable shared memory. One of the major issues is the unit of sharing. For hardware or operating system dependent approaches (i.e., using cache line or memory page) the major problem is false sharing while for application-dependent approaches the major problem is efficiency. In this project we are developing an application-dependent scalable, shared memory management scheme: (1) for minimizing false sharing at the intra-node level and (2) for reducing extra work due to data-migration at the inter-node level.

Our implementation is based on the extendible hash table (EHT) structure and method that originally was proposed in by Faginn in [18] for databases and recently was used for the efficient implementation of parallel adaptive finite difference solvers [11] and n-body simulations [10]. The EHT method in combination with an indexing scheme based on Space Filling Curves are used to implement pointer arithmetic that preserves spatial locality. Instead of storing the global hash key table on every single node as in [11] we use MOL's distributed directory and message forwarding protocols to implement efficiently a distributed global hash key table. The advantages of our approach are: (1) it avoids the replication of data and (2) it eliminates the overhead of global reduction operations required for the updates of the replicated global hash key table; these overheads can be extremely large for adaptive applications such as adaptive mesh refinement for crack propagation and mixed classical molecular and quantum dynamics.

Dynamic Load Balancing Load-balancing can be performed either explicitly (using domain-specific information) or implicitly by letting the runtime system move computations. Explicit methods [19] determine a partition of the mesh among the processors to balance computational load and reduce communication during the solution phase. Although these methods have been successful for static problems, many of them are computationally expensive for adaptive mesh refinement. Implicit methods move computations to balance load, but do so without any particular domain-specific information. In this project we are developing new *work–stealing* load balancing strategies [20] for adaptive mesh refinement that use data-dependencies to minimize communication in the solution phase. In contrast to *work–sharing* strategies such as diffusion methods or centralized methods [19], the work-stealing methods let the individual under-utilized processors request work and thus maximize processor utilization.

5   Parallel Adaptive Mesh Refinement

The efficient implementation of an unstructured mesh generator on distributed memory multiprocessors requires the maintenance of complex and dynamic distributed data structures for tolerating latency, minimizing communication and balancing processor workload. In this Section we describe an implementation of the parallel 2D Constrained Delaunay Triangulation (CDT) which uses the run-time system we described above. A parallel implementation for 3D domains is in progress.

Constrained Delaunay Triangulation. The mesh generator uses the Parallel Constrained Delaunay Triangulation method [21] to generate a guaranteed–quality mesh. Given a precomputed domain decomposition, each subdomain is refined independently of the other regions, except at the interfaces between regions. For 2D meshes, the extent of the refinement is defined by a "constrained" interface and boundary edges. If a boundary or interface edge is part of a triangle to be refined, that edge is split. Since interface edges are shared between regions, splitting an edge in one region causes the change to propagate to the region which shares the split edge. The target region is updated as if it had split the edge itself.

Load Balancing with the MOL. The input to the mesh generator is a decomposition of a domain into some number of regions, which are assigned to processors in a way that maximizes data locality. Each processor is responsible for managing multiple regions, since, in general, there will be many regions in the decomposition. Subsequently, imbalance can arise due to both unequal distribution of regions and large differences in computation (e.g. between high– and low–accuracy regions in the solution).

The "work–stealing" load balancing method we implement maintains a counter of the number of work–units that are currently waiting to be processed, and consults a threshold of work to determine when work should be requested from other processors. When the number of work–units falls below the threshold, a processor requests work from a neighborhood of processors [20] in order to maintain its level of work. The topology of the decomposition is used to define the adjacency-relationship between the regions. These relationships are used to compute at runtime the processors' neighborhoods. The regions can be viewed as the work-units or objects which the load balancer can migrate to re-balance the computation. Using the MOL, each region is viewed as a mobile object; we associate a mobile pointer with each region, which allows messages sent to migrated regions to be forwarded to the new locations.

Data Movement Using the MOL. The critical steps in the load balancing phase are the region migration and the updates for the edge split messages between regions. To move a region, the MOL requires a packing primitive (*mob_uninstallObj*) be called to update the sending processor's local directory to reflect the pending change in the region's location. Next, a programmer–supplied procedure is used to pack the region's data into a buffer, which must also contain the region's mobile pointer and the *MoveInfo* structure returned by *mob_uninstallObj* to track the region's migration. Then, a message–passing primitive (e.g. *mpi_send or dmcs_put*) is invoked to transport the buffer, and another user–supplied procedure unpacks and rebuilds the region in the new processor. After the region has been unpacked, another primitive *mob_installObj* must be called with the region's mobile pointer and the MoveInfo structure to update the new processor's directory.

Since the MOL is used to move data, standard message–passing primitives, like *mpi_send or dmcs_put*, it will not work to send split edge messages from one region to another, since regions can be migrated. Therefore, the MOL supplies a procedure, *mob_message*, which uses an object's mobile pointer to send messages to the object's last known location. The MOL will forward a split edge message sent with *mob_message* and it will update the sending processor's directory so that, unless the target region moves again, subsequent messages will be sent directly.

6   Preliminary Performance Data

In this Section we present three sets of performance experiments taken on an IBM RISC/6000 SP System, using the Active Messages implementation developed in [7]. The first set of performance experiments involves simple message passing using DMCS. These experiments indicate that DMCS's communication parameters are very close (within 10%) of the underlying active message layer. For example, one-way latency for a 0-byte message in DMCS is $29\mu s$, which represents an overhead of 10% over the underlying active message latency of $26\mu s$. This compares with a handler to handler latency of $79\mu s$ in Nexus, which represents an overhead of 80% over the native MPL one-way latency of $44\mu s$. Similarly, the bandwidth achieved in DMCS for bulk transfers is $33.6MBytes/sec$ for get operations, and $29MBytes/sec$ for put operations.

The second set of performance experiments involves: (1) the *mob_message*, which allows messages to be sent to a mobile object via a mobile pointer and (2) the *mob_request*, which directs messages of 1024 bytes or less to specific processors without explicitly requesting storage space on the target processor. The performance of both operations is very reasonable; the latency of *mob_request* is within about 11% of the latency of *am_store*, while *mob_message*'s latency is about 12% to 14% higher than *am_store*'s latency. Not surprisingly, the latency of the forwarded messages was about twice as high as that of the unforwarded messages. Figure 2 indicates that the overall (amortized) cost for forwarding messages is extremely small. This is because messages are forwarded immediately after an object moves but not after the updates have been received i.e., all subsequent messages until the next move are directly send to the object.

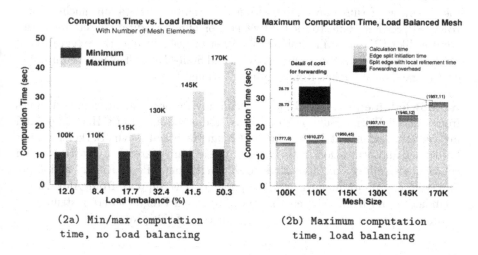

(2a) Min/max computation time, no load balancing

(2b) Maximum computation time, load balancing

Fig. 2. Parallel CDT mesh generation performance results.

Finally, the third set of experiments is gathered from the parallel meshing application described in Section 5. For these set of data instead of using the low-latency Active Messages we are using Nexus as the underline communication layer. Figure 2 depicts data for a parallel mesh with between $100,000$ and $170,000$ elements, and for load imbalances of between 8 and 50 percent (Figure 2a). Each of the four processors in the system started with 16 regions. Given above each bar is the number of elements generated in the mesh for that particular run. Figure 2b displays the maximum computation time for a series of load–balanced mesh computations which used the same initial mesh as the non–load balanced experiments. The tuple above each bar gives the number of split edge requests and the number of object migrations from the processor represented by the bar.

7 Summary and Future Work

We have presented a run-time system that integrates active messages with multi-threading. For certain numerical operations it overlaps computation and communication in the instruction level so that it can mask some of the communication overhead. It supports global address space, maintains automatically the validity of global pointers as data migrates from one processor to another, and it implements a correct and efficient message forwarding and communication mechanism between the migrating objects. Finally, it provides an efficient application-driven memory management and dynamic load balancing strategies that are required for the parallelization of adaptive applications.

Currently we are evaluating and fine-tuning this system in the context of two challenging applications: (1) 3D crack propagation simulations from fracture mechanics [22] and (2) mixed classical molecular and quantum dynamics simulations from reactions in microporous catalysts [23]. Also, we are studying the evolution of run-time systems from the traditional Petaflop machines like IBM's SP/SMP systems to the next generation of non-traditional Petaflop architectures like JPL's HTMT design [13]. Other projects involve the integration of MOL with parallel languages like EARTH and Split-C as well as meta-computing environments like Globus from Argonne National Labs.

Acknowledgments I would like to thank my former students and colleagues from Cornell Florian Sukup, Induprakas Kodukula, and Chris Hawblitzel and my current students at Notre Dame Demian Nave and Kevin Barker for their hard work and dedication in this project. Also, I am grateful to my friends and colleagues Geoffrey Fox, Keshav Pingali, Paul Chew, Peter Kogge, and Marc Snir whose support made this work possible. The support from Alex Nason Foundation and IBM's Shared University Research Program were instrumental in providing the resources critical to the success of this project.

References

1. Portable Run-Time Systems Consortium,
   http://www.cs.uoregon.edu/research/paracomp/ports
2. I. Foster, C. Kesselamn, S. Tuecke. The Nexus Task-parallel Runtime System, *Proc. 1st Intl Workshop on Parallel Processing*, 1994.
3. Pete Beckman and Dennis Gannon, Tulip: Parallel Run-time Support System for pC++, Computer Science, Indiana University, http://www.extreme.indiana.edu, 1996.

4. J.S. Chase , F.G. Amador, E.D. Lazowska, H.M. Levy and R.J. Littlefield. The Amber System: Parallel Programming on a Network of Multiprocessors SOSP12, pp 147–158, December, 1989.
5. R. Chandra, A. Gupta, J.L. Hennessy, COOL Parallel Programming Using C++ (eds. Wilson, G. and Lu, P.), The MIT Press, 1998.
6. L. Kale, S. Krishnan, Charm++, Parallel Programming Using C++ (eds. Wilson, G. and Lu, P.), The MIT Press, 1998.
7. C. Chang, G. Czajkowski, C. Hawblitzel, and T. von Eicken, Low-latency communication on the IBM RISC/6000 SP system, In Proceedings of Supercomputing'96.
8. Chris Hawblitzel and Nikos Chrisochoides Mobile Object Layer: A data migration framework for Active Messages Paradigm University of Notre Dame Department of Computer Science and Engineering TR 98-07, 1998.
9. D. Culler, J. Singh, A. Gupta Parallel Computer Architecture: A Hardware/Software Approach Morgan Kaufmann, 1998.
10. Warren, M. S. and Salmon J. K., A Parallel Hashed Oct-tree N-body Algorithm Proceedings, Supercomputing '93
11. Manish Parashar and James C. Browne. Distributed dynamic data-structures for parallel adaptive mesh refinement. In HiPC, 1995.
12. N. Chrisochoides, K. Pingali, I. Kodukula Data Movement and Control Substrate for Parallel Scientific Computing Lecture Notes in Computer Science, Springer-Verlag, Vol. 1199 pp 256–268, 1997
13. Thomas Sterling HTMT Tech Note: Notes from Notre Dame Architecture Workshop Technical Report, Serial No 016, Jet Propulsion Laboratory, 1998.
14. D. E. Culler, A. Dusseau, S. Goldstein, A. Krishnamurthy, S. Lumeta, T. von Eicken, and K. Yelick, Parallel Programming in Split-C, In Proceedings of Supercomputing, 1993.
15. M. Chandy and C. Kesselman, CC++: A Declarative Concurrent Object-Oriented Programming Notation, In Research Directions in Concurrent Object-Oriented Programming, MIT Press, 1993.
16. E. Arjomandi, W. O'Farrell, I. Kalas, G. Koblents, F. Ch. Eigler, and G. G. Gao, ABC++: Concurrency by Inheritance in C++, IBM Systems Journal, Vol. 34, No.1, pp. 120-137, 1995.
17. R. Fowler, The complexity of using forwarding addresses for decentralized object finding In Proceedings of the 5th annual ACM symposium on principles of distributed computing, 1986.
18. R. Faginn, Extendible Hasing– A Fast Access Method for Dynamic Files, ACM TODS, Vol 4, pp 315, 1979.
19. Fox, G., R. Williams and P. Messina Parallel Computing Works! Morgan Kaufmann Publishers, Inc. San Francisco, California, 1994.
20. N. Chrisochoides, Multithreaded Model for Load Balancing Parallel Adaptive Computations On Multicomputers, Journal of Applied Numerical Mathematics 6 pp 1–17, 1996.
21. P. Chew, N. Chrisochoides, and F. Sukup Parallel Constrained Delaunay Meshing In the proceedings of 1997 Joint ASME/ASCE/SES Summer Meeting, Special Symposium on Trends in Unstructured Mesh Generation, 1997.
22. P. Chew, N. Chrisochoides, G. Gao, T. Ingrafea, K. Pingali, S. Vavasis, Crack Propagation on Teraflop Computers CISE Challenge Proposal Report (unpublished), Computer Science, Cornell University, 1997.
23. H-C. Chang, N. Chrisochoides, S. Hammes-Schiffer, and E. Maginn Multi-scale Molecular and Quantum Dynamics Simulations of Zeolite Transport and Chemistry KDI Proposal Report (unpublished), University of Notre Dame, 1998.

Author Index

Lecture Notes in Computer Science

For information about Vols. 1–1371

please contact your bookseller or Springer-Verlag